AF676530

Insect Microbiome and Immunity

Insect Microbiome and Immunity

Editors

Hongyu Zhang
Xiaoxue Li
Yin Wang

MDPI • Basel • Beijing • Wuhan • Barcelona • Belgrade • Manchester • Tokyo • Cluj • Tianjin

Editors

Hongyu Zhang
Huazhong Agricultural University
Wuhan
China

Xiaoxue Li
Huazhong Agricultural University
Wuhan
China

Yin Wang
University of Georgia
Athens
USA

Editorial Office
MDPI
St. Alban-Anlage 66
4052 Basel, Switzerland

This is a reprint of articles from the Special Issue published online in the open access journal *Insects* (ISSN 2075-4450) (available at: https://www.mdpi.com/journal/insects/special_issues/Microbiome_Immunity).

For citation purposes, cite each article independently as indicated on the article page online and as indicated below:

LastName, A.A.; LastName, B.B.; LastName, C.C. Article Title. *Journal Name* **Year**, *Volume Number*, Page Range.

ISBN 978-3-0365-6868-3 (Hbk)
ISBN 978-3-0365-6869-0 (PDF)

© 2023 by the authors. Articles in this book are Open Access and distributed under the Creative Commons Attribution (CC BY) license, which allows users to download, copy and build upon published articles, as long as the author and publisher are properly credited, which ensures maximum dissemination and a wider impact of our publications.
The book as a whole is distributed by MDPI under the terms and conditions of the Creative Commons license CC BY-NC-ND.

Contents

About the Editors

Hongyu Zhang

Hongyu Zhang is a Professor in Huazhong Agricultural University, studies insect microbiome and molecular biology, as well as the green pest control technologies for horticultural and urban pests. He have achieved remarkable and innovative research results and published more than 110 papers in Nature Communications, Cell Reports, The ISME Journal, PLoS Pathogens, and PLoS Genetics etc.. He was awarded the 1st prize one time, 2nd twice times and 3rd prizes three times at the Hubei Science and Technology Progress Awards, as well as a 3rd prize at the Hubei Natural Science Awards. he has 16 Patents. he was editor-in-chief of Plant Pest Quarantine (the 1st–3rd edits) (Science Press), the Atlas of Agricultural Insect Pests Identification and Controls (Guangdong Science and Technology Press), Photographic Guide to Key Control Techniques for Citrus Disease and Insect Pests (Chinese Agricultural Press), Urban Entomology (Chinese Agricultural Press), Storage Technology of Grain and Seeds (Jindun Press), and The Tobacco Whitefly and its Control (Jindun Press). He has successively served as a scientific advisor of the International Foundation for science; a Council Member of Citrus Society of China, the Chinese Cereals and Oils Association respectively; a vice chairman of the professional committee of insect microbiome, a professional committee member of exotic species and quarantine at the Entomological Society of China; a professional committee member of horticultural pest control and a committee member of the biological invasion branch at China Plant Protection Society; and a vice chairman of the Hubei Plant Protection Society.

Xiaoxue Li

Dr. Xiaoxue Li, a co-editor of our Special Issue on Insects, is a professor of Plant Science and Technology at Huazhong Agricultural University. He received a PhD in Agricultural Entomology and Pest Control. He underwent Postdoc training at Bruno Lemaitre's lab in EPFL, Switzerland. Dr. Li has studied insect immunity and stress response using *Bactrocera dorsalis* and *Drosophila melanogaster*. His research primarily focuses on how insects cope with various biological and non-biological environmental challenges at a whole-organism level. He found that Drosophila purges its hemopymphatic lipids to prevent the lipid peroxidation caused by stress-posed ROS bursts, revealing a whole organismal way to prevent the tissue damage caused by ROS bursts. He is interested in how insects coordinate different tissues to accomplish stress and immune response and how these mechanisms contribute to the high adaptation of insects, especially economically important pests. He is currently working on how *B. dorsalis* successfully copes with different host plant and environmental challenges, such as temperature changes and invading microbes and microbiota functions. He is particularly interested in the role of the Malpighian tubule in maintaining whole-organism homeostasis by removing unwanted waste from hemolymphs and maintaining water and ion balance. He is also interested in the role of interorgan communication and its role in insect suitability.

Yin Wang

Dr. Yin Wang was a co-editor for our Special Issue of the Insects journal; he is currently working in the Entomology department at the University of Georgia after receiving his PhD at Leiden University in Netherlands. Since 2009, Dr. Y Wang has been studying insect–microbiota symbiosis using a variety of insect models, including termites; carrion beetles Nicrophoroines; and vector arthropods, such as kissing bugs Triatomine and mosquito species, among others. His research primarily focuses on insect–microbiome interactions from the standpoints of ecology and evolution. He has spent decades studying the effects of insect–microbiota interactions on host growth, immune response, and other physiological processes, such as social behavior, development and reproduction, diet, and host exploitation. His PhD dissertation, the Effects of Burying Beetle Nicrophorus vespilloides Social Behaviors on Microbial interactions, first shed light on the carrion insect species that exhibits a novel parental behavior, and then demonstrated how parental care evolved in this insect species, shaping host–bacterial communities and ultimately contributing to host fitness and resistance to pathogens. After that, he started a new project at UGA on vector arthropod systems. Using a Tn-seq method, he looked into how Triatomine species kissing bugs obtain obligate symbiont species and keep their gut community stable by sharing key genes and interacting metabolically (unpublished data). He also concentrated his efforts on blood-feeding mosquitoes, which transmit human diseases, such as malaria, yellow fever, and dengue. His recent research demonstrated that riboflavin deficiency is the key factor underlying the requirement of gut microbiota for mosquito development, and he stated that he will continue to investigate the mechanisms underlying the association between insects and the microbiome involved in host physiology.

Preface to "Insect Microbiome and Immunity"

Microbiome, a Swiss Knife to Insects' Fitness

Xiaoxue Li [1], Hongyu Zhang [1], Yin Wang [2]

1. National Key Laboratory for Germplasm Innovation & Utilization of Horticultural Crops, Hubei Hongshan Laboratory, China-Australia Joint Research Centre for Horticultural and Urban Pests, Institute of Urban and Horticultural Entomology, College of Plant Science and Technology, Huazhong Agricultural University, Wuhan 430070, China
2. College of Agricultural & Environmental Sciences, University of Georgia, 120 Cedar St., 413 Bio Science Bldg., Athens, GA 30602, USA

Insects are the most diverse group of organisms in nature, living in a wide variety of habitats. Insects, like all other organisms, encounter a diverse range of microbes. Recent studies proved that microbes could be both dangerous and beneficial to the hosts. Insects are good models for studying the nature of host–microbial ecology and exploring the mechanisms underlying host–microbiota coevolution. Furthermore, microbes harbored by insects could contribute to better crop management strategies and disease control. The innate immune system is the key controller of microbiome homeostasis. On one hand, microbiome homeostasis is regulated by the immune pathways and effectors. On the other hand, the microbiome also maintains host homeostasis and affects host physiology and overall fitness, including host immunity. Recent advances in technologies, such as high-throughput sequencing, single-microbe genomics, and microbiome/metagenome-wide association studies (MWAS) have greatly promoted our understanding of insect microbiome diversity and their potential functions in insect biology, elucidating host–microbe interactions from both the single-strain and whole-microbiome levels.

The diversity of insect microbiomes

Insects harbor a vast diversity of microorganisms. The cuticle, gut, and certain cell types are favorable for microorganism colonization [1]. Host–microbe interactions can be neutral, harmful, or have beneficial effects on the host, and can be designated as either commensalism, parasitism, or mutualism [2]. Many insects have been found to host a stable and specific microbiome, including the important model organism fruit fly *Drosophila melanogaster* [3,4], bee *Apis mellifera* [5], tick *Ixodes scapularis* [6], mosquito *Anopheles stephensi* [7], and silkworm *Bombyx mori* [8]. There is also increasing evidence proving the existence of a stable microbial community in tephratiedae fruit flies, such as *Bactrocera dorsalis* [9], *Bactrocera minax* [10,11], and *Ceratitis capitata* [12], as well as burying beetles, such as *Nicrophorus vespilloides* [13], and the scarab beetle *Holotrichia parallela* [14]. However, the microbiome is essentially extrinsic in some insects. For example, the gut microbial composition of crickets, cockroaches, and wasps is transitory, relying on the external environment and host diet [15,16].

The microbiota composition also varies extensively within insects. Lab-raised *D. melanogaster* strains are associated with bacteria belonging to *Acetobacter* and *Lactobacillus* [17]. The associations are also thought to be transient and strongly impacted by food condition. On contrary, wild-caught *D. melanogaster* establishes a stable and specific mutualistic interaction with *Acetobacter thailandicus* [18]. Unlike *Drosophila*, the endosymbiont *Wolbachia* is the dominant species in *Culex quinquefasciatus* and the Asian tiger mosquito *Aedes albopictus*. Apart from *Wolbachia*, some reports show that Proteobacteria are most prevalent in the microbiome in mosquitos, while others found that

Bacteroidetes or Acinetobacter phyla are the major components of the microbiome. This suggests that the microbiomes of mosquitoes can vary substantially between individuals, life stages, and geographically different strains [19,20]. In the Tephratidae fruit fly *B. dorsalis, Enterobacteriaceae* is the predominant species in its gut microbiome [9]. Exploring the diversity of insect microbiomes is the first step toward understanding how mutualism is established, maintained, and adapted under the intricate interactions between insect hosts and microbiota.

The biological function of the insect-associated microbiome

Scientists have spent centuries attempting to comprehend the significance of the insect microbiome to the biological functioning of the host. Recent studies have revealed the insect microbiome roles on host physiology and ecology. For example, the most well known function of the microbiome is to promote host growth and development. Germ-free Drosophila larvae grow slower and smaller on a poor diet. Monoassciation with either one of two commensal bacterium, namely Acetobacter pomorum or Lactobacillus plantarum, promote larvae growth through the TOR and insulin pathways [21–23]. Similar phenotypes were also observed in other insects, such as *Bactrocera dorsalis* [24], *A. mellifera* [25], and *Ceratitis capitate* [26]. In the beetle Holotrichia parallela, microbiota participate in host metabolism and provide nutrients by degrading cellulose and hemicellulose in host food [27–30]. Moreover, microbiome confers host tolerance or resistance to xenobiotic or other environmental stressors [31,32]. In *B. dorsalis,* a gut bacterial strain Klebsiella oxytoca (BD177) microbiota restores the ecological fitness decline caused by irradiation through improving food intake [33]. One interesting aspect of the microbiome is that they can regulate host behavior, such as mating, feeding, or parental care [34–36]. A recent study showed that the microbiome–gut–brain axis regulates host feeding behavior in response to the amino acid deficit in *Drosophila* [35]. Furthermore, the insect microbiome provides protection against pathogen infections via direct antimicrobial antagonisms or indirectly priming host immune response, including affecting the efficiency of disease transmission in Anopheles gambiae [37] or enhancing immune system formation and maturity in the Tsetse fly [38], protecting the host from pathogens in Anopheles stephensi [39]. Last but not the least, the insect microbiome can also assist intra- and inter-host species communication [10,40,41].

Insect immunity and microbiome homeostasis regulation

Although microbiota can be beneficial to the host under certain circumstances, they can also be harmful without proper control. Insects live in such complex environments with high evolvability and plasticity that insect microbiota can be affected by a variety of environmental and host factors, such as developmental stage, oxygen levels, and pH range, etc. [42–44]. Insect microbiota abundance and homeostasis, on the other hand, are mainly regulated by the immune system [45]. So far, most studies on the homeostasis regulation of insect gut microbiota have shown that physical defenses, reactive oxygen species (ROS), Imd signaling pathways, the Jak-STAT pathway, and intestinal symbiotic flora all play a role in the insect gut microbiome's homeostatic maintenance [42]. The major effector that regulates microbiota homeostasis is reactive oxygen species (ROS). ROS produced by dual oxidase (Duox) served as the primary bactericidal agent [46–49]. Moreover, scientists also revealed that the Imd pathway plays an important role in maintaining microbiota homeostasis [17,50–52]. Furthermore, from an evolutionary standpoint, insect microbiome homeostasis should rely on innate immunity, particularly because many insects are likely to have lost their adaptive immune systems due to their short life spans and simple body structures [53].

Insect microbiome applications

Insect microbiota have become a novel target for green pest control and human disease control. Although researchers face substantial challenges in manipulating the insect microbiome to improve the beneficial host effect and mitigate undesirable effects, targeting insect–microbe symbioses has a wide range of applications, including controlling insect pest populations and limiting vector-borne diseases through symbiosis disruption and incompatible insect technique (IIT), increasing the survival rate and the enhancing growth and development of beneficial insects as probiotics, and utilizing the symbionts for industrially important processes, such as symbiont-produced antimicrobial activity, a source of digestive enzymes that plays a possible role in bioremediation and detoxification [54,55]. Green pest control strategies targeting insect microbiota have been applied to control pests, such as *Rhodnius prolixus* [56], *Frankliniella occidentalis* [57], *Aedes albopictus* [58], and so on. Microbiota-based control strategies have been successfully adopted in mosquito control. Engineered nonpathogenic bacteria, such as AS1, can inhibit malaria parasite development in the midgut [59]. In addition to the biocontrol and other applications described above, microbial insect symbionts have been used for genetic engineering, such as expressing bio-pesticides and other functional genes [55]. Gene editing techniques have also been applied to insect symbionts, such as the bacteria Wolbachia [60]. Many potential applications of beneficial insect–microbe symbiosis have yet to be realized, despite all the genetic engineering advances.

This book examines insect models, including some important crop pest species, and their association with their intestinal microbial communities. The importance of studies is discussed in order to gain a better understanding of insect microbiomes and their composition, function, as well as host immunity and resistance. Within the scope of this book, we have seen numerous examples of functional investigations on insect microbiome association in terms of potential developmental support and immunity response in hosts, among other factors, which are critical in these fields.

Peer-reviewed research on insect–microbiome connections in agricultural pest species, including the fall armyworm, hairy fungal beetle, grasshoppers, Asian citrus psyllid, and Asian corn borer, are included in our book. Li et al. established that the fall armyworm's microbiome may be transformed by life features, such as developmental phases, diet change, and host habitat, and they also proposed that a diet-related microbiome change can result in the varied activity of host metabolic genes. It is easy to see how insect life traits can influence the structure of insect gut microbiota. Julius demonstrated that microbial communities differ significantly between lab-reared beetle and field-collected hairy fungal beetle *Typhaea stercorea*. This is also true for many insect species, which lack obligate symbionts and acquire gut communities horizontally from their diet and living environment, such as mosquitoes.

Another finding in Jing's study revealed that the dominant genus of four species of grasshoppers is significantly related to cellulose or hemicellulose digestibility; however, there is no significant difference in the gut microbiome of the four targeted grasshoppers, implying a food-digestion-derived microbiome. Other herbivore pests, such as Asian citrus psyllid, shifted the microbial diversity of host gut communities by feeding on various types of citrus plants. Based on our findings, we have learned that pest species microbiomes can be influenced by their hosts'developmental characteristics, eating behavior, and diet, and host gut microbiota have a significant effect on phenotypic change, implying a potential adaptation of host insects to the varied environmental factors contributed by their microbiomes.

Again, similar conclusions are supported in our book by not only focusing on pest insects but also by researching other insect species. Zheng's research determined that insect eating habits can result in a distinct microbial function in insect nutritional ecology, shaping the structure of insect gut microbiota in terms of diversity and functional specificity, implying a link between gut microbiota nutritional functions and insect eating habits. Zhang therefore concluded that parasitization by the wasp *Cotesia vestalis* modifies the gut microbiota of the host *Plutella xylostella*, resulting in a functional shift of gut microbiota in host nutrition metabolism and immunological control toward an environment conducive for parasitoid wasp development. Despite the fact that our book focuses on a small number of insect species, it also supports the notion that gut microbiota play an important role in insect nutritional ecology and that environmental factors, such as parasitization by a wasp, which may significantly alter the microbial community structure; therefore, further research into the function of insect microbiomes in host metabolism pathways and immunity response is warranted. For example, in our book, Wang et al. investigated the Asian corn borer microbiome and discovered that both Gram-positive and -negative bacteria can prime host immune response via peptidoglycan recognition systems.

Furthermore, the effect of fungal species in insect-associated microbiomes has recently gained attention. Our book published two articles written by Wu and Wang that report that fungal species can affect insect microbial ecology by either interfering with the insect beneficial symbionts community or preventing the insect host from pathogen infection. More research on insect fungal communities, we believe, will lead to more more contributions to the fields of insect immune systems and ecosystem functions, ultimately benefitting our understanding of the relationship between invertebrates and microbial communities.

Finally, our book also includes two papers regarding host immune response. In one paper, Wang and colleagues characterized the function of important peptidoglycan recognition proteins PGRP-A and PGRP-B in Ostrinia furnacalis. In the second paper, Gu and colleagues demonstrated that an IMD pathway negative regulator BdNub is essential for maintaining gut immune response, thus maintaining microbiota homeostasis in oriental fruit flies (Bactrocera dorsalis).

Thus, our book, which detailed research on insects and their associated microbiomes, will help researchers better appreciate the potential role of microbiomes in future investigations into insect nutrition and immunology. Last but not least, we thank all the contributions from our authors, editorial board, and all the researchers and reviewers who helped to review and improve the impact of our journal. We sincerely hope that our readers enjoy this issue and find it helpful in their research.

References

1. Douglas, A.E. Multiorganismal Insects: Diversity and Function of Resident Microorganisms. *Annu. Rev. Entomol.* **2015**, *60*, 17–34, doi:10.1146/annurev-ento-010814-020822.
2. Drew, G.C.; Stevens, E.J.; King, K.C. Microbial Evolution and Transitions along the Parasite–Mutualist Continuum. *Nat Rev Microbiol* **2021**, *19*, 623–638, doi:10.1038/s41579-021-00550-7.
3. Iatsenko, I.; Boquete, J.-P.; Lemaitre, B. Microbiota-Derived Lactate Activates Production of Reactive Oxygen Species by the Intestinal NADPH Oxidase Nox and Shortens Drosophila Lifespan. *Immunity* **2018**, *49*, 929-942.e5, doi:10.1016/j.immuni.2018.09.017.
4. Li, H.; Qi, Y.; Jasper, H. Preventing Age-Related Decline of Gut Compartmentalization Limits Microbiota Dysbiosis and Extends Lifespan. *Cell Host & Microbe* **2016**, *19*, 240–253, doi:10.1016/j.chom.2016.01.008.
5. Powell, J.E.; Martinson, V.G.; Urban-Mead, K.; Moran, N.A. Routes of Acquisition of the Gut Microbiota of the Honey Bee Apis Mellifera. *Appl Environ Microbiol* **2014**, *80*, 7378–7387, doi:10.1128/AEM.01861-14.

6. Narasimhan, S.; Rajeevan, N.; Liu, L.; Zhao, Y.O.; Heisig, J.; Pan, J.; Eppler-Epstein, R.; DePonte, K.; Fish, D.; Fikrig, E. Gut Microbiota of the Tick Vector Ixodes Scapularis Modulate Colonization of the Lyme Disease Spirochete. *Cell Host & Microbe* **2014**, *15*, 58–71, doi:10.1016/j.chom.2013.12.001.
7. Pike, A.; Dong, Y.; Dizaji, N.B.; Gacita, A.; Mongodin, E.F.; Dimopoulos, G. Changes in the Microbiota Cause Genetically Modified Anopheles to Spread in a Population. *Science* **2017**, *357*, 1396–1399, doi:10.1126/science.aak9691.
8. Chen, B.; Du, K.; Sun, C.; Vimalanathan, A.; Liang, X.; Li, Y.; Wang, B.; Lu, X.; Li, L.; Shao, Y. Gut Bacterial and Fungal Communities of the Domesticated Silkworm (Bombyx Mori) and Wild Mulberry-Feeding Relatives. *ISME J* **2018**, *12*, 2252–2262, doi:10.1038/s41396-018-0174-1.
9. Wang, H.; Jin, L.; Zhang, H. Comparison of the Diversity of the Bacterial Communities in the Intestinal Tract of Adult Bactrocera Dorsalis from Three Different Populations. *Journal of applied microbiology* **2011**, *110*, 1390–1401, doi:10.1111/j.1365-2672.2011.05001.x.
10. Wang, A.; Yao, Z.; Zheng, W.; Zhang, H. Bacterial Communities in the Gut and Reproductive Organs of Bactrocera Minax (Diptera: Tephritidae) Based on 454 Pyrosequencing. *PLoS ONE* **2014**, *9*, e106988, doi:10.1371/journal.pone.0106988.
11. Yao, Z.; Ma, Q.; Cai, Z.; Raza, M.F.; Bai, S.; Wang, Y.; Zhang, P.; Ma, H.; Zhang, H. Similar Shift Patterns in Gut Bacterial and Fungal Communities Across the Life Stages of Bactrocera Minax Larvae From Two Field Populations. *Frontiers in Microbiology* **2019**, *10*.
12. Behar, A.; Yuval, B.; Jurkevitch, E. Gut Bacterial Communities in the Mediterranean Fruit Fly (Ceratitis Capitata) and Their Impact on Host Longevity. *Journal of Insect Physiology* **2008**, *54*, 1377–1383, doi:10.1016/j.jinsphys.2008.07.011.
13. Zhang, H.; Jackson, T.A. Autochthonous Bacterial Flora Indicated by PCR-DGGE of 16S RRNA Gene Fragments from the Alimentary Tract of *Costelytra Zealandica* (Coleoptera: Scarabaeidae). *Journal of Applied Microbiology* **2008**, *105*, 1277–1285, doi:10.1111/j.1365-2672.2008.03867.x.
14. Huang, S.; Zhang, H. The Impact of Environmental Heterogeneity and Life Stage on the Hindgut Microbiota of Holotrichia Parallela Larvae (Coleoptera: Scarabaeidae). *PLoS ONE* **2013**, *8*, e57169, doi:10.1371/journal.pone.0057169.
15. Reeson, A.F.; Jankovic, T.; Kasper, M.L.; Rogers, S.; Austin, A.D. Application of 16S RDNA-DGGE to Examine the Microbial Ecology Associated with a Social Wasp Vespula Germanica. *Insect Mol Biol* **2003**, *12*, 85–91, doi:10.1046/j.1365-2583.2003.00390.x.
16. Kane, M.D.; Breznak, J.A. Effect of Host Diet on Production of Organic Acids and Methane by Cockroach Gut Bacteria. *Appl Environ Microbiol* **1991**, *57*, 2628–2634, doi:10.1128/aem.57.9.2628-2634.1991.
17. Ryu, J.-H.; Kim, S.-H.; Lee, H.-Y.; Bai, J.Y.; Nam, Y.-D.; Bae, J.-W.; Lee, D.G.; Shin, S.C.; Ha, E.-M.; Lee, W.-J. Innate Immune Homeostasis by the Homeobox Gene Caudal and Commensal-Gut Mutualism in Drosophila. *Science* **2008**, *319*, 777–782, doi:10.1126/science.1149357.
18. Pais, I.S.; Valente, R.S.; Sporniak, M.; Teixeira, L. Drosophila Melanogaster Establishes a Species-Specific Mutualistic Interaction with Stable Gut-Colonizing Bacteria. *PLoS Biol* **2018**, *16*, e2005710, doi:10.1371/journal.pbio.2005710.
19. Cansado-Utrilla, C.; Zhao, S.Y.; McCall, P.J.; Coon, K.L.; Hughes, G.L. The Microbiome and Mosquito Vectorial Capacity: Rich Potential for Discovery and Translation. *Microbiome* **2021**, *9*, 111, doi:10.1186/s40168-021-01073-2.
20. Hegde, S.; Khanipov, K.; Albayrak, L.; Golovko, G.; Pimenova, M.; Saldaña, M.A.; Rojas, M.M.; Hornett, E.A.; Motl, G.C.; Fredregill, C.L.; et al. Microbiome Interaction Networks and Community Structure From Laboratory-Reared and Field-Collected Aedes Aegypti, Aedes Albopictus, and Culex Quinquefasciatus Mosquito Vectors. *Front. Microbiol.* **2018**, *9*, 2160, doi:10.3389/fmicb.2018.02160.

21. Shin, S.C.; Kim, S.-H.; You, H.; Kim, B.; Kim, A.C.; Lee, K.-A.; Yoon, J.-H.; Ryu, J.-H.; Lee, W.-J. Drosophila Microbiome Modulates Host Developmental and Metabolic Homeostasis via Insulin Signaling. *Science* **2011**, *334*, 670–674, doi:10.1126/science.1212782.
22. Storelli, G.; Defaye, A.; Erkosar, B.; Hols, P.; Royet, J.; Leulier, F. Lactobacillus Plantarum Promotes Drosophila Systemic Growth by Modulating Hormonal Signals through TOR-Dependent Nutrient Sensing. *Cell Metabolism* **2011**, *14*, 403–414, doi:10.1016/j.cmet.2011.07.012.
23. Storelli, G.; Strigini, M.; Grenier, T.; Bozonnet, L.; Schwarzer, M.; Daniel, C.; Matos, R.; Leulier, F. Drosophila Perpetuates Nutritional Mutualism by Promoting the Fitness of Its Intestinal Symbiont Lactobacillus Plantarum. *Cell Metabolism* **2018**, *27*, 362-377.e8, doi:10.1016/j.cmet.2017.11.011.
24. Guo, Q.; Yao, Z.; Cai, Z.; Bai, S.; Zhang, H. Gut Fungal Community and Its Probiotic Effect on *Bactrocera Dorsalis*. *Insect Science* **2022**, *29*, 1145–1158, doi:10.1111/1744-7917.12986.
25. Zheng, H.; Powell, J.E.; Steele, M.I.; Dietrich, C.; Moran, N.A. Honeybee Gut Microbiota Promotes Host Weight Gain via Bacterial Metabolism and Hormonal Signaling. *Proc. Natl. Acad. Sci. U.S.A.* **2017**, *114*, 4775–4780, doi:10.1073/pnas.1701819114.
26. Behar, A.; Yuval, B.; Jurkevitch, E. Enterobacteria-Mediated Nitrogen Fixation in Natural Populations of the Fruit Fly Ceratitis Capitata. *Molecular Ecology* **2005**, *14*, 2637–2643, doi:10.1111/j.1365-294X.2005.02615.x.
27. Huang, S.-W.; Zhang, H.-Y.; Marshall, S.; Jackson, T.A. The Scarab Gut: A Potential Bioreactor for Bio-Fuel Production. *Insect Science* **2010**, *17*, 175–183, doi:10.1111/j.1744-7917.2010.01320.x.
28. Huang, S.; Sheng, P.; Zhang, H. Isolation and Identification of Cellulolytic Bacteria from the Gut of Holotrichia Parallela Larvae (Coleoptera: Scarabaeidae). *International Journal of Molecular Sciences* **2012**, *13*, 2563–2577, doi:10.3390/ijms13032563.
29. Sheng, P.; Huang, S.; Wang, Q.; Wang, A.; Zhang, H. Isolation, Screening, and Optimization of the Fermentation Conditions of Highly Cellulolytic Bacteria from the Hindgut of Holotrichia Parallela Larvae (Coleoptera: Scarabaeidae). *Appl Biochem Biotechnol* **2012**, *167*, 270–284, doi:10.1007/s12010-012-9670-3.
30. Sheng, P.; Xu, J.; Saccone, G.; Li, K.; Zhang, H. Discovery and Characterization of Endo-Xylanase and β-Xylosidase from a Highly Xylanolytic Bacterium in the Hindgut of Holotrichia Parallela Larvae. *Journal of Molecular Catalysis B: Enzymatic* **2014**, *105*, 33–40, doi:10.1016/j.molcatb.2014.03.019.
31. Raza, M.F.; Wang, Y.; Cai, Z.; Bai, S.; Yao, Z.; Awan, U.A.; Zhang, Z.; Zheng, W.; Zhang, H. Gut Microbiota Promotes Host Resistance to Low-Temperature Stress by Stimulating Its Arginine and Proline Metabolism Pathway in Adult Bactrocera Dorsalis. *PLOS Pathogens* **2020**, *16*, e1008441, doi:10.1371/journal.ppat.1008441.
32. Ceja-Navarro, J.A.; Vega, F.E.; Karaoz, U.; Hao, Z.; Jenkins, S.; Lim, H.C.; Kosina, P.; Infante, F.; Northen, T.R.; Brodie, E.L. Gut Microbiota Mediate Caffeine Detoxification in the Primary Insect Pest of Coffee. *Nat Commun* **2015**, *6*, 7618, doi:10.1038/ncomms8618.
33. Cai, Z.; Yao, Z.; Li, Y.; Xi, Z.; Bourtzis, K.; Zhao, Z.; Bai, S.; Zhang, H. Intestinal Probiotics Restore the Ecological Fitness Decline of Bactrocera Dorsalis by Irradiation. *Evol Appl* **2018**, *11*, 1946–1963, doi:10.1111/eva.12698.
34. Sharon, G.; Segal, D.; Ringo, J.M.; Hefetz, A.; Zilber-Rosenberg, I.; Rosenberg, E. Commensal Bacteria Play a Role in Mating Preference of *Drosophila Melanogaster*. *Proc. Natl. Acad. Sci. U.S.A.* **2010**, *107*, 20051–20056, doi:10.1073/pnas.1009906107.
35. Kim, B.; Kanai, M.I.; Oh, Y.; Kyung, M.; Kim, E.-K.; Jang, I.-H.; Lee, J.-H.; Kim, S.-G.; Suh, G.S.B.; Lee, W.-J. Response of the Microbiome–Gut–Brain Axis in Drosophila to Amino Acid Deficit. *Nature* **2021**, *593*, 570–574, doi:10.1038/s41586-021-03522-2.

36. Wang, Y.; Rozen, D.E. Gut Microbiota in the Burying Beetle, *Nicrophorus Vespilloides*, Provide Colonization Resistance against Larval Bacterial Pathogens. *Ecol Evol* **2018**, *8*, 1646–1654, doi:10.1002/ece3.3589.
37. Cirimotich, C.M.; Dong, Y.; Clayton, A.M.; Sandiford, S.L.; Souza-Neto, J.A.; Mulenga, M.; Dimopoulos, G. Natural Microbe-Mediated Refractoriness to Plasmodium Infection in Anopheles Gambiae. *Science* **2011**, *332*, 855–858, doi:10.1126/science.1201618.
38. Weiss, B.L.; Wang, J.; Aksoy, S. Tsetse Immune System Maturation Requires the Presence of Obligate Symbionts in Larvae. *PLoS Biology* **2011**, *9*, e1000619, doi:10.1371/journal.pbio.1000619.
39. Kalappa, D.M.; Subramani, P.A.; Basavanna, S.K.; Ghosh, S.K.; Sundaramurthy, V.; Uragayala, S.; Tiwari, S.; Anvikar, A.R.; Valecha, N. Influence of Midgut Microbiota in Anopheles Stephensi on Plasmodium Berghei Infections. *Malar J* **2018**, *17*, 385, doi:10.1186/s12936-018-2535-7.
40. Davis, T.S.; Crippen, T.L.; Hofstetter, R.W.; Tomberlin, J.K. Microbial Volatile Emissions as Insect Semiochemicals. *J Chem Ecol* **2013**, *39*, 840–859, doi:10.1007/s10886-013-0306-z.
41. Zhao, L.; Mota, M.; Vieira, P.; Butcher, R.A.; Sun, J. Interspecific Communication between Pinewood Nematode, Its Insect Vector, and Associated Microbes. *Trends in Parasitology* **2014**, *30*, 299–308, doi:10.1016/j.pt.2014.04.007.
42. Bai, S.; Yao, Z.; Raza, M.F.; Cai, Z.; Zhang, H. Regulatory Mechanisms of Microbial Homeostasis in Insect Gut. *Insect Science* **2021**, *28*, 286–301, doi:10.1111/1744-7917.12868.
43. Clark, R.I.; Salazar, A.; Yamada, R.; Fitz-Gibbon, S.; Morselli, M.; Alcaraz, J.; Rana, A.; Rera, M.; Pellegrini, M.; Ja, W.W.; et al. Distinct Shifts in Microbiota Composition during Drosophila Aging Impair Intestinal Function and Drive Mortality. *Cell Reports* **2015**, *12*, 1656–1667, doi:10.1016/j.celrep.2015.08.004.
44. Engel, P.; Moran, N.A. The Gut Microbiota of Insects – Diversity in Structure and Function. *FEMS Microbiol Rev* **2013**, *37*, 699–735, doi:10.1111/1574-6976.12025.
45. Buchon, N.; Silverman, N.; Cherry, S. Immunity in Drosophila Melanogaster from Microbial Recognition to Whole-Organism Physiology. *Nature Reviews Immunology* **2014**, doi:10.1038/nri3763.
46. Ha, E.-M.; Oh, C.-T.; Bae, Y.S.; Lee, W.-J. A Direct Role for Dual Oxidase in Drosophila Gut Immunity. *Science* **2005**, *310*, 847–850, doi:10.1126/science.1117311.
47. Lee, K.-A.; Kim, S.-H.; Kim, E.-K.; Ha, E.-M.; You, H.; Kim, B.; Kim, M.-J.; Kwon, Y.; Ryu, J.-H.; Lee, W.-J. Bacterial-Derived Uracil as a Modulator of Mucosal Immunity and Gut-Microbe Homeostasis in Drosophila. *Cell* **2013**, *153*, 797–811, doi:10.1016/j.cell.2013.04.009.
48. Xiao, X.; Yang, L.; Pang, X.; Zhang, R.; Zhu, Y.; Wang, P.; Gao, G.; Cheng, G. A Mesh-Duox Pathway Regulates Homeostasis in the Insect Gut. *Nature Microbiology* **2017**, *2*, 17020, doi:10.1038/nmicrobiol.2017.20.
49. Yao, Z.; Wang, A.; Li, Y.; Cai, Z.; Lemaitre, B.; Zhang, H. The Dual Oxidase Gene BdDuox Regulates the Intestinal Bacterial Community Homeostasis of Bactrocera Dorsalis. *The ISME Journal* **2016**, *10*, 1037–1050, doi:10.1038/ismej.2015.202.
50. Broderick, N.A.; Buchon, N.; Lemaitre, B. Microbiota-Induced Changes in Drosophila Melanogaster Host Gene Expression and Gut Morphology. *mBio* **2014**, *5*, e01117-14-e01117-14, doi:10.1128/mBio.01117-14.
51. Guo, L.; Karpac, J.; Tran, S.L.; Jasper, H. PGRP-SC2 Promotes Gut Immune Homeostasis to Limit Commensal Dysbiosis and Extend Lifespan. *Cell* **2014**, *156*, 109–122, doi:10.1016/j.cell.2013.12.018.
52. Yao, Z.; Cai, Z.; Ma, Q.; Bai, S.; Wang, Y.; Zhang, P.; Guo, Q.; Gu, J.; Lemaitre, B.; Zhang, H. Compartmentalized PGRP Expression along the Dipteran Bactrocera Dorsalis Gut Forms a Zone of Protection for Symbiotic Bacteria. *Cell Reports* **2022**, *41*, 111523, doi:10.1016/j.celrep.2022.111523.

53. Buchon, N.; Broderick, N.A.; Lemaitre, B. Gut Homeostasis in a Microbial World: Insights from Drosophila Melanogaster. *Nature Reviews Microbiology* **2013**, *11*, 615–626, doi:10.1038/nrmicro3074.
54. Berasategui, A.; Shukla, S.; Salem, H.; Kaltenpoth, M. Potential Applications of Insect Symbionts in Biotechnology. *Appl Microbiol Biotechnol* **2016**, *100*, 1567–1577, doi:10.1007/s00253-015-7186-9.
55. Arora, A.K.; Douglas, A.E. Hype or Opportunity? Using Microbial Symbionts in Novel Strategies for Insect Pest Control. *Journal of Insect Physiology* **2017**, *103*, 10–17, doi:10.1016/j.jinsphys.2017.09.011.
56. Taracena, M.L.; Oliveira, P.L.; Almendares, O.; Umaña, C.; Lowenberger, C.; Dotson, E.M.; Paiva-Silva, G.O.; Pennington, P.M. Genetically Modifying the Insect Gut Microbiota to Control Chagas Disease Vectors through Systemic RNAi. *PLoS Negl Trop Dis* **2015**, *9*, e0003358, doi:10.1371/journal.pntd.0003358.
57. Whitten, M.M.A.; Facey, P.D.; Del Sol, R.; Fernández-Martínez, L.T.; Evans, M.C.; Mitchell, J.J.; Bodger, O.G.; Dyson, P.J. Symbiont-Mediated RNA Interference in Insects. *Proc. R. Soc. B.* **2016**, *283*, 20160042, doi:10.1098/rspb.2016.0042.
58. Zheng, X.; Zhang, D.; Li, Y.; Yang, C.; Wu, Y.; Liang, X.; Liang, Y.; Pan, X.; Hu, L.; Sun, Q.; et al. Incompatible and Sterile Insect Techniques Combined Eliminate Mosquitoes. *Nature* **2019**, *572*, 56–61, doi:10.1038/s41586-019-1407-9.
59. Wang, S.; Dos-Santos, A.L.A.; Huang, W.; Liu, K.C.; Oshaghi, M.A.; Wei, G.; Agre, P.; Jacobs-Lorena, M. Driving Mosquito Refractoriness to emphPlasmodium Falciparum with Engineered Symbiotic Bacteria. *Science* **2017**, *357*, 1399–1402, doi:10.1126/science.aan5478.
60. Wang, G.-H.; Gamez, S.; Raban, R.R.; Marshall, J.M.; Alphey, L.; Li, M.; Rasgon, J.L.; Akbari, O.S. Combating Mosquito-Borne Diseases Using Genetic Control Technologies. *Nat Commun* **2021**, *12*, 4388, doi:10.1038/s41467-021-24654-z.

Hongyu Zhang, Xiaoxue Li, and Yin Wang
Editors

insects

MDPI

Review

The Role of Insect Symbiotic Bacteria in Metabolizing Phytochemicals and Agrochemicals

Man Zhao, Xingyu Lin and Xianru Guo *

Henan International Laboratory for Green Pest Control, College of Plant Protection, Henan Agricultural University, Zhengzhou 450002, China; zhaoman821@henau.edu.cn (M.Z.); xingyulin666666@163.com (X.L.)
* Correspondence: guoxianru@126.com; Tel.: +86-0371-63558170

Simple Summary: To counter plant chemical defenses and exposure to agrochemicals, herbivorous insects have developed several adaptive strategies to guard against the ingested detrimental substances, including enhancing detoxifying enzyme activities, avoidance behavior, amino acid mutation of target sites, and lower penetration through a thicker cuticle. Insect microbiota play important roles in many aspects of insect biology and physiology. To better understand the role of insect symbiotic bacteria in metabolizing these detrimental substances, we summarize the research progress on the function of insect bacteria in metabolizing phytochemicals and agrochemicals, and describe their future potential application in pest management and protection of beneficial insects.

Abstract: The diversity and high adaptability of insects are heavily associated with their symbiotic microbes, which include bacteria, fungi, viruses, protozoa, and archaea. These microbes play important roles in many aspects of the biology and physiology of insects, such as helping the host insects with food digestion, nutrition absorption, strengthening immunity and confronting plant defenses. To maintain normal development and population reproduction, herbivorous insects have developed strategies to detoxify the substances to which they may be exposed in the living habitat, such as the detoxifying enzymes carboxylesterase, glutathione-S-transferases (GSTs), and cytochrome P450 monooxygenases (CYP450s). Additionally, insect symbiotic bacteria can act as an important factor to modulate the adaptability of insects to the exposed detrimental substances. This review summarizes the current research progress on the role of insect symbiotic bacteria in metabolizing phytochemicals and agrochemicals (insecticides and herbicides). Given the importance of insect microbiota, more functional symbiotic bacteria that modulate the adaptability of insects to the detrimental substances to which they are exposed should be identified, and the underlying mechanisms should also be further studied, facilitating the development of microbial-resource-based pest control approaches or protective methods for beneficial insects.

Keywords: insect microbiota; plant secondary substance; insecticide resistance; detoxifying enzymes; insect immune system

Citation: Zhao, M.; Lin, X.; Guo, X. The Role of Insect Symbiotic Bacteria in Metabolizing Phytochemicals and Agrochemicals. *Insects* **2022**, *13*, 583. https://doi.org/10.3390/insects13070583

Academic Editors: Hongyu Zhang, Yin Wang and Xiaoxue Li

Received: 25 April 2022
Accepted: 23 June 2022
Published: 26 June 2022

Publisher's Note: MDPI stays neutral with regard to jurisdictional claims in published maps and institutional affiliations.

Copyright: © 2022 by the authors. Licensee MDPI, Basel, Switzerland. This article is an open access article distributed under the terms and conditions of the Creative Commons Attribution (CC BY) license (https://creativecommons.org/licenses/by/4.0/).

1. Introduction

Insects, which are the most abundant and widely distributed species in the animal kingdom, can survive and reproduce under various conditions [1,2]. The diversity and adaptability of insects are closely related to their symbiotic microbes, including bacteria, fungi, viruses, protozoa and archaea [3]. In insects, these microbes inhabit the exoskeleton, gut, blood cavity, salivary gland, and other organs, as well as individual cells, accounting for 1–10% of insect biomass and playing critical roles in many aspects of the biology and physiology of insects [4–7].

During the interaction between microbial symbionts and insects, the insects provide the habitat and nutrition for microbes, and in return, these symbionts help the host insects with food digestion, nutrition absorption, defense responses to pathogens, and xenobiotic metabolism, while also promoting insect development and reproduction [8,9]. For example, the fungal-yeast-like symbiotes in planthoppers and aphids are vital for the synthesis of essential amino acids and for maintaining the vitamin supply in the insect host [10–12]. The polydnavirus from parasitoid wasps can disorder the immune system of host insects to ensure the survival of wasp offspring, and three partiti-like viruses identified from the African armyworm (*Spodoptera exempta*) can enhance the resistance of *S. exempta* to nucleopolyhedrovirus [13,14]. For wood-feeding lower termites, they rely on symbiotic flagellates to decompose the lignocelluloses in their plant diet, and methanogenic archaea to produce methane [15,16]. In terms of symbiotic bacteria, these comprise the most abundant microorganism species in insects and are mainly distributes in the gut, including Proteobacteria, Bacteroidetes, Firmicutes, Clostridia, Actinomycetes, and others, and contribute to the development, behavior, communication, and adaptation of host insects [3,17–22]. In addition, the composition of bacterial communities in insects can be influenced by food resources, environmental factors, pathogenic microbials, or the detrimental substances to which they are exposed [23–25].

Early studies on the symbiotic microbes of insects have mainly relied on traditional isolation and culture methods, but a major limitation of these methods is that many microbes are uncultured, and their functional roles cannot be studied in vivo. In recent years, the rapid development of high-throughput metagenomic sequencing technology and methods for rearing germ-free insects has promoted research on functional characterization of the microorganisms in insects, especially the symbiotic bacteria [26]. For some insect species linked to agriculture (such as pests, pollinators, and parasitic enemies), their development, learning behavior, and resistance evolution are highly relevant to gut bacteria [4,24,27]. For insect vectors transmitting human diseases (such as mosquitoes), some symbiotic bacteria, influencing the vector transmitting efficiency or reproduction of mosquitoes, can be targets for potential public disease control [28,29]. In past decades, extensive studies have been conducted on insect bacterial community diversity and interactions of bacteria with host insects (Figure 1). This review focuses on the research progress of insect symbiotic bacteria in metabolizing phytochemicals and agrochemicals (insecticides and herbicides), which are two main kinds of substances insects encounter in their life histories. Finally, the microbe-based pest control approaches, pest resistance management strategies, and protective methods of natural enemy insects that may apply in the future should be examined.

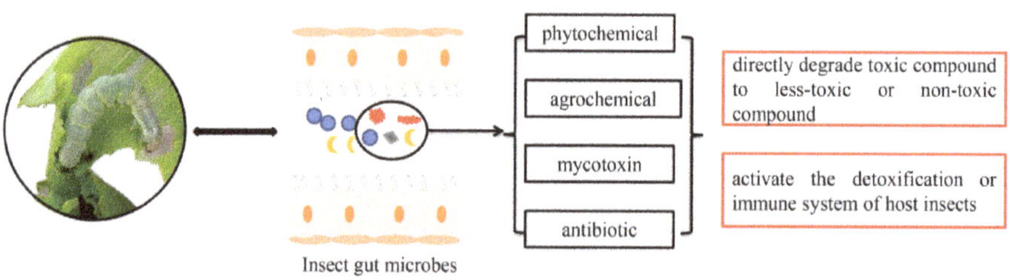

Figure 1. An overview of symbiont-mediated detoxification in insects.

2. Insect Bacteria Confer Resistance to Phytochemicals

In nature, more than half of insects are herbivores, which damage different kinds of crops and even cause economical losses [30,31]. To defend themselves from attack by insect herbivory, plants have evolved various defensive mechanisms, including production of phytochemicals such as alkaloids, terpenoids, phenols, and some other secondary substances that show detrimental effects on the growth and survival of insects or attract the natural enemies of herbivores [32]. To cope, herbivorous insects have developed several strategies to detoxify the ingested phytochemicals, including the concerned biochemical counteradaptations [33].

In addition to biochemical responses, insect symbiotic bacteria play key roles in countering plant defenses [7,34]. Before feeding, the oral secretions of some insect herbivores contain a few effector molecules that suppress the antiherbivore defenses, and some bacteria (belonging to the genera *Stenotrophomonas*, *Pseudomonas*, and *Enterobacter*), identified from oral secretions of *Leptinotarsa decemlineata*, are also responsible for plant defense suppression [35–37]. After ingestion, the consumed plant tissue enters the digestive tract of the insects, and the gut bacterial community is able to help hosts with food digestion, nutrition absorption, and countering the toxic or harmful phytochemicals from the plant diet [9,34]. For generalist insects, to some extent, their polyphagous habits rely on several symbiotic bacteria to adapt to phytochemicals from different host plants [38]. For example, when fed with *Arabidopsis thaliana*, the gut bacteria of *Trichoplusia ni* were dominated by *Shinella*, *Terribacillus* and *Propionibacterium*, which are known to have the ability to degrade the plant allelochemical glucosinolate; when feeding on *Solanum lycopersicum*, the relative abundances of *Agrobacterium* and *Rhizobium* able to degrade alkaloids were significantly increased [39]. However, specialist insects may need specific bacteria to degrade the toxic compounds in their host plants, such as *Enterococcus* sp. from *Hyles euphorbiae* and *Brithys crini*, which have the ability to tolerate alkaloid and latex [40].

Terpenes are a class of toxic phytochemicals that are highly present in coniferous plants. To overcome these toxic compounds, the pests that colonize these plants metabolize the toxic compounds with the aid of symbiotic bacteria. For example, the gut bacteria *Serritia*, *Pseudomonas*, and *Rahnella* from *Dendroctonus ponderosae* have a strong ability to degrade monoterpenes and diterpene acids because these genera contain the majority of the genes that participate in terpene degradation [41,42]. For another mountain pine beetle, *Dendroctonus valens*, its gallery lengths and body weight were significantly suppressed when fed on a diet containing α-pinene at 6 and 12 mg/mL, and three bacterial strains (*Serratia* sp., *Pseudomonas* sp., and *Rahnella aquatilis*) degraded 20–50% of α-pinene [43]. However, the role of these bacteria in degrading terpene has not been verified in the two pine beetles in vivo. A further study on the gut microbiota of the pine weevil (*Hylobius abietis*) found that the weevil can degrade substantial amounts of diterpene in its plant diet, and this degradability was significantly reduced after eliminating gut microbes with antibiotics and then restored again after supplying a normal gut microbial community. When inoculating the gut bacterial community with dehydroabietic acid for five days, the amount of bacteria significantly reduced, and the metagenomic analysis results showed that beetles fed on Norway spruce contained 10 degradation genes (*dit*), which were almost eliminated after treating with antibiotics [44]. In another weevil (*Curculio chinensis*), the gut bacteria from the genus *Acinetobacter* degraded tea saponin and used it as source of carbon and nitrogen [45]. Moreover, some insects, such as *Rhodnius prolixus*, counteract the toxic effects of azadirachtin (a triterpenoid compound of terpenes) by promoting the gene of equivalent NF-kB transcription factor (*RpDorsal*) and antimicrobial peptide (*defC AMP*), as well as the abundance of the gut bacterium *Serratia marcescens* [46].

Alkaloids, a kind of plant phytochemical, are neurotoxic to a wide range of insects, and most of them have been used as botanical agrochemicals for pest control [47]. Although alkaloids exhibit toxicity to most insects, a few species still show high tolerance to these substances, such as *Hypothenemus hampei*, which can consume coffee beans rich in the alkaloid caffeine. Later researchers found that the tolerance of *H. hampei* to caffeine is underpinned by its gut microbiota. After eliminating the gut microorganisms with antibiotics, the fitness of *H. hampei*, fed on a caffeine-treated diet, declined and showed no decrease in caffeine concentration in their frass. Through a culture-dependent approach, a gut bacterium, *Pseudomonas fulva*, was isolated, which processed a gene coding one subunit of a caffeine demethylase, and the reinstatement of *P. fulva* in germ-free *H. hampei* recovered its capacity to degrade caffeine [48]. As another important phytochemical, phenols inhibit herbivorous insects by inducing reactive oxygen production. When feeding on unripe olives, the olive fly *Bactrocera olea* requires the gut bacterium *Erwinia dacicola* to overcome the toxic phenolic glycoside in unripe olives [49]. Metagenomic analysis revealed that the bacterium *Novosphingobium* sp. in *D. valens* possesses putative genes involved in the degradation of naringenin, and the survival rate of *D. valens* on a naringenin-treated diet significantly increased when supplied with *Novosphingobium* sp. [50]. In addition, the gut bacteria *Acinetobacter* sp. in *Lymantria dispar* also use condensed tannins as a carbon source [51]. In the cabbage stem flea beetle *Psylliodes chrysocephala*, when its gut bacteria were removed with antibiotics, the beetles accumulated 11.3-fold higher levels of unmetabolized isothiocyanates compared to control beetles, and the isothiocyanate degradation ability was restored when reintroducing the bacteria *Pantoea* sp. Pc8 in antibiotic-fed beetles [52]. For the phytochemical oxalate, the gut bacterium *Ishikawaella capsulata* in stinkbug *Megacopta punctatissima* encodes genes for oxalate decarboxylase, suggesting the possible role of the bacterium in oxalate detoxification [53]. In human, calcium oxalate is formed if the food-derived oxalate cannot be metabolized, which can result in kidney stone disease, so the identification of insect bacteria able to degrade oxalate may act as a novel treatment for kidney stone patients [54] (Table 1).

Table 1. Symbiont-mediated detoxification of phytochemicals.

Plant Allelochemical		Functional Bacteria and Host	Description	Reference
Terpenoid	Monoterpene	*Serritia marcescens*, *Pseudomonas mandelii*, and *Rahnella aquatilis* from *Dendroctonus ponderosae*	*S. marcescens* reduced 49–79% of 3-carene and (−)-β-pinene, and *P. mandelii* decreased concentrations of all monoterpenes by 15–24%, while *R. aquatilis* decreased (−)-α-pinene (38%) and (+)-α-pinene (46%) by 40% and 45% (by GC-MS), respectively	[41]
		Pseudomonas, *Rahnella*, *Serratia*, and *Burkholderia* in *D. ponderosae*	Genera contained most genes involved in terpene degradation (by metagenomics)	[42]
		Serratia sp., *Pseudomonas* sp., and *Rahnella aquatilis* in *Dendroctonus valens*	Degraded 20–50% of α-pinene (by GC-MS)	[43]
	Diterpene	gut microbiota of *Hylobius abietis*	Gut bacterial community of *H. abietis* reduced most diterpenes, and metagenomic analysis results showed gut community contained 10 degradation genes (*dit*) (by metagenome sequencing and GC-MS)	[44]
	Saponin	*Acinetobacter* sp. in *Curculio chinensis*	*Acinetobacter* sp. in *C. chinensis* enriched after treating with saponin, and when incubating bacteria with saponin for 72 h, saponin content significantly decreased from 4.054 to 1.867 mg/mL (by 16S rRNA metagenome sequencing and HPLC)	[45]
	Azadirachtin	*Serratia marcescens* in *Rhodnius prolixus*	*S. marcescens* load in *R. prolixus* increased when fed diet containing azadirachtin at 1 μg/mL (by qRT-PCR)	[46]

Table 1. *Cont.*

Plant Allelochemical		Functional Bacteria and Host	Description	Reference
Alkaloid	Caffeine	*Pseudomonas fulva* in *Hypothenemus hampei*	*P. fulva* processed gene coding one subunit of caffeine demethylase, and reinstatement of *P. fulva* in germ-free *H. hampei* degraded all caffeine consumed (by 16S rRNA gene sequencing and GC-MS)	[48]
	Aconitine, nicotine	entire gut bacteria of *Dendrolimus superans* and *Lymantria dispar*	Abundance of genus *Pseudomonas* in *D. superans* larvae increased, but *Serratia* and *Enterobacter* decreased, and *L. dispar* larvae fed on aconitine-treated diet and nicotine-treated diet shared dominant bacteria *Enterococcus* (by 16S rRNA gene sequencing)	[55]
Phenol	Phenolic glycoside	*Erwinia dacicola* in *Bactrocera olea*	Larvae developed in unripe olive harbored more *E. dacicola* (by 16S rRNA gene sequencing)	[49]
	Phenolic naringenin	*Novosphingobium* sp. in *D. valens*	*Novosphingobium* sp. possesses putative genes involved in degradation of naringenin, and *D. valens* supplied with *Novosphingobium* sp. acquired protection against naringenin (by metagenomic analysis)	[50]
	Tannins	*Acinetobacter* sp. in *Lymantria dispar*	Condensed tannins improved growth of *Acinetobacter* sp. by 15% (by measuring the optical density)	[51]
Glucosinolate		*Pantoea* sp. Pc8 in *Psylliodes chrysocephala*	Laboratory-reared and field-collected *P. chrysocephala* all contained three core genera *Pantoea*, *Acinetobacter* and *Pseudomonas*, and reintroduction of *Pantoea* sp. Pc8 in antibiotic-fed beetles restored isothiocyanate degradation ability in vivo (by 16S rRNA gene sequencing and LC-MS)	[52]
Oxalate		*Ishikawaella capsulata* in *Megacopta punctatissima*	Encodes genes of oxalate decarboxylase (by whole-genome shotgun sequencing)	[53]

Apart from detoxification roles, some insect bacteria can convert phytochemicals into pheromone compounds and thus influence the chemical communication of host insects [56]. For instance, the gut bacteria *Pantoea agglomerans*, *Klebsiella pneumonia*, and *Enterobacter cloacae* of *Schistocerca gregaria* can use the plant-derived vanillic acid to produce guaiacol and phenol, which are two main components of the locust cohesion pheromone [20]. In the mine beetle *Chrysolina herbacea*, its gut bacteria has the ability to metabolize terpenoids into pheromone compounds [57]. In addition to phytochemicals, the *Bacillus* species isolated from the male rectum of *Bactrocera dorsalis* can directly produce sex pheromone components (2,3,5-trimethylpyrazine and 2,3,5,6-tetramethylpyrazine) by using glucose and threonine as the substrates. After treating male flies with antibiotics, the levels of the two components were significantly reduced [58]. These findings suggest that some insect bacteria may be an ideal choice for microbe-based pest control because their products can disorder the normal aggregation or mating behavior of pests.

3. Association between Gut Bacteria and Insects' Adaptation to Agrochemicals

To promote crop yield and quality, many agrochemicals are applied on fields to control the dominant economic pests, but frequent application of these chemicals has also resulted in severe health and environmental issues, as well as the resistance evolution of pests to these widely used chemicals, and nontarget toxicity to natural enemies or pollinators [59–63]. To find alternatives with novel modes of action against pests, genetically modified (GM) crops that express insecticidal proteins derived from *Bacillus thuringiensis* (Bt) have been developed and commercially planted since 1996, but resistant populations of target pests were also recorded after several years [64–66].

Amino acid mutation of target sites and upregulation of detoxification enzymes or transporters mainly confer the resistance evolution of insects to these agrochemicals [67–69], but recently, insect-associated bacteria have also been reported to directly or indirectly participate in the adaptability of insects to agrochemicals (Table 2).

3.1. Symbionts Directly Degrade Agrochemicals

When exposed to agrochemicals, the survival, development, behavior, as well as the composition and abundance of gut bacteria in target insects are affected [70]. However, under long-term high selection pressure of agrochemicals, the target insects also evolve resistance to the exposed agrochemicals, and, in some cases, the diversity and abundance of gut microbiota between resistant insect populations and susceptible insect populations are significantly different [71–73]. Compared with susceptible insect strains, the uniquely enriched gut bacteria in resistant insects should receive more attention, because these bacteria may participate in conferring insect resistance to some agrochemicals [74]. In *Aedes albopictus*, an important urban pest that can transmit viruses such as dengue, Zika, and chikungunya, the 16S rRNA sequencing results of intestinal bacteria between deltamethrin-resistant and -sensitive strains showed that the bacteria *Serratia oryzae* and *Acinetobacter junii* had higher abundance in resistance strains, and these strains may help *Ae. albopictus* develop resistance to deltamethrin, but their roles have not been verified in vitro or in vivo [72]. In deltamethrin-resistant *Spodoptera frugiperda*, the isolated bacterium *Arthrobacter nicotinovorans* grew better in the selective media and cleared 54.9% of deltamethrin [75]. Similarly, the gut symbionts *Burkholderia* from *Riptortus pedestris* and *Cletus punctiger* metabolize fenitrothion (an organophosphorus agrochemical) into nontoxic substances and use them as the available carbon source, thus promoting the development of host insects and conferring their resistance to fenitrothion. These bacteria are also present in the soil, and when treating field soil with fenitrothion for one month, the bacterial community increased to 10^7 to 10^8 CFU/g, of which >80% showed fenitrothion-degrading activities, suggesting that the insects may acquire fenitrothion-degrading bacteria from the soil [76,77]. Furthermore, in *Blatta orientalis*, the degradation rates of bacteria *Pseudomonas aeruginosa* G1, *Stenotrophomonas maltophilia* G2, and *Acinetobacter lwoffii* G5 to α-endosulfan were all >80%, which may facilitate insecticide resistance evolution and make cockroaches difficult to control [78]. In *Anopheles gambiae*, the gut bacteria *Sphingobacterium*, *Lysinibacillus*, *Streptococcus*, and *Rubrobacter* are highly associated with its resistance to permethrin [79]. Apart from insecticides, the insect gut bacterium *Acetobacter tropicalis*, isolated from *Drosophila melanogaster*, is also responsible for atrazine detoxification (one herbicide), and the restoration of *A. tropicalis* in germ-free flies reduces atrazine toxicity. Genome sequencing results showed that this bacterium contains candidate genes *atzA*, *atzB*, and *atzC*, which are involved in atrazine metabolism [80]. Furthermore, the gut bacteria *Serratia marcescens* and *Pseudomonas protegens* in *Nasonia vitripennis* also confer atrazine resistance. When exposed to atrazine for several generations, the bacterial densities of *S. marcescens* and *P. protegens* in *N. vitripennis* significantly increased. The degradation rates of these strains to atrazine were 20% and 10%, respectively, and whole-genome sequencing results also indicated the possession of the atrazine metabolism genes [24].

During the interaction of insect gut microbes with agrochemicals, some detoxification enzymes, encoded by the genes of symbionts, also play important roles in the metabolism of agrochemicals. The results of comparative genomics analysis showed that the gut symbiont *Citrobacter* sp. of *Bactrocera dorsalis* encodes genes of phosphatase hydrolase, and the gene expression levels are higher when exposed to trichlorphon. When antibiotic-treated flies were supplied with *Citrobacter* sp., the hosts obtained insecticide resistance to trichlorphon [81]. The bacterial esterase and carboxylesterase facilitated the degradation of indoxacarb in *Plutella xylostella* [82]. The above findings suggest that the degradation effects of insect gut bacteria directly mediate insect resistance to agrochemicals.

3.2. Indirect Regulation of Insect Resistance by Gut Bacteria

In addition to direct degradation, insect microbes can regulate insect resistance to agrochemicals by activating the detoxification the enzyme or immune system in hosts [83,84]. For instance, after treatment with polymyxin B, the survival rate of *Bombyx mori* exposed to chlorpyrifos was significantly lower, and 16S rRNA gene sequencing results showed that the abundances of the genera *Stenotrophomonas* and *Enterococcus* were decreased. When supplying germ-free silkworms with *S. maltophilia*, the host resistance to chlorpyrifos was enhanced. However, this bacterium cannot directly degrade chlorpyrifos in the gut, but by promoting the activity levels of acetylcholinesterase in hosts [85]. In *Culex pipiens*, the abundance of the intestinal bacterium *Aeromonas hydrophila* in deltamethrin-resistant populations was found to be much higher. After eliminating the gut bacteria of the resistant strains with antibiotics, its resistance level was reduced by 66%, while the enzyme activity of cytochrome P450 monooxygenases (CYP450s) in the hosts was reduced by 58%. Supplying *A. hydrophila* restored the resistance and enzyme activity of CYP450s, indicating that *A. hydrophila* increases the resistance of hosts to deltamethrin by enhancing the activity of CYP450s [86]. In addition, the *Enterococcus* sp. isolated from the guts of *Plutella xylostella* enhance insecticide resistance to chlorpyrifos by regulating the expression of an antimicrobial peptide named gloverin [87]. After exposure to imidacloprid, the abundance of *Wolbachia* in *Nilaparvata lugens* increased, and removing this bacterium reduced the enzyme activity of CYP450s, while the transcript level of NlCYP4CE1 also significantly decreased. This result suggested that *Wolbachia* enhances the resistance of hosts to imidacloprid by promoting the expression of NlCYP4CE1 [88]. For pollinators such as the honeybee (*Apis mellifera*), the gut microbiota promotes the expression of some immune-related genes (hymenoptaecin, defensin1) and detoxification-related genes (*CYP450s*, *GST*, and *catalase*), and thus increase honeybee tolerance to thiacloprid, tau-fluvalinate, or flumethrin [89,90].

Table 2. Symbiont-mediated insect resistance to agrochemicals.

Bacteria and Insect Host	Target Agrochemical	Description	Reference
Serratia oryzae and *Acinetobacter junii* in *Aedes albopictus*	Deltamethrin	*S. oryzae* and *A. junii* had higher abundance in deltamethrin-resistant strain (by 16S rRNA sequencing)	[72]
Arthrobacter nicotinovorans in *Spodoptera frugiperda*		Cleared 54.9% of deltamethrin (by LC-MS)	[75]
Burkholderia strains in *Riptortus pedestris* and *Cletus punctiger*	Fenitrothion	Bacteria metabolized fenitrothion into nontoxic substance, and insects infected with fenitrothion-degrading *Burkholderia* strains had higher survival rate and larger body size (by HPLC).	[76,77]
Pseudomonas aeruginosa G1, *Stenotrophomonas maltophilia* G2, and *Acinetobacter lwoffii* G5 in *Blatta orientalis*	α-endosulfan	Degradation rates of *P. aeruginosa* G1, *S. maltophilia* G2, and *A. lwoffii* G5 to α-endosulfan were 88.5%, 85.5%, and 80.2%, respectively (by HPLC)	[78]
Sphingobacterium, and *Lysinibacillus Streptococcus* and *Rubrobacter* in *Anopheles gambiae*	Pyrethroid	*Sphingobacterium*, *Lysinibacillus*, *Streptococcus*, and *Rubrobacter* significantly more abundant in resistant mosquitoes (by 16S rRNA gene sequencing)	[79]

Table 2. *Cont.*

Bacteria and Insect Host	Target Agrochemical	Description	Reference
Acetobacter tropicalis in *Drosophila melanogaster*	Atrazine	Atrazine exposure reduced relative abundance of *Acetobacter*, and restoration of *A. tropicalis* in germ-free flies reduced atrazine toxicity bacterium contained genes involved in atrazine metabolism (by 16S rRNA gene sequencing)	[80]
Serratia marcescens and *Pseudomonas protegens* in *Nasonia vitripennis*		Bacterial densities of *S. marcescens* and *P. protegens* in atrazine-fed *N. vitripennis* significantly increased, and degradation rates to atrazine were 20% and 10%, respectively; both contained genes involved in atrazine metabolism (by 16S rRNA gene sequencing, HPLC, whole-genome sequencing)	[24]
Stenotrophomonas maltophilia in *Bombyx mori*	Chlorpyrifos	Enhanced host resistance to chlorpyrifos by increasing activities of acetylcholinesterase (by 16S rRNA gene sequencing, qRT-PCR, GC-MS)	[85]
Aeromonas hydrophila in *Culex pipiens*	Deltamethrin	Increased the resistance of hosts to deltamethrin by enhancing activities of CYP450s (measurement of activity levels of enzyme)	[86]
Enterococcus sp. in *Plutella xylostella*	Chlorpyrifos	Enhanced insecticide resistance to chlorpyrifos by regulating expression of antimicrobial peptide named gloverin (by using a UV spectrophotometer at 293 nm absorbance and qRT-PCR)	[87]
Wolbachia in *Nilaparvata lugens*	Imidacloprid	Enhanced resistance of hosts to imidacloprid by promoting expression of NlCYP4CE1 (by 16S rRNA gene sequencing, qRT-PCR, measurement of activity levels of enzyme)	[88]
gut bacteria in *Apis mellifera*	Thiacloprid, tau-fluvalinate and flumethrin	E=Enhanced insecticide resistance of hosts by promoting expression of immune-related genes and detoxification-related genes (by 16S rRNA gene sequencing, qRT-PCR, HPLC)	[89,90]

4. Degradation of Other Detrimental Substances by Insect Bacteria

As the main secondary metabolites produced by mycotoxigenic fungi, mycotoxins have been found in nearly all agricultural goods, and they can cause severe human health problems and economic losses during livestock production [91]. To prevent the contamination of agricultural commodities by mycotoxins, many strategies have been recommended; there has recently been increasing interest in detoxification methods involving functional microbes isolated from natural samples [92–94]. Under natural conditions, some herbivorous insects co-occur with mycotoxigenic fungi [95]. Accordingly, they must be able to tolerate exposure to these mycotoxins to ensure that they normally develop and reproduce. Thus, they may be useful sources of functional microbes capable of detoxifying mycotoxins. To date, most of the reported mycotoxin-degrading microorganisms were isolated from noninsect systems (such as soil, water, or contaminated crops), with only one study demonstrating that *Symbiotaphrina kochii*, which is a symbiont in the tobacco beetle *Lasioderma serricorne*, can detoxify mycotoxins, including deoxynivalenol, ochratoxin A, and sterigmatocystin [96]. Future studies should identify and isolate additional functional microbes in insects that are highly tolerant to mycotoxins [97].

The overuse and abuse of antibiotics in livestock production and the treatment of human disease have resulted in severe problems associated with antibiotic resistance and antibiotic residues [98]. The gut microbes of *Musca domestica* and *Hermetia illucens* can efficiently degrade oxytetracycline (>54.5%), implying that insect gut microorganisms may be useful for eliminating antibiotic residues [99–101]. Some insect bacteria can produce antimicrobial compounds that contribute to protection from pathogens. For example, the gut bacterium *Enterococcus mundtii* in *Spodoptera littoralis* can secrete an antimicrobial (mundticin KS) against the invading bacteria, and the purified mundticin can cure larvae infected with *E. faecalis* [21]. Furthermore, cockroaches also carry bacteria that can produce metabolites or proteins with potential industrial applications, such as the antibiotic-producing *Streptomyces* strain, *Bacillus* strain, *Enterococcus* strain, and *Pseudomonas* species, all of which may be suitable for development as pharmaceuticals or plant protection products and provide opportunities for biotechnological application [102].

5. Conclusions and Future Perspectives

Insect microbiota are critical for metabolizing diverse detrimental substances. Future research on beneficial insects, including pollinators and natural enemies of pests, should consider the utility of microorganisms as biocontrol agents that can provide protection from the effects of toxic substances. Regarding pests, the role of their microbial partners should be monitored when developing new strategies for controlling pests or decreasing the vector competence of pests (e.g., the death of male insects and parthenogenesis caused by *Wolbachia* and *Rickettsia* species), but this may require genetic modifications. Furthermore, identifying microbes in insects able to detoxify harmful compounds may have important implications for bioremediation or for limiting the toxicity of xenobiotics.

Author Contributions: X.G. conceived the ideas of this review. X.G., M.Z. and X.L. contributed to the writing and revising of the manuscript. All authors have read and agreed to the published version of the manuscript.

Funding: This review manuscript was supported by the National Natural Science Foundation of China (grant no. 31801735).

Institutional Review Board Statement: Not applicable.

Informed Consent Statement: Not applicable.

Data Availability Statement: Not applicable.

Conflicts of Interest: All authors declare no conflict interest.

References

1. Engel, M.S. Insect evolution. *Curr. Biol.* **2015**, *25*, 868–872. [CrossRef]
2. Zou, Y.; Feng, J.C.; Xue, D.Y.; Sang, W.G.; Axmacher, J. Insect diversity: Addressing an important but strongly neglected research topic in China. *J. Resour. Ecol.* **2011**, *4*, 380–384.
3. Engel, P.; Moran, N.A. The gut microbiota of insects—Diversity in structure and function. *FEMS Microbiol. Rev.* **2013**, *37*, 699–735. [CrossRef] [PubMed]
4. Qiao, H.L.; Keesey, L.W.; Hansson, B.S.; Knaden, M. Gut microbiota affects development and olfactory behavior in *Drosophila melanogaster*. *J. Exp. Biol.* **2019**, *222*, 1242. [CrossRef] [PubMed]
5. Lizé, A.; Mckay, R.; Lewis, Z. Kin recognition in *Drosophila*: The importance of ecology and gut microbiota. *ISME J.* **2014**, *8*, 469–477. [CrossRef] [PubMed]
6. Farine, J.P.; Habbachi, W.; Cortot, J.; Roche, S.; Ferveur, J.F. Maternally-transmitted microbiota affects odor emission and preference in Drosophila larva. *Sci. Rep.* **2017**, *7*, 6062. [CrossRef]
7. Douglas, A.E. Multiorganismal insects: Diversity and function of resident microorganisms. *Annu. Rev. Entomol.* **2015**, *60*, 17–34. [CrossRef]
8. Wang, H.; Xian, X.Q.; Gu, Y.J.; Castane, C.; Arno, J.; Wu, S.; Wan, F.H.; Liu, W.X.; Zhang, G.F.; Zhang, Y.B. Similar bacterial communities among different populations of a newly emerging invasive species, *Tuta absoluta* (Meyrick). *Insects* **2022**, *13*, 252. [CrossRef]
9. Yang, H.; Huang, Y.P. Insect microbiome: As guardians of insect health and adaptation. *Acta Microbiol. Sinica.* **2018**, *58*, 961–962.
10. Dong, S.Z.; Pang, K.; Bai, X.; Yu, X.P.; Hao, P.Y. Identification of two species of yeast-like symbiotes in the brown planthopper, *Nilaparvata lugens*. *Curr. Microbiol.* **2011**, *62*, 1133–1138. [CrossRef]
11. Vogel, K.J.; Moran, N.A. Functional and evolutionary analysis of the genome of an obligate fungal symbiont. *Genome Biol Evol.* **2013**, *5*, 891–904. [CrossRef] [PubMed]

12. Pang, K.; Dong, S.Z.; Hao, P.Y.; Chen, T.T.; Wang, X.L.; Yu, X.P.; Lin, H.F. Fungicides reduce the abundance of yeast-like symbionts and survival of white-backed planthopper *Sogatelle furcifera* (Homoptera: Delphacidae). *Insects* **2020**, *11*, 209. [CrossRef] [PubMed]
13. Strand, M.R.; Burke, G.R. Polydnavirus-wasp associations: Evolution, genome organization, and function. *Curr. Opin. Virol.* **2013**, *3*, 587–594. [CrossRef]
14. Xu, P.J.; Yang, L.Y.; Yang, X.M.; Li, T.; Graham, R.I.; Wu, K.M.; Wilson, K. Novel partiti-like viruses are conditional mutualistic symbionts in their normal lepidopteran host, African armyworm, but parasitic in a novel host, Fall armyworm. *PLoS Pathog.* **2020**, *16*, 1008467. [CrossRef] [PubMed]
15. Ohkuma, M. Symbioses of flagellates and prokaryotes in the gut of lower termites. *Trends Microbiol.* **2008**, *16*, 345–352. [CrossRef]
16. Shi, Y.; Huang, Z.; Han, S.; Fan, S.; Yang, H. Phylogenetic diversity of Archaea in the intestinal tract of termites from different lineages. *J. Basic Microbiol.* **2015**, *55*, 1021–1028. [CrossRef] [PubMed]
17. Engel, P.; Martinson, V.G.; Moran, N.A. Functional diversity within the simple gut microbiota of the honey bee. *Proc. Natl. Acad. Sci. USA* **2012**, *109*, 11002–11007. [CrossRef] [PubMed]
18. Engel, P.; Moran, N.A. Functional and evolutionary insights into the simple yet specific gut microbiota of the honey bee from metagenomic analysis. *Gut Microbes.* **2013**, *4*, 60–65. [CrossRef]
19. Dantur, K.I.; Enrique, R.; Welin, B.; Castagnaro, A.P. Isolation of cellulolytic bacteria from the intestine of *Diatraea saccharalis* larvae and evaluation of their capacity to degrade sugarcane biomass. *AMB Express.* **2015**, *5*, 15. [CrossRef]
20. Dillon, R.; Charnley, K. Mutualism between the desert locust *Schistocerca gregaria* and its gut microbiota. *Res. Microbiol.* **2002**, *153*, 503–509. [CrossRef]
21. Shao, Y.Q.; Chen, B.S.; Sun, C.; Ishida, K.; Hertweck, C.; Boland, W. Symbiont-derived antimicrobials contribute to the control of the lepidopteran gut microbiota. *Cell. Chem. Biol.* **2017**, *24*, 66–75. [CrossRef] [PubMed]
22. Zheng, H.; Powell, J.E.; Steele, M.I.; Dietrich, C.; Moran, N.A. Honeybee gut microbiota promotes host weight gain via bacterial metabolism and hormonal signaling. *Proc. Natl. Acad. Sci. USA* **2017**, *114*, 4775–4780. [CrossRef] [PubMed]
23. Gong, Q.; Cao, L.J.; Sun, L.N.; Chen, J.C.; Gong, Y.J.; Pu, D.Q.; Huang, Q.; Hoffmann, A.A.; Wei, S.J. Similar gut bacterial microbiota in two fruit-feeding moth pests collected from different host species and locations. *Insects* **2020**, *11*, 840. [CrossRef] [PubMed]
24. Wang, G.H.; Berdy, B.M.; Velasquez, O.; Jovanovic, N.; Alkhalifa, S.; Minbiole, K.P.C.; Brucker, R.M. Changes in microbiome confer multigenerational host resistance after sub-toxic pesticide exposure. *Cell Host Microbe.* **2020**, *27*, 213–224. [CrossRef] [PubMed]
25. Shapira, M. Gut microbiotas and host evolution: Scaling up symbiosis. *Trends Ecol. Evol.* **2016**, *31*, 539–549. [CrossRef]
26. Cao, L.; Ning, K. Metagenomics of the insect gut: The frontier of microbial big data. *Acta Microbiol. Sinica.* **2018**, *58*, 964–984.
27. Zhang, Z.J.; Mu, X.H.; Cao, Q.N.; Shi, Y.; Hu, X.S.; Zheng, H. Honeybee gut *Lactobacillus* modulates host learning and memory behaviors via regulating tryptophan metalolism. *Nat. Commun.* **2022**, *13*, 2037. [CrossRef]
28. Bai, L.; Wang, L.L.; Vega-Rodriguez, J.; Wang, G.D.; Wang, S.B. A gut symbiotic bacterium *Serratia marcescens* renders mosquito resistance to *Plasmodium* infection through activation of mosquito immune responses. *Front. Microbiol.* **2019**, *10*, 1580. [CrossRef]
29. Kaur, R.; Shropshire, J.D.; Cross, K.L.; Leigh, B.; Mansueto, A.J.; Stewart, V.; Bordenstein, S.R.; Bordenstein, S.R. Living in the endosymbiotic world of *Wolbachia*: A centennial review. *Cell Host Microbe.* **2021**, *29*, 879–893. [CrossRef]
30. Jiang, Y.J.; Zhang, C.X.; Chen, R.Z.; He, S.Y. Challenging battles of plants with phloem-feeding insects and prokaryotic pathogens. *Proc. Natl. Acad. Sci. USA* **2019**, *116*, 23390–23397. [CrossRef]
31. Oliveira, C.M.; Auad, A.M.; Mendes, S.M.; Frizzas, M.R. Crop losses and the economic impact of insect pests on Brazilian agriculture. *Crop. Prot.* **2014**, *56*, 50–54. [CrossRef]
32. Kessler, A.; Baldwin, I.T. Plant responses to insect herbivory: The emerging molecular analysis. *Annu. Rev. Plant Biol.* **2002**, *53*, 299–328. [CrossRef] [PubMed]
33. Alyokhin, A.; Chen, Y.H. Adaptation to toxic hosts as a factor in the evolution of insecticide resistance. *Curr. Opin. Insect Sci.* **2017**, *21*, 33–38. [CrossRef]
34. Hammer, T.J.; Bowers, M.D. Gut microbes may facilitate insect herbivory of chemically defended plants. *Oecologia* **2015**, *179*, 1–14. [CrossRef] [PubMed]
35. Consales, F.; Schweizer, F.; Erb, M.; Gouhier-Darimont, C.; Bodenhausen, N.; Bruessow, F.; Sobhy, I.; Reymond, P. Insect oral secretions suppress wound-induced responses in *Arabidopsis*. *J. Exp. Bot.* **2012**, *63*, 727–737. [CrossRef] [PubMed]
36. Atamian, H.S.; Chaudhary, R.; Cin, V.D.; Bao, E.; Girke, T.; Kaloshian, I. In planta expression or delivery of potato aphid *Macrosiphum euphorbiae* effectors *Me10* and *Me23* enhances aphid fecundity. *Mol. Plant Microbe Interact.* **2013**, *26*, 67–74. [CrossRef] [PubMed]
37. Chung, S.H.; Rosa, C.; Scully, E.D.; Peiffer, M.; Tooker, J.F.; Hoover, K.; Luthe, D.S.; Felton, G.W. Herbivore exploits orally secreted bacteria to suppress plant defenses. *Proc. Natl. Acad. Sci. USA* **2013**, *110*, 15728–15733. [CrossRef] [PubMed]
38. Santos-Garcia, D.; Mestre-Rincon, N.; Zchori-Fein, E.; Morin, S. Inside out: Microbiota dynamics during host-plant adaptation of whiteflies. *IMSE J.* **2020**, *14*, 847–856. [CrossRef]
39. Leite-Mondin, M.; Dilegge, M.J.; Manter, D.K.; Weir, T.L.; Silva-Filho, M.C.; Vivanco, J.M. The gut microbiota composition of *Trichoplusia ni* is altered by diet and may influence its polyphagous behavior. *Sci. Rep.* **2021**, *11*, 5786. [CrossRef]
40. Vilanova, C.; Baixeras, J.; Latorre, A.; Porcar, M. The generalist inside the specialist: Gut bacterial communities of two insect species feeding on toxic plants are dominated by *Enterococcus* sp. *Front. Microbiol.* **2016**, *7*, 1005. [CrossRef]

41. Boone, K.C.; Keefover-Ring, K.; Mapes, A.C.; Adams, A.S.; Bohlmann, J.; Raffa, K.F. Bacteria associated with a tree-killing insect reduce concentrations of plant defense compounds. *J. Chem. Ecol.* **2013**, *39*, 1003–1006. [CrossRef] [PubMed]
42. Adams, A.S.; Aylward, F.O.; Adams, S.M.; Erbilgin, N.; Aukema, B.H.; Currie, C.R.; Suen, G.; Raffa, K.F. Mountain pine beetles colonizing historical and naïve host trees are associated with a bacterial community highly enriched in genes contributing to terpene metabolism. *Appl. Environ. Microb.* **2013**, *79*, 3468–3475. [CrossRef] [PubMed]
43. Xu, L.T.; Lu, M.; Sun, J.H. Invasive bark beetle-associated microbes degrade a host defensive monoterpene. *Insect Sci.* **2016**, *23*, 183–190. [CrossRef] [PubMed]
44. Berasategui, A.; Salem, H.; Paetz, C.; Santoro, M.; Gershenzon, J.; Kaltenpoth, M.; Schmidt, A. Gut microbiota of the pine weevil degrades conifer diterpenes and increases insect fitness. *Mol. Ecol.* **2017**, *26*, 4099–4110. [CrossRef] [PubMed]
45. Zhang, S.K.; Shu, J.P.; Xue, H.J.; Zhang, W.; Zhang, Y.B.; Liu, Y.N.; Fang, L.X.; Wang, Y.D.; Wang, H.J. The gut microbiota in camellia weevils are influenced by plant secondary metabolites and contribute to saponin degradation. *Msystems* **2020**, *5*, e00692. [CrossRef] [PubMed]
46. Vieira, C.S.; Figueiredo, M.B.; Moraes, C.S.; Pereira, S.B.; Dyson, P.; Mello, C.B.; Castro, D.P.; Azambuja, P. Azadirachtin interferes with basal immunity and microbial homeostasis in the *Rhodnius prolixus* midgut. *Dev. Comp. Immunol.* **2021**, *114*, 103864. [CrossRef]
47. Nuringtyas, T.R.; Verpoorte, R.; Klinkhamer, P.G.L.; van Oers, M.M.; Leiss, K.A. Toxicity of pyrrolizidine alkaloids to *Spodoptera exigua* using insect cell lines and injection bioassays. *J. Chem. Ecol.* **2014**, *40*, 609–616. [CrossRef]
48. Ceja-Navarro, J.A.; Vega, F.E.; Karaoz, U.; Hao, Z.; Jenkins, S.; Lim, H.C.; Kosina, P.; Infante, F.; Northen, T.R.; Brodie, E.L. Gut microbiota mediate caffeine detoxification in the primary insect pest of coffee. *Nat. Commun.* **2015**, *6*, 7618. [CrossRef]
49. Ben-Yosef, M.; Pasternak, Z.; Jurkevitch, E.; Yuval, B. Symbiotic bacteria enable olive fly larvae to overcome host defences. *R. Soc. Open Sci.* **2015**, *2*, 150170. [CrossRef]
50. Cheng, C.H.; Wickham, J.D.; Chen, L.; Xu, D.D.; Lu, M.; Sun, J.H. Bacterial microbiota protect an invasive bark beetle from a pine defensive compound. *Microbiome* **2018**, *6*, 132. [CrossRef]
51. Mason, C.J.; Lowe-Power, T.M.; Rubert-Nason, K.F.; Lindroth, R.L.; Raffa, K.F. Interactions between bacteria and aspen defense chemicals at the phyllosphere-herbivore interface. *J. Chem. Ecol.* **2016**, *42*, 193–201. [CrossRef] [PubMed]
52. Shukla, S.P.; Beran, F. Gut microbiota degrades toxic isothiocyanates in a flea beetle pest. *Mol. Ecol.* **2020**, *29*, 4692–4705. [CrossRef] [PubMed]
53. Nikoh, N.; Hosokawa, T.; Oshima, K.; Hattori, M.; Fukatsu, T. Reductive evolution of bacterial genome in insect gut environment. *Genome Biol. Evol.* **2011**, *3*, 702–714. [CrossRef] [PubMed]
54. Al, K.F.; Daisley, B.A.; Chanyi, R.M.; Bjazevic, J.; Razvi, H.; Reid, G.; Burton, J.P. Oxalate-degrading *Bacillus subtilis* mitigates urolithiasis in a *Drosophila melanogaster* model. *Msphere* **2020**, *5*, e00498. [CrossRef]
55. Zeng, J.Y.; Wu, D.D.; Shi, Z.B.; Yang, J.; Zhang, G.C.; Zhang, J. Influence of dietary aconitine and nicotine on the gut microbiota of two lepidopteran herbivores. *Arch. Insect Biochem. Physiol.* **2020**, *104*, e21676. [CrossRef]
56. Engl, T.; Kaltenpoth, M. Influence of microbial symbionts on insect pheromones. *Nac. Prod. Rep.* **2018**, *35*, 386–397. [CrossRef]
57. Pizzolante, G.; Cordero, C.; Tredici, S.M.; Vergara, D.; Pontieri, P.; Giudice, L.D.; Capuzzo, A.; Rubiolo, P.; Kanchiswamy, C.N.; Zebelo, S.A.; et al. Cultivable gut bacteria provide a pathway for adaptation of *Chrysolina herbacea* to *Mentha aquatic* volatiles. *BMC Plant Biol.* **2017**, *17*, 30. [CrossRef]
58. Ren, L.; Ma, Y.G.; Xie, M.X.; Lu, Y.Y.; Cheng, D.F. Rectal bacteria produce sex pheromones in the male oriental fruit fly. *Curr. Biol.* **2021**, *31*, 2220–2226. [CrossRef]
59. Xu, L.; Wang, J.H.; Mei, Y.; Li, D.Z. Research progress on the molecular mechanisms of insecticides resistance mediated by detoxification enzymes and transporters. *Chin. J. Pestic. Sci.* **2020**, *22*, 1–10.
60. Mallinger, R.E.; Werts, P.; Gratton, C. Pesticide use within a pollinator-dependent crop has negative effects on the abundance and species richness of sweat bees, *Lasioglossum* spp., and on bumble bee colony growth. *J. Insect Conserv.* **2015**, *19*, 999–1010. [CrossRef]
61. Obregon, D.; Guerrero, O.R.; Stashenko, E.; Poveda, K. Natural habitat partially mitigates negative pesticide effects on tropical pollinator communities. *Glob. Ecol. Conserv.* **2021**, *28*, e01668. [CrossRef]
62. Mills, N.J.; Beers, E.H.; Shearer, P.W.; Unruh, T.R.; Amarasekare, K.G. Comparative analysis of pesticide effects on natural enemies in western orchards: A synthesis of laboratory bioassay data. *Biol. Control.* **2016**, *102*, 17–25. [CrossRef]
63. Pestana, D.; Teixeira, D.; Faria, A.; Domingues, V.; Monteiro, R.; Calhau, C. Effects of the environmental pesticide DDT and its metabolites on the human breast cancer cell line MCF-7. *Toxicol. Lett.* **2010**, *196*, 180. [CrossRef]
64. Edgerton, M.D.; Fridgen, J.; Anderson, J.R., Jr.; Ahlgrim, J.; Criswell, M.; Dhungana, P.; Gocken, T.; Li, Z.; Mariappan, S.; Pilcher, C.D.; et al. Transgenic insect resistance traits increase corn yield and yield stability. *Nat. Biotechnol.* **2012**, *30*, 493–496. [CrossRef] [PubMed]
65. Wu, Y.D. Detection and mechanisms of resistance evolved in insects to Cry toxins from *Bacillus thuringiensis*. *Adv. Insect Physiol.* **2014**, *47*, 297–342.
66. Jakka, S.R.K.; Gong, L.; Hasler, J.; Banerjee, R.; Sheets, J.J.; Narva, K.; Blanco, C.A.; Jurat-Fuentes, J.L. Field-evolved mode 1 fall armyworm resistance to Bt corn associated with reduced Cry1Fa toxin binding and midgut alkaline phosphatase expression. *Appl. Environ. Microb.* **2015**, *82*, 02871-15.

67. Guo, L.; Wang, Y.; Zhou, X.G.; Li, Z.Y.; Liu, S.Z.; Pei, L.; Gao, X.W. Functional analysis of a point mutation in the ryanodine receptor of *Plutella xylostella* (L.) associated with resistance to chlorantraniliprole. *Pest Manag. Sci.* **2014**, *70*, 1083–1089. [CrossRef]
68. Wang, X.L.; Cao, X.W.; Jiang, D.; Yang, Y.H.; Wu, Y.D. CRISPR/Cas9 mediated ryanodine receptor I4790M knockin confers unequal resistance to diamides in *Plutella xylostella*. *Insect Biochem. Mol. Biol.* **2020**, *125*, 103453. [CrossRef]
69. Li, X.X.; Li, R.; Zhu, B.; Gao, X.W.; Liang, P. Overexpression of cytochrome P450 CYP6BG1 may contribute to chlorantraniliprole resistance in *Plutella xylostella* (L.). *Pest Manag. Sci.* **2018**, *74*, 1386–1393. [CrossRef]
70. Hou, J.Y.; Yu, J.Z.; Qin, Z.H.; Liu, X.J.; Zhao, X.P.; Hu, X.Q.; Yu, R.X.; Wang, Q.; Yang, J.Y.; Shi, Y.; et al. Guadipyr, a new insecticide, induces microbiota dysbiosis and immune disorders in the midgut of silkworms (*Bombyx mori*). *Environ. Pollut.* **2021**, *286*, 117531. [CrossRef]
71. Wang, Y.T.; Shen, R.X.; Xing, D.; Zhao, C.P.; Gao, H.T.; Wu, J.H.; Zhang, N.; Zhang, H.D.; Chen, Y.; Zhao, T.Y.; et al. Metagenome sequencing reveals the midgut microbiota makeup of *Culex pipiens quinquefasciatus* and its possible relationship with insecticide resistance. *Front. Microbiol.* **2021**, *12*, 625539. [CrossRef] [PubMed]
72. Wang, H.Y.; Zhang, C.X.; Cheng, P.; Wang, Y.; Liu, H.M.; Wang, H.F.; Wang, H.W.; Gong, M.Q. Differences in the intestinal microbiota between insecticide-resistant and -sensitive *Aedes albopictus* based on full-length 16S rRNA sequencing. *MicrobiologyOpen* **2021**, *10*, 1177. [CrossRef] [PubMed]
73. Arévalo-Cortés, A.; Mejia-Jaramillo, A.M.; Granada, Y.; Coatsworth, H.; Lowenberger, C.; Triana-Chavez, O. The midgut microbiota of colombian *Aedes aegypti* populations with different levels of resistance to the insecticide lambda-cyhalothrin. *Insects* **2020**, *11*, 584. [CrossRef] [PubMed]
74. Gressel, J. Microbiome facilitated pest resistance: Potential problems and uses. *Pest Manag. Sci.* **2018**, *74*, 511–515. [CrossRef] [PubMed]
75. de Almeida, L.G.; de Moraes, L.A.B.; Trigo, J.R.; Omoto, C.; Consoli, F.L. The gut microbiota of insecticide-resistant insects houses insecticide-degrading bacteria: A potential source for biotechnological exploitation. *PLoS ONE* **2017**, *12*, e0174754. [CrossRef]
76. Kikuchi, Y.; Hayatsu, M.; Hosokawa, T.; Nagayama, A.; Tago, K.; Fukatsu, T. Symbiont-mediated insecticide resistance. *Proc. Natl. Acad. Sci. USA* **2012**, *109*, 8618–8622. [CrossRef]
77. Ishigami, K.; Jang, S.; Itoh, H.; Kikuchi, Y. Insecticide resistance governed by gut symbiosis in a rice pest, *Cletus punctiger*, under laboratory conditions. *Biol. Lett.* **2021**, *17*, 20200780. [CrossRef]
78. Ozdal, M.; Ozdal, O.G.; Alguri, O.F. Isolation and characterization of alpha-endosulfan degrading bacteria from the microflora of cockroaches. *Pol. J. Microbiol.* **2016**, *65*, 63–68. [CrossRef]
79. Omoke, D.; Kipsum, M.; Otieno, S.; Esalimba, E.; Sheth, M.; Lenhart, A.; Njeru, E.M.; Ochomo, E.; Dada, N. Western Kenyan *Anopheles gambiae* showing intense permethrin resistance harbour distinct microbiota. *Malar. J.* **2021**, *20*, 77. [CrossRef]
80. Brown, J.B.; Langley, S.A.; Snijders, A.M.; Wan, K.H.; Morris, S.N.S.; Booth, B.W.; Fisher, W.W.; Hammonds, A.S.; Park, S.; Weiszmann, R.; et al. An integrated host-microbiome response to atrazine exposure mediates toxicity in *Drosophila*. *Commun. Biol.* **2021**, *4*, 1324. [CrossRef]
81. Cheng, D.F.; Guo, Z.J.; Riegler, M.; Xi, Z.Y.; Liang, G.W.; Xu, Y.J. Gut symbiont enhances insecticide resistance in a significant pest, the oriental fruit fly *Bactrocera dorsalis* (Hendel). *Microbiome* **2017**, *5*, 13. [CrossRef] [PubMed]
82. Ramya, S.L.; Venkatesan, T.; Murthy, K.S.; Jalali, S.K.; Verghese, A. Detection of carboxylesterase and esterase activity in culturable gut bacterial flora isolated from diamondback moth, *Plutella xylostella* (Linnaeus), from India and its possible role in indoxacarb degradation. *Braz. J. Microbiol.* **2016**, *47*, 327–336. [CrossRef] [PubMed]
83. Wang, Z.Y.; Wang, W.F.; Lu, Y.J. Symbiotic microbiota and insecticide resistancein insects. *Chin. J. Appl. Entomol.* **2021**, *58*, 265–276.
84. Liu, X.D.; Guo, H.F. Importance of endosymbionts Wolbachia and Rickettsia in insect resistance development. *Curr. Opin. Insect Sci.* **2019**, *33*, 84–90. [CrossRef] [PubMed]
85. Chen, B.S.; Zhang, N.; Xie, S.; Zhang, X.C.; He, J.T.; Muhammad, A.; Sun, C.; Lu, X.M.; Shao, Y.Q. Gut bacteria of the silkworm *Bombyx mori* facilitate host resistance against the toxic effects of organophosphate insecticides. *Environ. Int.* **2020**, *143*, 105886. [CrossRef] [PubMed]
86. Xing, Y.F.; Liu, Z.H.; Zhang, R.M.; Zhou, D.; Sun, Y.; Ma, L.; Shen, B. Effect of the midgut symbiotic *Aeromonas hydrophila* on the deltamethrin resistance of Culex pipiens pallens. *J. Pathog. Biol.* **2021**, *16*, 661–666.
87. Xia, X.F.; Sun, B.T.; Gurr, G.M.; Vasseur, L.; Xue, M.Q.; You, M.S. Gut microbiota mediate insecticide resistance in the diamondback moth, *Plutella xylostella* (L.). *Front. Microbiol.* **2018**, *9*, 00025. [CrossRef] [PubMed]
88. Cai, T.W.; Zhang, Y.H.; Liu, Y.; Deng, X.Q.; He, S.; Li, J.H.; Wang, H. *Wolbachia* enhances expression of *NlCYP4CE1* in *Nilaparvata lugens* in response to imidacloprid stress. *Insect Sci.* **2021**, *28*, 355–362. [CrossRef]
89. Wu, Y.Q.; Zheng, Y.F.; Chen, Y.N.; Wang, S.; Chen, Y.P.; Hu, F.L.; Zheng, H.Q. Honey bee (*Apis mellifera*) gut microbiota promotes host endogenous detoxification capability via regulation of P450 gene expression in the digestive tract. *Microb. Biotechnol.* **2020**, *13*, 1201–1212. [CrossRef]
90. Yu, L.T.; Yang, H.Y.; Cheng, F.P.; Wu, Z.H.; Huang, Q.; He, X.J.; Yan, W.Y.; Zhang, L.Z.; Wu, X.B. Honey bee *Apis mellifera* larvae gut microbial and immune, detoxication responses towards flumethrin stress. *Environ. Pollut.* **2021**, *290*, 118107. [CrossRef]
91. Zhu, Y.; Hassan, Y.I.; Lepp, D.; Shao, S.Q.; Zhou, T. Strategies and methodologies for developing microbial detoxification systems to mitigate mycotoxins. *Toxins* **2017**, *9*, 130. [CrossRef] [PubMed]
92. He, J.W.; Zhou, T.; Young, J.C.; Boland, G.J.; Scott, P.M. Chemical and biological transformations for detoxification of trichothecene mycotoxins in human and animal food chains: A review. *Trends Food Sci. Technol.* **2010**, *21*, 67–76. [CrossRef]

93. Zhu, Y.; Hassan, Y.I.; Watts, C.; Zhou, T. Innovative technologies for the mitigation of mycotoxins in animal feed and ingredients—A review of recent patents. *Anim. Feed Sci. Technol.* **2016**, *216*, 19–29. [CrossRef]
94. He, J.W.; Hassan, Y.I.; Perilla, N.; Li, X.Z.; Boland, G.J.; Zhou, T. Bacterial epimerization as a route for deoxynivalenol detoxification: The influence of growth and environmental conditions. *Front. Microbiol.* **2016**, *7*, 572. [CrossRef]
95. Liu, Y.; Li, R.R.; He, K.L.; Bai, S.X.; Zhang, T.T.; Cong, B.; Wang, Z.Y. Effects of *Conogethes punctiferalis* (Lepidopteran: Grambidae) infestation on the occurrence of *Fusarium* ear rot and the yield loss of spring corn. *Acta Entomol. Sinica.* **2017**, *60*, 576–581.
96. Shen, S.K.; Dowd, P.F. Detoxification spectrum of the cigarette beetle symbiont *Symbiotaphrina kochii* in culture. *Entomol. Exp. Appl.* **1991**, *60*, 51–59. [CrossRef]
97. Bosch, G.; Fels-Klerx, H.J.; Rijk, T.C.; Oonincx, D.G. Aflatoxin B_1 tolerance and accumulation in black soldier fly larvae (*Hermetia illucens*) and yellow mealworms (*Tenebrio molitor*). *Toxins* **2017**, *9*, 185. [CrossRef] [PubMed]
98. Huang, Y.Y.; Cong, Y.L. Global health ethical reflection on antibiotics abuse. *Chin. Med. Ethics* **2017**, *30*, 412–416.
99. Liu, C.C.; Yao, H.Y.; Chapman, S.J.; Su, J.Q.; Wang, C.W. Changes in gut bacterial communities and the incidence of antibiotic resistance genes during degradation of antibiotics by black soldier fly larvae. *Environ. Int.* **2020**, *142*, 105834. [CrossRef]
100. Liu, C.C.; Yao, H.Y.; Wang, C.W. Black soldier fly larvae can effectively degrade oxytetracycline bacterial residue by means of the gut bacterial community. *Front. Microbiol.* **2021**, *12*, 663972. [CrossRef]
101. Jiang, C.L.; Li, H.Y.; Wang, X.; Teng, C.Y.; Feng, S.Y.; Lou, L.P.; Zhang, Z.J. Effects of fly maggot gut microbiota on the degradation of residual antibiotics in pig manure and its resistance genes. *Acta Microbiol. Sin.* **2018**, *58*, 1103–1115.
102. Guzman, J.; Vilcinskas, A. Bacteria associated with cockroaches: Health risk or biotechnological opportunity? *Appl. Microbiol. Biotechnol.* **2020**, *104*, 10369–10387. [CrossRef] [PubMed]

insects

Article

Fall Armyworm Gut Bacterial Diversity Associated with Different Developmental Stages, Environmental Habitats, and Diets

Dan-Dan Li, Jin-Yang Li, Zu-Qing Hu, Tong-Xian Liu and Shi-Ze Zhang *

State Key Laboratory of Crop Stress Biology for Arid Areas, College of Plant Protection, Northwest A&F University, Xianyang 712100, China
* Correspondence: shzzhang@nwafu.edu.cn; Tel.: +86-29-8708-2350

Simple Summary: Microorganisms play a crucial role during the growth and development of insects. However, as a major invasive pest, the diversity and dynamics of gut microbes with different developmental stages, environmental habitats, and diets in *Spodoptera frugiperda* remain unclear. The abundant gut microbes of *S. frugiperda* may be beneficial for its abilities of invasiveness and adaptation. Therefore, it is of great importance to systematically understand the microbial dynamics of *S. frugiperda*. This study systematically explored the changes of microorganisms of *S. frugiperda* at each developmental stage. Furthermore, the differences in gut microorganisms of *S. frugiperda* in different living environments (field and laboratory) and different foods (corn and artificial diet) were also explored. Our results suggest that *S. frugiperda* gut microbes vary greatly at different developmental stages and demonstrate vertical transmission of bacteria in *S. frugiperda*. Furthermore, environment and diet can also alter gut microbes. We performed a detailed investigation of the microbial community of *S. frugiperda* that provides a basis for future research. Since the plasticity of insect gut microbes helps insects utilize different foods and enhances insect fitness, a comprehensive understanding of *S. frugiperda*'s gut microbiome will help develop novel pest control strategies for this invasive pest prevention.

Abstract: The fall armyworm, *Spodoptera frugiperda* (Lepidoptera: Noctuidae), is a major invasive pest that seriously threatens world agricultural production and food security. Microorganisms play a crucial role in the growth and development of insects. However, the diversity and dynamics of gut microbes with different developmental stages, environmental habitats, and diets in *S. frugiperda* remain unclear. In this study, we found the changes of the microbiome of *S. frugiperda* across their life stages, and the bacteria were dominated by Firmicutes and Proteobacteria. The community composition of the egg stage was quite different from other developmental stages, which had the highest community diversity and community richness, and was dominated by Proteobacteria. The bacterial community compositions of male and female adults were similar to those of early larvae stage (L1–L2), and operational taxonomic units (OTUs) with abundant content were *Enterococcus* and Enterobacteriaceae bacteria, including *Enterobacteria*, *Klebsiella*, *Pantoea*, and *Escherichia*. The third instar larvae (L3) mainly consist of *Enterococcus*. The late stage larvae (L4–L6) harbored high proportions of *Enterococcus*, *Rhodococcus*, and *Ralstonia*. There was no significant difference in gut microbial composition between field populations and laboratory populations in a short period of rearing time. However, after long-term laboratory feeding, the gut microbial diversity of *S. frugiperda* was significantly reduced. *Enterococcus* and *Rhodococcccus* of *S. frugiperda* feeding on maize showed higher relative proportion, while the microbial community of *S. frugiperda* feeding on artificial diet was composed mainly of *Enterococcus*, with a total of 98% of the gut microbiota. The gene functions such as metabolism, cell growth and death, transport and catabolism, and environmental adaptation were more active in *S. frugiperda* feeding on corn than those feeding on artificial diet. In short, these results indicate that developmental stage, habitat, and diet can alter the gut bacteria of *S. frugiperda*, and suggest a vertical transmission route of bacteria in *S. frugiperda*. A comprehensive understanding of gut microbiome of *S. frugiperda* will help develop novel pest control strategies to manage this pest.

Citation: Li, D.-D.; Li, J.-Y.; Hu, Z.-Q.; Liu, T.-X.; Zhang, S.-Z. Fall Armyworm Gut Bacterial Diversity Associated with Different Developmental Stages, Environmental Habitats, and Diets. *Insects* **2022**, *13*, 762. https://doi.org/10.3390/insects13090762

Academic Editors: Hongyu Zhang, Yin Wang and Xiaoxue Li

Received: 10 July 2022
Accepted: 23 August 2022
Published: 24 August 2022

Publisher's Note: MDPI stays neutral with regard to jurisdictional claims in published maps and institutional affiliations.

Copyright: © 2022 by the authors. Licensee MDPI, Basel, Switzerland. This article is an open access article distributed under the terms and conditions of the Creative Commons Attribution (CC BY) license (https://creativecommons.org/licenses/by/4.0/).

Keywords: *Spodoptera frugiperda*; gut microbiota; developmental stage; host diet; environmental habitat; 16S rRNA

1. Introduction

Animal-microbial symbiosis is extremely important to the ecosystem [1]. Microbial symbionts are especially ubiquitous in insects, and they exist in insect exoskeletons, gut, and even within insect cells, which are usually beneficial or necessary for survival of insect hosts [2]. Insects can use microorganisms to enhance their life performance and adaptation to the various environmental changes [3]. Many insect-related microorganisms can not only provide specific nutrients that insects cannot synthesize themselves, such as essential amino acids [4,5] and B vitamins [6,7], but also protect their insect hosts against other invasive organisms, such as pathogens, parasitoids or predators [8–10]. In addition, symbiotic microorganisms can also enhance the resistance of insects to pesticides [11,12].

Many factors, including diet, life stage, and host habitat affect the structure of the gut microbial community [13–16]. In order to adapt to the different environmental changes, insects have evolved different composition of symbiotic microorganisms in the different developmental stages [14]. In principle, diet can influence the gut microbiota directly and indirectly [2,17]. For example, protein can lead to an increase in the abundance of specific microbiota in *Blattella germanica* [18]. Microbial communities of isogenic *Drosophila melanogaster* fed on different diets are different, but three distantly related Drosophilids fed on the same medium have similar bacterial microbiome [19].

The fall armyworm, *Spodoptera frugiperda* (Lepidoptera: Noctuidae) is a serious invasive insect pest. Due to its overeating major crops such as corn and rice, long-distance super migration and spreading ability, *S. frugiperda* was listed as one of the top 10 hazardous plant pests in the world by the CAB International (CABI) in 2017 (https://www.cabi.org/isc/fallarmyworm (accessed on 10 January 2021)). Moreover, *S. frugiperda* is posing a serious threat for potential economic losses to other staple crops such as wheat, soybean, cotton, tomato, and cabbage [20]. It is well known that insects have abundant and diverse gut microbes, and the microbiomes not only provide important nutrients for their insect hosts but also assist in the food digestion, immune defense, detoxification, and adaptation to changing environments [21,22]. The gut microbes Archaea and Bacteria play an important role in the nutritional requirement of the fifth instar larvae of *S. frugiperda* [23]. The analysis of the gut microbiota of susceptible, insecticide-resistant strains and field populations of *S. frugiperda* indicates that the gut microbes have a high diversity and the ability to metabolize insecticides in field populations of *S. frugiperda* [24]. Recently, Lv et al. [25] reported that the gut microbial community of the fifth instar larvae of *S. frugiperda* is significantly affected by different host species. However, the previous studies on the gut microbial community of *S. frugiperda* were limited to a certain stage of the host's development and rarely investigated the changes of microorganisms throughout the complete life stages. Thus, diversity and dynamics of the bacterial community across different developmental stages of *S. frugiperda* are still unclear.

It is well known that altering the insect gut microbiome can influence insect behavior, which may lead to new approaches to pest control, but these depend largely on a detailed understanding of insect-associated microorganisms [2]. For example, the elimination of the symbiont *Symbiotaphrina kochi* in *Lasioderma serricorne* beetles depresses larval development [26]. Insect gut microbes are able to interact with the host, and then the high abundance bacteria are more likely to play an important role in host adaptation. In addition, previous studies have shown that the diet consumed, living environment, and developmental stage of insects may lead to the differences of gut microbial communities and dynamics [18,27,28]. As one of the 10 most notorious plant pests in the world, systematic study on the interaction between *S. frugiperda* and gut microbiome can not only provide a basis for in-depth understanding of its rapid adaptation in migration area, but also could provide a theoretical basis for the development of new control strategies and technology. However, to date, limited data have been available on *S. frugiperda* microbiota. We hypothesized that the rapid adaptation of *S. frugiperda* in the invasive areas may be related to the abundance of its gut microbiome. Therefore, in this study, we systematically explored the changes of microorganisms of *S. frugiperda* at each developmental stage, and the differences of gut microorganisms of *S. frugiperda* in different living environments (field and laboratory) and different foods (corn and artificial diet). The present work not only provides valuable information for a comprehensive understanding of gut microbiome across the life history of *S. frugiperda*, but also assists the development of novel pest control strategies for prevention of this invasive pest.

2. Materials and Methods

2.1. Rearing of *S. frugiperda*

Maize (*Zea mays* L. var. Shandan 636) seeds were purchased from Yangling Agricultural High-Tech Development Co., Ltd. (Yangling, China), and sown in plastic pots with a 3:1:1 mixture of commercial peat moss (Pindstrup Mosebrug A/S, Ryomgaard, Denmark), perlite and vermiculite in an artificial climate room (25–30 °C, 50–80% RH and a photoperiod of 16L:8D). The plants of 14 days old were used for the experiments. The *S. frugiperda* larvae were collected from maize field (34°17′37.01″ N, 108°01′03.34″ E) in Yangling, Shaanxi Province, in July 2019, and individually put into plastic boxes (4 × 3 × 3 cm) and then brought back to the lab for rearing with maize seedlings in climatic chambers (LRH-400A-G3, Zhujiang®, Guangdong THK Scientific Instrument Co., Ltd., Shaoguan, Guangdong, China) at 25 ± 1 °C, 50–80% relative humidity and a photoperiod of 16:8 h (L:D).

2.2. Experimental Design

For field populations (Field), the 5th instar larvae of *S. frugiperda* were collected in the field and then brought back to the lab for dissection of the whole gut. The lab population of *S. frugiperda* (Lab0) was collected from the same field and was reared with maize seedlings under laboratory conditions. Field and lab populations were used to compare the differences of gut microbiome of *S. frugiperda* in different environments. The Lab0 population was continuously raised for 10 generations (Lab10) under laboratory conditions to validate the shaping of the gut microbiome by the environment. The artificial diet (DF) and maize leaves (MF) were used in rearing the *S. frugiperda* to test the effect of the diet on the gut bacterial composition, respectively. Artificial diet was improved according to Prasanna et al. [29] and the main ingredients are as follows: 180 mL distilled water, 15 g soybean powder, 12 g wheat bran, 2 g casein, 4 g yeast powder, 1.2 g ascorbic acid, 4 g agar, 150 mg choline chloride, 300 mg sorbic acid, 35 mg inositol, 30 mg cholesterol, 750 mg methyl-p-hydroxybenzoate, 0.1 mL formaldehyde.

2.3. Collection of Tissue Samples and DNA Extraction

The surface of *S. frugiperda* larvae and adults was washed with 0.5% NaClO for 2 min, 75% ethanol for 1 min and rinsed three times with sterilized-deionized water [30]. Previous studies have shown that the entire gut can provide a more accurate assessment of gut microbial composition [31], so the whole gut of *S. frugiperda* larvae was used in this study. The gut tissue was dissected in 0.01 M phosphate-buffered solution (PBS; PH7.4) under a dissecting microscope (Nanjing Jiangnan Novel Optics Co., Ltd., Nanjing, China). Due to the small size of the early instar larvae and eggs of *S. frugiperda*, a large number of samples were required for sequencing. Gut tissue collection for each replication at different developmental stages: the first instar larvae (L1) sample contained 500 individuals, the second instar larvae (L2) sample contained 300 individuals, the third instar larvae (L3) sample contained 100 individuals, the fourth instar larvae (L4) sample contained 50 individuals, the fifth instar larvae (L5) and the sixth instar larvae (L6) sample contained 5 individuals, respectively, the male (Male) and female (Female) adult sample contained 20 adults, respectively. The whole egg was used for sampling and each replicate contained

500 eggs. In addition, the fifth instars of lab (Lab0 and Lab10) and field (Field) populations, and artificial diet-feeding (DF) and maize leaf-feeding (MF) populations were used to collect the gut tissue. Each treatment included 3 replicates. The dissected gut tissue samples were collected into the 1.5 mL tube and were immediately flash frozen in liquid nitrogen and stored at −80 °C. The total nucleic acid was extracted using the FastDNA® SPIN Kit for Soil (MP Biomedicals, Qbiogene Inc., Carlsbad, CA, USA) following the manufacturer's protocol. The sterile PBS without insect tissue was used as a negative control both in DNA extraction and PCR amplification to detect reagents and environmental contamination [31]. The integrity and quality of the extracted DNA were evaluated on 1% agarose gel electrophoresis and a NanoDrop® ND-2000 spectrophotometer (Thermo Fisher Scientific, Wilmington, DE, USA), respectively [32].

2.4. Sequencing of 16S rRNA Gene

Targeted amplicons of the V3−V4 region of 16S rRNA gene were generated with primers 338F and 806R [33]. A 20 µL PCR reaction mixture contained 4 µL of 5× FastPfu Buffer, 2 µL of 2.5 mM dNTPs, 0.8 µL of Forward Primer (5 µM), 0.8 µL of Reverse Primer (5 µM), 0.4 µL of FastPfu Polymerase, 0.2 µL of BSA, and 10 ng of Template DNA. PCR amplification was conducted in ABI GeneAmp® 9700 following the conditions: 3 min at 95 °C, followed by 30 cycles of 30 s at 95 °C; 30 s at 50 °C; 45 s at 72 °C, and 10 min at 72 °C. All samples were amplified in triplicate. The PCR product was extracted from 2% agarose gel and purified using the AxyPrep DNA Gel Extraction Kit (Axygen Biosciences, Union City, CA, USA) according to manufacturer's instructions and quantified using Quantus™ Fluorometer (Promega, Madison, Madison, USA). Sequencing libraries were generated with TruSeq™ DNA Sample Prep Kit (New England Biolabs, Ipswich, SD, USA) and were sequenced on the Illumina MiSeq PE300 platform (Illumina, San Diego, CA, USA). Sequencing was performed by the Shanghai Majorbio Bio-pharm Technology Co., Ltd. (Shanghai, China) Thirty-nine DNA samples were sequenced successfully.

2.5. Microbiome Analyses

The PE reads obtained by Miseq sequencing were spliced according to the overlap relationship, and then quality-filtered by fastp version 0.19.6 and merged by FLASH [34]. The number of mismatches allowed by barcode was 0, and the maximum number of primer mismatches was 2. Raw data of the sequence were analyzed using QIIME. Reads that could not be assembled were discarded. Sequences with 97% similarity were clustered as operational taxonomic units (OTUs) using UPARSE. Manually filter the OTU table, i.e., remove chloroplast and mitochondria sequences in all samples. To minimize the effects of sequencing depth on alpha and beta diversity measure, the number of 16S rRNA gene sequences from each sample was rarefied, which still yielded an average Good's coverage of 99.09%, respectively. The classification of representative sequences for each OTU were analyzed using RDP Classifier against a 16S rRNA gene database (Silva v138) using a confidence threshold of 0.7. The microbiome function was predicted by PICRUSt2 based on OTU representative sequences. Bioinformatic analysis of the gut microbiota was carried out using the Majorbio Cloud platform (https://cloud.majorbio.com (accessed on 15 May 2022)). Based on taxonomic information, statistical analysis of community structure was performed at each classification level. On the basis of the above analysis, a series of in-depth statistical and visual analyses such as multivariate analysis and difference significance test were performed on the community composition and phylogenetic information of multiple samples. Alpha diversity including Chao1 richness, Ace index, Shannon index, and Simpson index were calculated with Mothur to investigate community diversity and community richness. The Unifrac distance matrices were constructed and visualized in principal coordinate analysis (PCoA). More details about the tools used are listed on Table S1.

2.6. Statistical Analysis

The PERMANOVA test was used to assess the percentage of variation explained by the treatment along with its statistical significance using Vegan v2.5–3 package. Statistical test of significance was performed for multiple (one-way ANOVA, LSD post hoc test) and two-group (Student's *t*-test, $p < 0.05$) treatments to detect statistical changes in community structure between treatments. These differences were considered significant at $p < 0.05$ level. Data were analyzed by using statistical software package SPSS 20.0 (SPSS Inc., Chicago, IL, USA). A similarity analysis (ANOSIM) was performed on bacterial communities at different developmental stages and different treatment groups.

3. Results

3.1. Sequencing Data of 16S rRNA

Negative controls are key to identify potential contamination. In this study, no bacteria were detected in the negative control, and the contamination of environmental and reagent microorganisms was excluded. Data sequencing and analysis of 39 samples for studying diversity were completed, and a total of 1,697,034 optimized sequences and 719,118,584 bases were obtained, with an average sequence length of 423 bp. Sequencing data statistics of all samples are shown on Table S2. The rarefaction curves of all samples reached a plateau stage, indicating that the sample numbers of all samples were sufficient (Figure S1). At the phylum level, Firmicutes, Proteobacteria, Actinobacteriota, Cyanobacteria, and Chloroflexi were the top five phyla.

3.2. Gut Microbiota Composition of *S. frugiperda* across Different Developmental Stages

To investigate the variability of *S. frugiperda* bacterial communities at different developmental stages, we collected the samples of egg, L1–L6, and adult (male and female). Our results showed that the microbial diversity in the egg stage was the highest, and the microbial diversity decreased dramatically after the eggs hatched into larvae; in the larval stage, L6 had the highest microbial diversity; the adult stage had the lowest community richness (Figure S2). Firmicutes were the most abundant bacterial community of the larval stage; the dominant bacterial phylum in the egg and adult stages was Proteobacteria, followed by Firmicutes (Figure 1A). At the genus level, *Ralstonia* was the most abundant bacterium in the egg stage, followed by Enterobacteriaceae, including *Enterobacteria*, *Klebsiella*, *Pantoea*, and *Escherichia*; the bacterial community composition of male and female adults was similar to that of early larvae stage (L1–L2), and OTUs with abundant content were *Enterococcus* and Enterobacteriaceae bacteria, including *Enterobacteria*, *Klebsiella*, *Pantoea*, and *Escherichia*; the bacterial community of L3 mainly consisted of *Enterococcus*; the community composition of the late larvae (L4–L6) harbored high proportions of *Enterococcus*, *Rhodococcus*, and *Ralstonia* (Figure 1B).

The community heatmap analysis at family level allowed us to view the community composition in more details (Figure 1D). During the egg stage, the most abundant OTUs were Enterobacteriaceae and Burkholderiaceae. The bacterial community composition of male and female adults was similar to early larvae stage (L1–L2), and OTUs with abundant content were Enterococcaceae and Enterobacteriaceae. The dominant OTUs in the L3 were Enterococcaceae. The community composition of the late larvae (L4–L6) was similar, and the abundant OTUs were Enterococcaceae and Enterobacteriaceae, followed by Burkholderiaceae and Nocardiaceae. Among them, Enterococcaceae had a higher abundance at all developmental stages. PCoA based on the weighted unifrac distance showed that the samples from male and female adults were the most uniform, sharing similarities (Figure 2). The similarity analysis results indicated that there were significant differences in the bacterial community of *S. frugiperda* across developmental stages (ANOSIM: $R = 0.533$, $p = 0.001$; PERMANOVA: $R = 0.061$, $p = 0.001$).

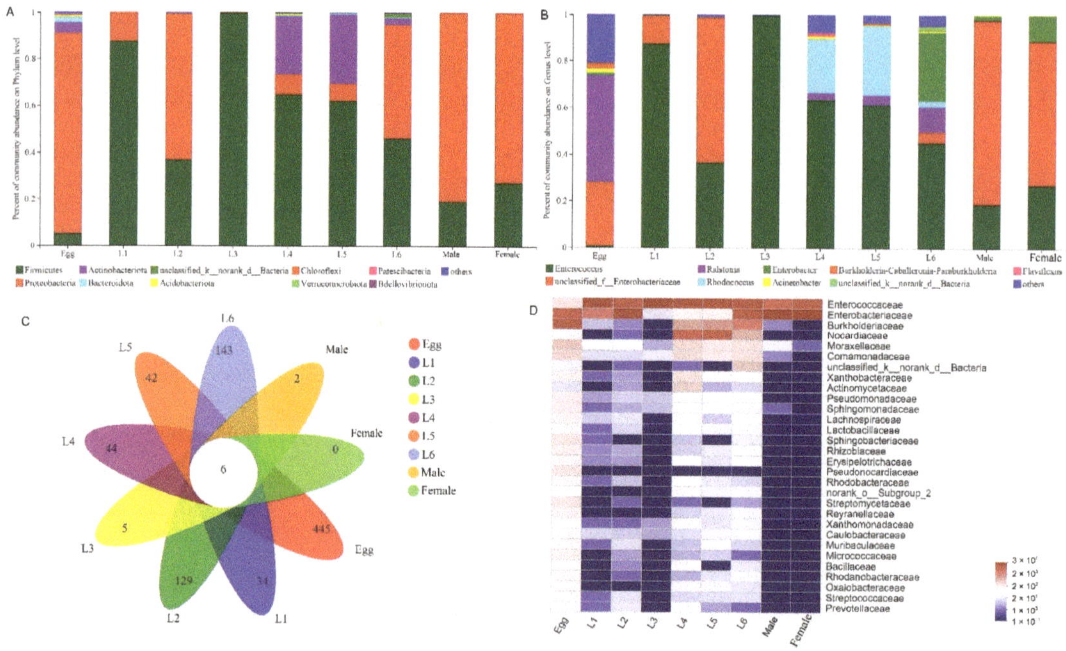

Figure 1. Gut bacterial community dynamics in different developmental life stages of *S. frugiperda*. (**A**) Gut bacteria composition at the phylum level; (**B**) Gut bacteria composition at the genus level; (**C**) Venn plot of OTUs in different developmental life stages; (**D**) Heatmap of the top 30 abundant families showing the relative abundance of the bacteria taxa assigned to a family level.

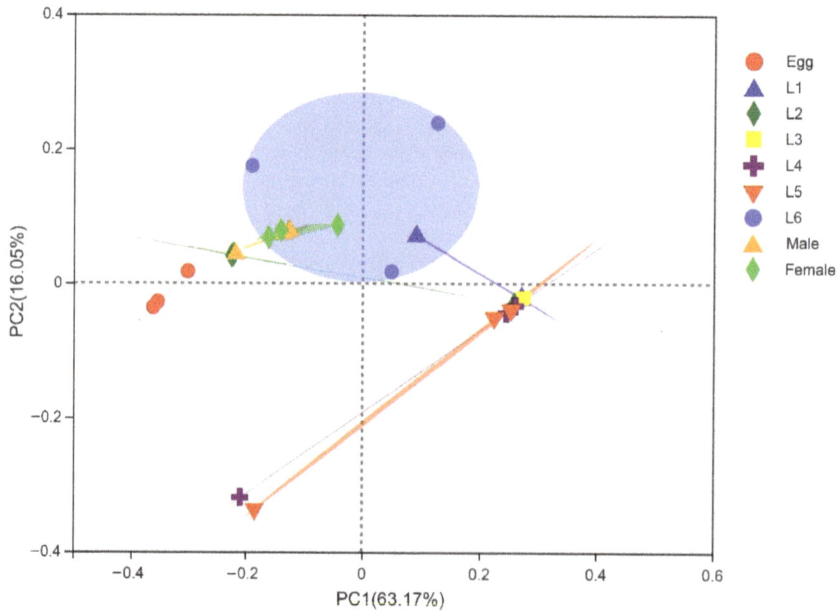

Figure 2. Principal coordinate analysis (PCoA) of community structure of different developmental life stages of *S. frugiperda*. Each symbol represents a sample.

3.3. Common and Unique Microbes among All Developmental Stages of S. frugiperda

Six OTUs, i.e., OTU478 (Proteobacteria, Enterobacteriaceae), OTU956 (Actinobacteriota, Corynebacteriaceae), OTU877 (Proteobacteria, Enterobacteriaceae), OTU346 (Firmicutes, Enterococcaceae), OTU784 (Firmicutes, Enterococcaceae), OTU884 (Proteobacteria, Moraxellaceae) were stable in different developmental stages of *S. frugiperda* (Figures 1C and S3A; Table S3). The microbiomes of female adults had no additional OTUs, but male adults had two unique OTUs, i.e., OTU1446 (Rhodocyclaceae) and OTU1260 (Rikenellaceae); the egg stage had the largest number and diversity of unique OTUs, with the highest proportions being Desulfitobacteriaceae (10.75%), Clostridia (8.36%) and Thermoanaerobacteraceae (5.97%) (Figure S3B); the most abundant unique OTUs in L1 were Dojkabacteria (22.22%), 37–13 (16.67%) and Run-SP154 (16.67%) (Figure S3C); in L2, the most abundant unique OTUs were Cyanobiaceae (10.91%), Subgroup_7 (10.91%) and Pirellulaceae (9.09%) (Figure S3D); in L3, unique OTUs consisted of Desulfomicrobiaceae (75%) and Calditrichaceae (25%) (Figure S3E); in L4, the most abundant unique OTUs were 11–24 (34.24%), PHOS-HE36 (25.76%), and Magnetospirillaceae (17.97%) (Figure S3F); in L5, the most abundant unique OTUs were Petrotogaceae (49.45%), Marinobacteraceae (10.99%), and Desulfuromonadia (10.99%) (Figure S3G); in L6, the most abundant unique OTUs were Proteobacteria (17.97%), Hymenobacteraceae (15.63%), and Leptospirillaceae (10.16%) (Figure S3H).

PICRUST analysis predicted that "Metabolic pathways" and "Biosynthesis of secondary metabolites" were abundant in all developmental stages of *S. frugiperda*. Phosphotransferase system (PTS) was more abundant in larvae and adults than in eggs. PTS mainly phosphorylates various sugars and their derivatives through the phosphorylation cascade and then transports them into the cell. Starch and sucrose metabolism were more abundant in larvae than in adults and eggs (Figure S4).

3.4. Comparison of Gut Bacterial Communities of S. frugiperda Associated with Different Environmental Habitats of Host

The more abundant common OTUs associated with laboratory and field populations of *S. frugiperda* were Moraxellaceae (23.57%), Microtrichaceae (5.25%), Nocardiaceae (4.87%), and Enterococcaceae (4.25%) (Figure S5A). Among the OTUs unique to the laboratory population, the higher contents were Dermatophilaceae (8.33%), Eggerthellaceae (6.25%), Spirochaetaceae (5%), and GEKB124 (4.17%) (Figure S5B). However, among the OTUs unique to the field population, the higher contents were Thermomicrobiaceae (16.49%), Syntrophomonadaceae (7.45%), Neisseriaceae (6.91%), and Cytophagaceae (4.26%) (Figure S5C). We employed Alpha diversity (Shannon's diversity, Simpson, Chao1, Ace) to estimate the diversity of the microbial community associated with laboratory and field populations of *S. frugiperda*. Alpha diversity analysis showed that there was no significant difference in microbial abundance and diversity between field and laboratory populations (Figure S6). Since the laboratory rearing conditions were stable without various adverse conditions, we explored the changes of gut microbiota in *S. frugiperda* when it was raised in laboratory conditions for more than 10 generations (about one year). The Alpha diversity index showed that both the gut microbiota diversity and community richness of the Lab0 generation were higher than those of the Lab10 generation (Figure S6). PCoA with similar degrees of bacterial communities showed that samples from laboratory populations clustered relatively tightly, but there were significant differences among field population samples (Figure 3). The gut microbial community of Lab0 generation was diverse, while Lab10 generation showed the higher relative proportion of *Enterococcus* and *Rhodococcus* (Figure 4). The similarity analysis results indicated that there were significant differences in the bacterial community of *S. frugiperda* associated with host environment habitat (ANOSIM: R = 0.449, $p = 0.044$; PERMANOVA: R = 0.566, $p = 0.007$).

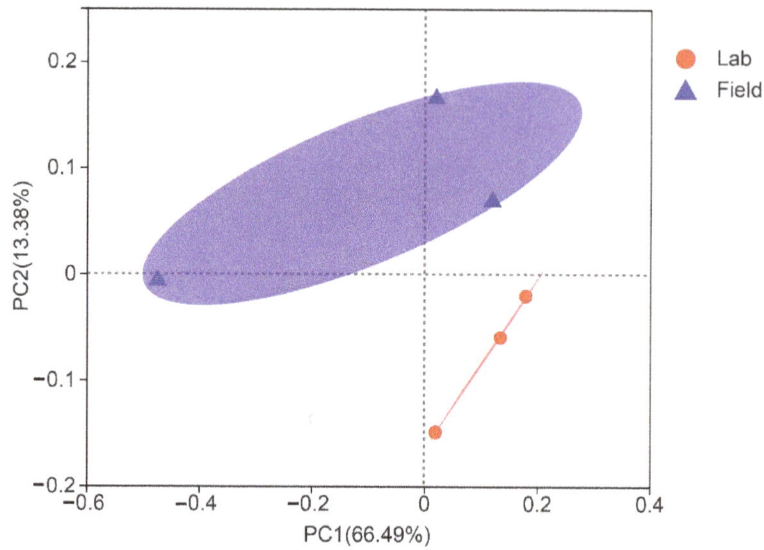

Figure 3. Principal coordinate analysis (PCoA) of community structure from lab and field groups of *S. frugiperda*.

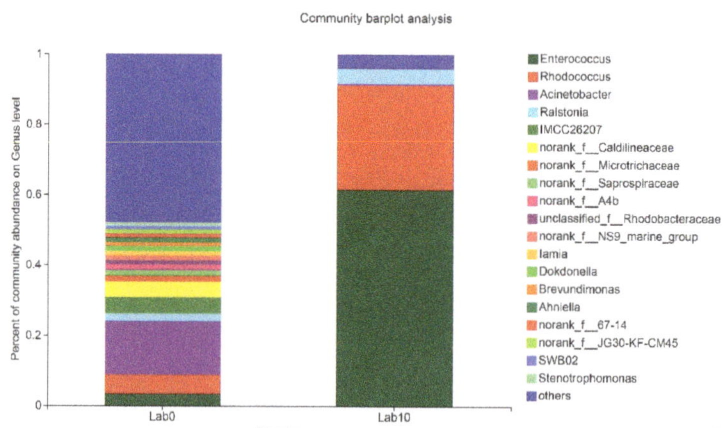

Figure 4. Relative abundance of bacterial composition of *S. frugiperda* after one year of laboratory rearing at the genus level.

3.5. Comparison of Gut Microbiota of S. frugiperda Fed Maize and Artificial Diet

PCoA analysis using Bray–Curtis indicated that the samples from feeding on artificial diet (DF) were the most uniform, while the samples from feeding on maize (MF) showed higher variation within groups (Figure 5). The Shannon and Simpson indices of the gut microbiota diversity of *S. frugiperda* fed with maize were higher than those fed with artificial diet. The Chao and Ace index suggested a higher community richness of *S. frugiperda* fed on maize compared with that fed on artificial diet (Figure S7). *Enterococcus* and *Rhodococccus* of *S. frugiperda* fed on maize showed the higher relative proportion, while the microbial community of *S. frugiperda* fed on artificial diet was composed mainly of *Enterococcus*, with a total of 98% of the gut microbiota (Figure S8). The similarity analysis results indicated that there were no significant differences in the bacterial community of *S. frugiperda* fed on maize and artificial diet (ANOSIM: R = 0.444, $p = 0.098$; PERMANOVA: R = 0.209, $p = 0.2$).

The function of gut microbiota was predicted using the KEGG level 2 and level 3, and the functions such as metabolism, cell growth and death, transport and catabolism, and environmental adaptation were more active in *S. frugiperda* fed on maize (Figure 6).

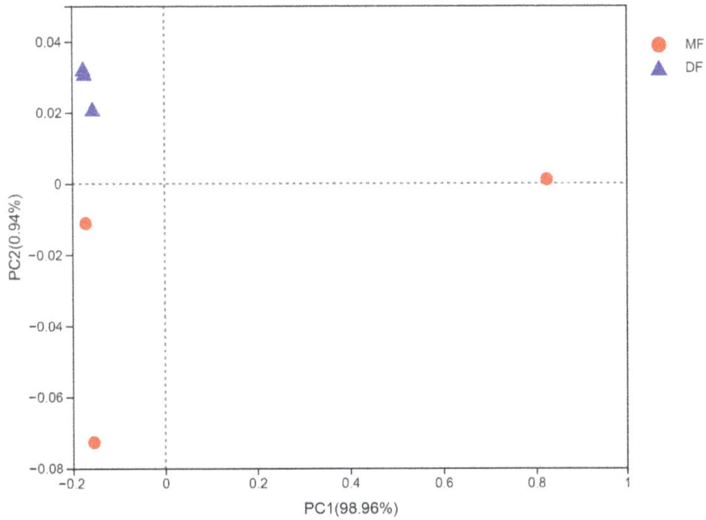

Figure 5. Principal coordinate analysis (PCoA) using the Bray–Curtis dissimilarity measurement comparing the alpha diversity of the bacterial community. MF, *S. frugiperda* fed on maize leaves; DF, *S. frugiperda* fed on artificial diet.

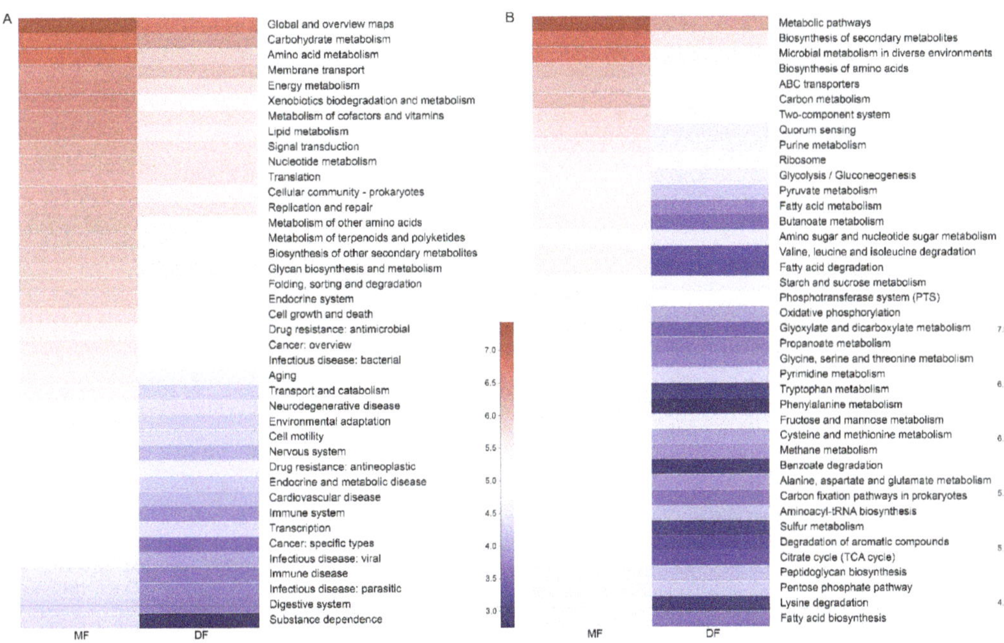

Figure 6. Functional gut microbiota profiles of *S. frugiperda* fed different diets at (**A**) KEGG-level 2 and (**B**) KEGG-level 3.

4. Discussion

Systematically analyzing the diversity of microbial communities is challenging due to the high complexity of sampling volume, sampling method, and sampling stage. For example, due to the small size of the eggs and early instar larvae of *S. frugiperda*, a large number of samples is required for sequencing. *S. frugiperda* is a major invasive pest with great reproduction and strong adaptability, which may rely on a variety of microbiota to quickly adapt to different environmental conditions, and such differences may provide a model for investigating and comparing microbial population dynamics. Although microbiomes associated with *S. frugiperda* have been reported in previous studies [23,24], few have investigated the dynamics of microorganisms. In this study, we found support for our hypotheses that *S. frugiperda* utilizes abundant gut microbial community to help it quickly adapt to the environment of the invasion site. Our results indicate that the bacteria in *S. frugiperda* were dominated by Firmicutes and Proteobacteria at the phylum level, which is consistent with previous studies in Lepidopterans [13,35–39]. However, we also found significant differences in the bacterial communities of *S. frugiperda*, which depend on the developmental stages (egg, larvae, and adults), diets, and environmental habitats.

In the present study, we found that *S. frugiperda* differed considerably in the microbial compositions across different life stages. The microbiota diversity was the highest in the egg stage. We speculated that this may be related to the lack of sterilization on the egg surface. When the eggs were sterilized, not enough microorganisms were extracted for sequencing. Therefore, the egg microorganisms might include two parts: most of them were carried by the egg itself, and a few might be the microorganisms in the environment when the egg contacted the environment. The larval gut microbiome was mainly composed of Firmicutes. The results were consistent with the findings of Chen et al. [36] in *Spodoptera littoralis* and Gomes et al. [24] in *S. frugiperda*. Since the food intake of the late larval instars (L4–L6) of *S. frugiperda* was significantly increased compared with that of early larval instars (L1–L3) and the body size grew faster, the changes in the gut microbiota were associated with the growth and development of the host insects, which was consistent with previous reports in *Bombyx mori* [40]. Many studies have shown that early larval stages are more sensitive to environmental changes, which are related to their body sizes and the development of their immune systems [41]. Therefore, the differences in gut microbes between early and late larval stages may also be related to host immunity. *S. frugiperda* is a holometabolous insect, and the gut of adults and larvae have a huge difference. The dynamics of insect gut microbiota can be determined by gut morphology and physicochemical conditions, such as pH and oxygen availability [21,42]. As insects go through their life cycle, gut morphology changes dramatically due to metamorphosis, and gut shape may affect oxygen availability [43,44]. These different gut conditions may lead to changes in the host-specific gut microbiota in insects. Our results showed that gut microbes also were detected in non-feeding adults that had just emerged for one day. Whether these microbes remain before the pupation or exist stably on the gut tissue of *S. frugiperda* requires further research.

Although there were differences in the gut microbiota of *S. frugiperda* during different developmental stages, Firmicutes and Proteobacteria were the dominant bacteria throughout the various developmental stages. The results were consistent with the findings of Broderick et al. [45] in *Lymantria dispar*, Priya et al. [46] in *Helicoverpa armigera*, Xia et al. [35] in *Plutella xylostella* and Chen et al. [40] in *Bombyx mori*. It is known that Proteobacteria and Firmicutes symbionts are involved in the digestion and nutritional utilization of a series of polysaccharides, including cellulose and hemicellulose [47–49]. Our results showed that Enterobacteriaceae and Enterococcaceae existed in the whole developmental stage of *S. frugiperda*. Similar to our results, Gomes et al. [24] reported that the dominant bacterium of *S. frugiperda* is Enterococcaceae in five Brazilian states. Enterobacteriaceae contributes to the synthesis of vitamins and pheromones and the degradation of plant compounds, and involves the process of nitrogen fixation and cellulose catabolism [50–52]. Enterococcaceae is reported in other Lepidopterous insects such as *Spodoptera litura*, *Manduca sexta*, and *H. armigera* [13,53,54]. *Enterococcus* within Enterococcaceae is able to degrade alkaloids and/or latex, suggesting that *Enterococcus* has a putative role in insect tolerance to their toxic diet [55]. The results obtained above indicate that these conservative bacterial communities could help herbivorous insects adapt to the host and play an important role in physiological metabolism.

Some studies have shown that diet and environment can greatly influence the structure of the host microbiota [56,57]. Our results showed that the diversity of the gut microbes of the laboratory-raised *S. frugiperda* was lower than that directly collected from the field. Correspondingly, the gut microbial diversity of *S. frugiperda* was also reduced after one year of continuous laboratory rearing. The environment of the field is more complex and variable than that of the laboratory, so the *S. frugiperda* may need more symbiotic microorganisms to defend against adverse environments or pathogens. In addition, the leaf microbiome of host plants can be enriched by the environmental microbiome, e.g., by rain splash or wind [58]. Previous studies have shown large differences in microbial titers between field and greenhouse-grown maize leaves [59], which may contribute to differences in gut microbes that were introduced into the gut of *S. frugiperda* through diet consumed.

Previous reports have shown that changing diet can dramatically alter the gut microbiome of the host insect [18,60]. Mason et al. [59] demonstrated that different diets affect the proliferation of gut microbes of *S. frugiperda* by counting colony forming units. Our results by 16S rRNA sequencing suggest that the gut microbiota of *S. frugiperda* fed with maize leaves and artificial diets is differs greatly. On the one hand, since the nutritional components of corn leaves and artificial diets are different, the differences in gut microbial composition of *S. frugiperda* may be related to different nutrient metabolism. A dynamic gut microbiome facilitates adaptation of herbivores to a new diet [61]. On the other hand, maize leaves contain microbes but the artificial diets are sterile, so differences in microbes introduced during feeding may lead to differences in gut microbes. Finally, plant tissues contain large amounts of indigestible and toxic compounds, so herbivorous insects have evolved a range of plant-adaptive strategies, including symbiosis with microbes to adapt to host plants.

In recent years, insect gut microbes have shown great application potential in the development of novel pest biological control strategies, such as *Bacillus thuringiensis* and *Pseudomonas protegens* species [62,63]. Luo et al. [64] reported that *Enterobacter*, *Providencia* and *Serratia* are highly attractive to *Bactrocera tau* adults, which provides a basis for the development of odor attractants made by microorganisms. The invasion of *P. protegens* type strain CHA0 leads to significant changes in gut microbes of *Pieris brassicae*, which eventually results in the death of insect hosts [65]. Therefore, the detailed characterization of the gut microbes of *S. frugiperda* may help to develop novel pest biological control strategies through the elimination of important symbiotic microorganisms or the discovery of entomopathogenic microorganisms.

5. Conclusions

The abundant gut microbes of *S. frugiperda* may be beneficial for its abilities of invasion and adaptation. In this study, we collected different *S. frugiperda* gut samples and performed 16S rRNA sequencing. Our results showed that *S. frugiperda* gut microbes vary greatly at different developmental stages and suggest vertical transmission of bacteria in *S. frugiperda*. Furthermore, we found that different environmental conditions and diets can also alter gut microbes. The detailed investigation of the gut microbiota of *S. frugiperda* provides a basis for future research. Since the plasticity of insect gut microbes helps insects utilize different foods and enhances adaptation of insects, a comprehensive understanding of *S. frugiperda*'s gut microbiome will help the development of novel pest control strategies for preventing this invasive pest.

Supplementary Materials: The following supporting information can be downloaded at: https://www.mdpi.com/article/10.3390/insects13090762/s1, Figure S1. Rarefaction curves of each sample based on Miseq sequencing. Figure S2. Comparison of Alpha diversity of gut microbiota across different life stages of *S. frugiperda*. Different letters indicate statistical significance (one-way ANOVA, LSD post hoc test, $p < 0.05$). (**A**) Shannon index; (**B**) Simpson index; (**C**) Chao index; (**D**) Ace index. The larger the Shannon value, the higher the community diversity. The larger the Simpson index value, the lower the community diversity. The larger the Chao and Ace index values, the higher the community richness. Figure S3. Bacterial communities of *S. frugiperda* among different developmental stages. (**A**) Shared bacteria communities between developmental stages; unique bacteria communities to egg (**B**), L1 (**C**), L2 (**D**), L3 (**E**), L4 (**F**), L5 (**G**), L6 (**H**). Figure S4. Gut microbiota of functional profiles of *S. frugiperda* across different life stages at KEGG-level 3. Figure S5. Shared and unique bacterial communities of *S. frugiperda* associated with laboratory and field populations. (**A**) shared bacterial communities of laboratory and field populations; (**B**) unique bacterial communities to laboratory populations and (**C**) field populations. Figure S6. Alpha diversity index of lab and field population of *S. frugiperda*. Different letters indicate statistical significance (one-way ANOVA, LSD post hoc test, $p < 0.05$). (**A**) Shannon index; (**B**) Simpson index; (**C**) Chao index; (**D**) Ace index. Figure S7. Alpha diversity index of *S. frugiperda* fed on different diets (Student's *t*-test, $p < 0.05$). (**A**) Shannon index; (**B**) Simpson index; (**C**) Chao index; (**D**) Ace index. MF: *S. frugiperda* was reared by maize leaves; DF: *S. frugiperda* was reared by artificial diet. Figure S8. Relative abundance of gut bacterial community of *S. frugiperda* related to different diets at the genus level. Table S1. Primers used in this study. Table S2. Sequencing data statistics of all samples. Table S3. Bacteria identified persist throughout different stages of development.

Author Contributions: Conceived and designed the experiments, S.-Z.Z. and D.-D.L.; analyzed the data, D.-D.L. and S.-Z.Z.; performed the experiments, D.-D.L. and J.-Y.L.; wrote the first draft, D.-D.L.; improved, S.-Z.Z., Z.-Q.H. and T.-X.L.; funding acquisition, S.-Z.Z. All authors have read and agreed to the published version of the manuscript.

Funding: This work was supported by Key Research and Development Program of Shaanxi (2021NY-038), National Key R&D Program of China (2017YFD0201006).

Institutional Review Board Statement: Not applicable.

Informed Consent Statement: Not applicable.

Data Availability Statement: Raw sequencing data were deposited in the NCBI Short Read Archive (SRA) BioProject PRJNA790707.

Acknowledgments: We are grateful for the assistance of all staff members and students in the Key Laboratory of Applied Entomology, Northwest A&F University at Yangling, Shaanxi, China.

Conflicts of Interest: The authors declare no conflict of interest.

References

1. Lu, M.; Hulcr, J.; Sun, J.H. The role of symbiotic microbes in insect invasions. *Annu. Rev. Ecol. Evol. Syst.* **2016**, *47*, 487–505. [CrossRef]
2. Douglas, A.E. Multiorganismal insects: Diversity and function of resident microorganisms. *Annu. Rev. Entomol.* **2015**, *60*, 17–34. [CrossRef] [PubMed]
3. Vivero, R.J.; Jaramillo, N.G.; Cadavid-Restrepo, G.; Soto, S.I.; Herrera, C.X. Structural differences in gut bacteria communities in developmental stages of natural populations of *Lutzomyia evansi* from Colombia's Caribbean coast. *Parasites Vectors* **2016**, *9*, 496. [CrossRef] [PubMed]
4. Feldhaar, H.; Straka, J.; Krischke, M.; Berthold, K.; Stoll, S.; Mueller, M.J.; Gross, R. Nutritional upgrading for omnivorous carpenter ants by the endosymbiont Blochmannia. *BMC Biol.* **2007**, *5*, 48. [CrossRef] [PubMed]
5. Tokuda, G.; Elbourne, L.D.; Kinjo, Y.; Saitoh, S.; Sabree, Z.; Hojo, M.; Yamada, A.; Hayashi, Y.; Shigenobu, S.; Bandi, C.; et al. Maintenance of essential amino acid synthesis pathways in the *Blattabacterium cuenoti* symbiont of a wood-feeding cockroach. *Biol. Lett.* **2013**, *9*, 20121153. [CrossRef]
6. Shigenobu, S.; Watanabe, H.; Hattori, M.; Sakaki, Y.; Ishikawa, H. Genome sequence of the endocellular bacterial symbiont of aphids *Buchnera* sp. APS. *Nature* **2000**, *407*, 81–86. [CrossRef]
7. McCutcheon, J.P.; Moran, N.A. Parallel genomic evolution and metabolic interdependence in an ancient symbiosis. *Proc. Natl. Acad. Sci. USA* **2007**, *104*, 19392–19397. [CrossRef]

8. Oliver, K.M.; Moran, N.A.; Hunter, M.S. Variation in resistance to parasitism in aphids is due to symbionts not host genotype. *Proc. Natl. Acad. Sci. USA* **2005**, *102*, 12795–12800. [CrossRef]
9. Cardoza, Y.J.; Klepzig, K.D.; Raffa, K.F. Bacteria in oral secretions of an endophytic insect inhibit antagonistic fungi. *Ecol. Entomol.* **2006**, *31*, 636–645. [CrossRef]
10. Florez, L.V.; Biedermann, P.H.; Engl, T.; Kaltenpoth, M. Defensive symbioses of animals with prokaryotic and eukaryotic microorganisms. *Nat. Prod. Rep.* **2015**, *32*, 904–936. [CrossRef]
11. Chen, B.S.; Zhang, N.; Xie, S.; Zhang, X.; He, J.; Muhammad, A.; Sun, C.; Lu, X.; Shao, Y. Gut bacteria of the silkworm *Bombyx mori* facilitate host resistance against the toxic effects of organophosphate insecticides. *Environ. Int.* **2020**, *143*, 105886. [CrossRef]
12. Wang, G.-H.; Dittmer, J.; Douglas, B.; Huang, L.; Brucker, R.M. Coadaptation between host genome and microbiome under long-term xenobiotic-induced selection. *Sci. Adv.* **2021**, *7*, eabd4473. [CrossRef]
13. Xiang, H.; Wei, G.F.; Jia, S.; Huang, J.; Miao, X.X.; Zhou, Z.; Zhao, L.-P.; Huang, Y.-P. Microbial communities in the larval midgut of laboratory and field populations of cotton bollworm (*Helicoverpa armigera*). *Can. J. Microbiol.* **2006**, *52*, 1085–1092. [CrossRef] [PubMed]
14. Adams, A.S.; Currie, C.R.; Cardoza, Y.; Klepzig, K.D.; Raffa, K.F. Effects of symbiotic bacteria and tree chemistry on the growth and reproduction of bark beetle fungal symbionts. *Can. J. For. Res.* **2009**, *39*, 1133–1147. [CrossRef]
15. Yun, J.H.; Roh, S.W.; Whon, T.W.; Jung, M.J.; Kim, M.S.; Park, D.S.; Yoon, C.; Nam, Y.-D.; Kim, Y.-J.; Choi, J.-H.; et al. Insect gut bacterial diversity determined by environmental habitat, diet, developmental stage, and phylogeny of host. *Appl. Environ. Microbiol.* **2014**, *80*, 5254–5264. [CrossRef] [PubMed]
16. Xue, Z.J.; Zhang, J.L.; Zhang, R.L.; Huang, Z.D.; Wan, Q.; Zhang, Z. Comparative analysis of gut bacterial communities in housefly larvae fed different diets using a high-throughput sequencing approach. *FEMS Microbiol. Lett.* **2019**, *366*, fnz126. [CrossRef] [PubMed]
17. Liu, Y.J.; Shen, Z.J.; Yu, J.M.; Li, Z.; Liu, X.X.; Xu, H.L. Comparison of gut bacterial communities and their associations with host diets in four fruit borers. *Pest Manag. Sci.* **2020**, *76*, 1353–1362. [CrossRef]
18. Perez-Cobas, A.E.; Maiques, E.; Angelova, A.; Carrasco, P.; Moya, A.; Latorre, A. Diet shapes the gut microbiota of the omnivorous cockroach *Blattella germanica*. *FEMS Microbiol. Ecol.* **2015**, *91*, fiv022. [CrossRef]
19. Chandler, J.A.; Lang, J.M.; Bhatnagar, S.; Eisen, J.A.; Kopp, A. Bacterial communities of diverse Drosophila species: Ecological context of a host-microbe model system. *PLoS Genet.* **2011**, *7*, e1002272. [CrossRef]
20. Wang, W.W.; He, P.Y.; Zhang, Y.Y.; Liu, T.X.; Jing, X.F.; Zhang, S.Z. The Population Growth of *Spodoptera frugiperda* on Six Cash Crop Species and Implications for Its Occurrence and Damage Potential in China. *Insects* **2020**, *11*, 639. [CrossRef]
21. Engel, P.; Moran, N.A. The gut microbiota of insects—Diversity in structure and function. *FEMS Microbiol. Rev.* **2013**, *37*, 699–735. [CrossRef] [PubMed]
22. Berasategui, A.; Salem, H.; Paetz, C.; Santoro, M.; Gershenzon, J.; Kaltenpoth, M.; Schmidt, A. Gut microbiota of the pine weevil degrades conifer diterpenes and increases insect fitness. *Mol. Ecol.* **2017**, *26*, 4099–4110. [CrossRef]
23. Rozadilla, G.; Cabrera, N.A.; Virla, E.G.; Greco, N.M.; McCarthy, C.B. Gut microbiota of *Spodoptera frugiperda* (J.E. Smith) larvae as revealed by metatranscriptomic analysis. *J. Appl. Entomol.* **2020**, *144*, 351–363. [CrossRef]
24. Gomes, A.F.F.; Omoto, C.; Cônsoli, F.L. Gut bacteria of field-collected larvae of *Spodoptera frugiperda* undergo selection and are more diverse and active in metabolizing multiple insecticides than laboratory-selected resistant strains. *J. Pest Sci.* **2020**, *93*, 833–851. [CrossRef]
25. Lv, D.B.; Liu, X.Y.; Dong, Y.L.; Yan, Z.Z.; Zhang, X.; Wang, P.; Yuan, X.Q.; Li, Y.P. Comparison of gut bacterial communities of fall armyworm (*Spodoptera frugiperda*) reared on different host plants. *Int. J. Mol. Sci.* **2021**, *22*, 11266. [CrossRef]
26. Shen, S.K.; Dowd, P.F. Detoxification spectrum of the cigarette beetle symbiont Symbiotaphrina kochii in culture. *Entomol. Exp. Appl.* **1991**, *60*, 51–59. [CrossRef]
27. Zhao, X.; Zhang, X.; Chen, Z.; Wang, Z.; Lu, Y.; Cheng, D. The divergence in bacterial components associated with *Bactrocera dorsalis* across developmental stages. *Front. Microbiol.* **2018**, *9*, 114. [CrossRef]
28. Aguirre, L.M.; Scully, E.D.; Trick, H.N.; Zhu, K.Y.; Smith, C.M. Comparative analyses of transcriptional responses of *Dectes texanus* LeConte (Coleoptera: Cerambycidae) larvae fed on three different host plants and artificial diet. *Sci. Rep.* **2021**, *11*, 11448. [CrossRef]
29. Prasanna, B.M.; Huesing, J.E.; Eddy, R.; Peschke, V.M. *Fall Armyworm in Africa: A Guide for Integrated Pest Management*; International Maize and Wheat Improvement Center (CIMMYT): Ciudad de Mexico, Mexico, 2018; pp. 51–54.
30. Chen, B.; Yu, T.; Xie, S.; Du, K.; Liang, X.; Lan, Y.; Sun, C.; Lu, X.; Shao, Y. Comparative shotgun metagenomic data of the silkworm *Bombyx mori* gut microbiome. *Sci. Data* **2018**, *5*, 180285. [CrossRef]
31. Zhang, N.; He, J.T.; Shen, X.Q.; Sun, C.; Muhammad, A.; Shao, Y.Q. Contribution of sample processing to gut microbiome analysis in the model Lepidoptera, silkworm *Bombyx mori*. *Comput. Struct. Biotechnol.* **2021**, *19*, 4658–4668. [CrossRef]
32. Muhammad, A.; He, J.T.; Yu, T.; Sun, C.; Shi, D.; Jiang, Y.; Xianyu, Y.; Shao, Y. Dietary exposure of copper and zinc oxides nanoparticles affect the fitness, enzyme activity, and microbial community of the model insect, silkworm *Bombyx mori*. *Sci. Total Environ.* **2022**, *813*, 152608. [CrossRef] [PubMed]
33. Xu, N.; Tan, G.C.; Wang, H.Y.; Gai, X.P. Effect of biochar additions to soil on nitrogen leaching, microbial biomass and bacterial community structure. *Eur. J. Soil Biol.* **2016**, *74*, 1–8. [CrossRef]

34. Magoc, T.; Salzberg, S.L. FLASH: Fast length adjustment of short reads to improve genome assemblies. *Bioinformatics* **2011**, *27*, 2957–2963. [CrossRef] [PubMed]
35. Xia, X.; Zheng, D.; Zhong, H.; Qin, B.; Gurr, G.; Vasseur, L.; Lin, H.; Bai, J.; He, W.; You, M. DNA sequencing reveals the midgut microbiota of diamondback moth, *Plutella xylostella* (L.) and a possible relationship with insecticide resistance. *PLoS ONE* **2013**, *8*, e68852. [CrossRef] [PubMed]
36. Chen, B.S.; Teh, B.S.; Sun, C.; Hu, S.R.; Lu, X.M.; Boland, W.; Shao, Y. Biodiversity and activity of the gut microbiota across the life history of the insect herbivore *Spodoptera littoralis*. *Sci. Rep.* **2016**, *6*, 29505. [CrossRef]
37. Snyman, M.; Gupta, A.K.; Bezuidenhout, C.C.; Claassens, S.; van den Berg, J. Gut microbiota of *Busseola fusca* (Lepidoptera: Noctuidae). *World, J. Microbiol. Biotechnol.* **2016**, *32*, 115. [CrossRef]
38. Bapatla, K.G.; Singh, A.; Yeddula, S.; Patil, R.H. Annotation of gut bacterial taxonomic and functional diversity in *Spodoptera litura* and *Spilosoma obliqua*. *J. Basic. Microbiol.* **2018**, *58*, 217–226. [CrossRef]
39. Xia, X.F.; Sun, B.T.; Gurr, G.M.; Vasseur, L.; Xue, M.Q.; You, M.S. Gut microbiota mediate insecticide resistance in the diamondback moth, *Plutella xylostella* (L.). *Front. Microbiol.* **2018**, *9*, 25. [CrossRef]
40. Chen, B.S.; Du, K.Q.; Sun, C.; Vimalanathan, A.; Liang, X.L.; Li, Y.; Wang, B.; Lu, X.; Li, L.; Shao, Y. Gut bacterial and fungal communities of the domesticated silkworm (*Bombyx mori*) and wild mulberry-feeding relatives. *ISME J.* **2018**, *12*, 2252–2262. [CrossRef]
41. Kyi, A.; Zalucki, M.P.; Titmarsh, I.J. An experimental study of early stage survival of Helicoverpa armigera (Lepidoptera: Noctuidae) on cotton. *Bull. Entomol. Res.* **1991**, *81*, 263–271. [CrossRef]
42. Dillon, R.J.; Dillon, V.M. The gut bacteria of insects: Nonpathogenic interactions. *Annu. Rev. Entomol.* **2004**, *49*, 71–92. [CrossRef] [PubMed]
43. Moll, R.M.; Romoser, W.S.; Modrzakowski, M.C.; Moncayo, A.C.; Lerdthusnee, K. Meconial peritrophic membranes and the fate of midgut bacteria during mosquito (Diptera: Culicidae) metamorphosis. *J. Med. Entomol.* **2001**, *38*, 29–32. [CrossRef] [PubMed]
44. Ke, J.; Sun, J.Z.; Nguyen, H.D.; Singh, D.; Lee, K.C.; Beyenal, H.; Chen, S.-L. In-situ oxygen profiling and lignin modification in guts of wood-feeding termites. *Insect Sci.* **2010**, *17*, 277–290. [CrossRef]
45. Broderick, N.A.; Raffa, K.F.; Goodman, R.M.; Handelsman, J. Census of the bacterial community of the gypsy moth larval midgut by using culturing and culture-independent methods. *Appl. Environ. Microbiol.* **2004**, *70*, 293–300. [CrossRef]
46. Priya, N.G.; Ojha, A.; Kajla, M.K.; Raj, A.; Rajagopal, R. Host plant induced variation in gut bacteria of *Helicoverpa armigera*. *PLoS ONE* **2012**, *7*, e30768. [CrossRef]
47. Engel, P.; Martinson, V.G.; Moran, N.A. Functional diversity within the simple gut microbiota of the honey bee. *Proc. Natl. Acad. Sci. USA* **2012**, *109*, 11002–11007. [CrossRef]
48. Anand, A.A.; Vennison, S.J.; Sankar, S.G.; Prabhu, D.; Vasan, P.T.; Raghuraman, T.; Geoffrey, C.J.; Vendan, S.E. Isolation and characterization of bacteria from the gut of bombyx mori that degrade cellulose, xylan, pectin and starch and their impact on digestion. *J. Insect Sci.* **2010**, *10*, 20. [CrossRef]
49. Warnecke, F.; Luginbühl, P.; Ivanova, N.; Ghassemian, M.; Richardson, T.H.; Stege, J.T.; Cayouette, M.; McHardy, A.C.; Djordjevic, G.; Aboushadi, N.; et al. Metagenomic and functional analysis of hindgut microbiota of a wood-feeding higher termite. *Nature* **2007**, *450*, 560–565. [CrossRef]
50. Lilburn, T.G.; Kim, K.S.; Ostrom, N.E.; Byzek, K.R.; Leadbetter, J.R.; Breznak, J.A. Nitrogen fixation by symbiotic and free-living spirochetes. *Science* **2001**, *292*, 2495–2498. [CrossRef]
51. Xu, J.; Gordon, J.I. Honor thy symbionts. *Proc. Natl. Acad. Sci. USA* **2003**, *100*, 10452–10459. [CrossRef]
52. Morales-Jimenez, J.; Zuniga, G.; Ramirez-Saad, H.C.; Hernandez-Rodriguez, C. Gut-associated bacteria throughout the life cycle of the bark beetle *Dendroctonus rhizophagus* Thomas and Bright (Curculionidae: Scolytinae) and their cellulolytic activities. *Microb. Ecol.* **2012**, *64*, 268–278. [CrossRef] [PubMed]
53. Brinkmann, N.; Martens, R.; Tebbe, C.C. Origin and diversity of metabolically active gut bacteria from laboratory-bred larvae of *Manduca sexta* (Sphingidae, Lepidoptera, Insecta). *Appl. Environ. Microbiol.* **2008**, *74*, 7189–7196. [CrossRef] [PubMed]
54. Thakur, A.; Dhammi, P.; Saini, H.S.; Kaur, S. Pathogenicity of bacteria isolated from gut of *Spodoptera litura* (Lepidoptera: Noctuidae) and fitness costs of insect associated with consumption of bacteria. *J. Invertebr. Pathol.* **2015**, *127*, 38–46. [CrossRef]
55. Vilanova, C.; Baixeras, J.; Latorre, A.; Porcar, M. The generalist inside the specialist: Gut bacterial communities of two insect species feeding on toxic plants are dominated by *Enterococcus* sp. *Front. Microbiol.* **2016**, *7*, 1005. [CrossRef] [PubMed]
56. Egert, M.; Marhan, S.; Wagner, B.; Scheu, S.; Friedrich, M.W. Molecular profiling of 16S rRNA genes reveals diet-related differences of microbial communities in soil, gut, and casts of *Lumbricus terrestris* L. (Oligochaeta: Lumbricidae). *FEMS Microbiol. Ecol.* **2004**, *48*, 187–197. [CrossRef]
57. Antwis, R.E.; Haworth, R.L.; Engelmoer, D.J.; Ogilvy, V.; Fidgett, A.L.; Preziosi, R.F. Ex situ diet influences the bacterial community associated with the skin of red-eyed tree frogs (*Agalychnis callidryas*). *PLoS ONE* **2014**, *9*, e85563. [CrossRef]
58. Zarraonaindia, I.; Owens, S.M.; Weisenhorn, P.; West, K.; Hampton-Marcell, J.; Lax, S.; Bokulich, N.A.; Mills, D.A.; Martin, G.; Taghavi, S.; et al. The soil microbiome influences grapevine-associated microbiota. *mBio* **2015**, *6*, e02527-14. [CrossRef]
59. Mason, C.J.; Clair, A.S.; Peiffer, M.; Gomez, E.; Jones, A.G.; Felton, G.W.; Hoover, K. Diet influences proliferation and stability of gut bacterial populations in herbivorous lepidopteran larvae. *PLoS ONE* **2020**, *15*, e0229848. [CrossRef]
60. Erkosar, B.; Yashiro, E.; Zajitschek, F.; Friberg, U.; Maklakov, A.A.; van der Meer, J.R.; Kawecki, T.J. Host diet mediates a negative relationship between abundance and diversity of Drosophila gut microbiota. *Ecol. Evol.* **2018**, *8*, 9491–9502. [CrossRef]

61. McMurdie, P.J.; Holmes, S. Phyloseq: An R package for reproducible interactive analysis and graphics of microbiome census data. *PLoS ONE* **2013**, *8*, e61217. [CrossRef]
62. Franzosa, E.A.; Hsu, T.; Sirota-Madi, A.; Shafquat, A.; Abu-Ali, G.; Morgan, X.C.; Huttenhower, C. Sequencing and beyond: Integrating molecular 'omics' for microbial community profiling. *Nat. Rev. Microbiol.* **2015**, *13*, 360–372. [CrossRef] [PubMed]
63. Xie, S.; Lan, Y.; Sun, C.; Shao, Y. Insect microbial symbionts as a novel source for biotechnology. *World J. Microbiol. Biotechnol.* **2019**, *35*, 25. [CrossRef] [PubMed]
64. Luo, M.; Zhang, H.; Du, Y.; Idrees, A.; He, L.; Chen, J.; Ji, Q.E. Molecular identification of cultivable bacteria in the gut of adult *Bactrocera tau* (Walker) and their trapping effect. *Pest Manag. Sci.* **2018**, *74*, 2842–2850. [CrossRef] [PubMed]
65. Vacheron, J.; Pechy-Tarr, M.; Brochet, S.; Heiman, C.M.; Stojiljkovic, M.; Maurhofer, M.; Keel, C. T6SS contributes to gut microbiome invasion and killing of an herbivorous pest insect by plant-beneficial *Pseudomonas protegens*. *ISME J.* **2019**, *13*, 1318–1329. [CrossRef]

insects

Article

Characterization of Microbial Communities from the Alimentary Canal of *Typhaea stercorea* (L.) (Coleoptera: Mycetophagidae)

Julius Eason and Linda Mason *

Department of Entomology, Purdue University, 901 West State Street, West Lafayette, IN 47907, USA; eason2@purdue.edu
* Correspondence: lmason@purdue.edu

Simple Summary: Hairy fungus beetle, *Typhaea stercorea,* is a secondary post-harvest pest of stored grains that thrives by feeding on mytoxigenic fungi. Bacterial communities residing in the alimentary canal of most insects contribute to their host's development. While there are many examples, little is known about the role of bacterial communities in the alimentary canal of *T. stercorea.* The objectives of this study were to (1) characterize the microbial communities residing in *T. stercorea* and (2) compare the microbial compositions of field-collected and laboratory-reared populations. In this study, we were able to identify bacterial communities that possess mycolytic properties and track mark changes in the microbiota profiles associated with development. The genus *Pseudomonas* was enriched in *T. stercorea* larvae compared to adults. Furthermore, field-collected *T. sterocrea* adults had a lower species richness than both larva and adult laboratory-reared *T. sterocrea.* Moreover, the gut microbial compositions of field-collected and laboratory-reared populations were vastly different. Overall, our results suggest that the environment and physiology can shift the microbial composition in the alimentary canal of *T. stercorea.*

Abstract: The gut microbiomes of symbiotic insects typically mediate essential functions lacking in their hosts. Here, we describe the composition of microbes residing in the alimentary canal of the hairy fungus beetle, *Typhaea stercorea* (L.), at various life stages. This beetle is a post-harvest pest of stored grains that feeds on fungi and serves as a vector of mycotoxigenic fungi. It has been reported that the bacterial communities found in most insects' alimentary canals contribute to nutrition, immune defenses, and protection from pathogens. Hence, bacterial symbionts may play a key role in the digestive system of *T. stercorea.* Using 16S rRNA amplicon sequencing, we examined the microbiota of *T. stercorea.* We found no difference in bacterial species richness between larvae and adults, but there were compositional differences across life stages (PERMANOVA:pseudo-$F_{(8,2)}$ = 8.22; p = 0.026). The three most abundant bacteria found in the alimentary canal of the larvae and adults included *Pseudomonas* (47.67% and 0.21%, respectively), an unspecified genus of the Enterobacteriaceae family (46.60 % and 90.97%, respectively), and *Enterobacter* (3.89% and 5.75%, respectively). Furthermore, *Pseudomonas* spp. are the predominant bacteria in the larval stage. Our data indicated that field-collected *T. stercorea* tended to have lower species richness than laboratory-reared beetles (Shannon: H = 5.72; p = 0.057). Furthermore, the microbial communities of laboratory-reared insects resembled one another, whereas field-collected adults exhibited variability (PERMANOVA:pseudo-$F_{(10,3)}$ = 4.41; p = 0.006). We provide evidence that the environment and physiology can shift the microbial composition in the alimentary canal of *T. stercorea.*

Keywords: *Typhaea stercorea*; fungivore; alimentary canal; 16S rRNA amplicon sequencing; bacterial symbionts

Citation: Eason, J.; Mason, L. Characterization of Microbial Communities from the Alimentary Canal of *Typhaea stercorea* (L.) (Coleoptera: Mycetophagidae). *Insects* **2022**, *13*, 685. https://doi.org/10.3390/insects13080685

Academic Editors: Hongyu Zhang and Xiaoxue Li

Received: 21 June 2022
Accepted: 24 July 2022
Published: 29 July 2022

Publisher's Note: MDPI stays neutral with regard to jurisdictional claims in published maps and institutional affiliations.

Copyright: © 2022 by the authors. Licensee MDPI, Basel, Switzerland. This article is an open access article distributed under the terms and conditions of the Creative Commons Attribution (CC BY) license (https://creativecommons.org/licenses/by/4.0/).

1. Introduction

Mycotoxins produced by pathogenic fungi have an economic impact on stored grains and pose a serious threat to food security. The mycotoxins of most concern are produced by certain species of *Aspergillus* and *Penicillium* that are commonly associated with stored grain products during processing and storage [1]. These fungal infestations are influenced by storage, environmental, and ecological conditions [2], which can affect the fungal growth on or in stored grains. Fungal contamination accounts for a large percentage ($\leq$25%) of direct losses in grain production [3,4] and can trigger secondary insect infestations [5]. The most prevalent fungal feeding insects associated with stored grain are *Ahasverus advena* (Waltl) (Coleoptera: Silvanidae), *Cryptolestes ferrugineus* (Stephens) (Coleoptera: Laemophloeidae), and *Typhaea stercorea* (L.) (Coleoptera: Mycetophagidae) [6], of which *T. stercorea* is the most common species found throughout fungal infested storage structures [7].

Typhaea stercorea, also known as the hairy fungus beetle, feeds on an array of fungi growing on stored grains and vectors of mycotoxigenic fungi throughout storage structures [8]. The fungi eaten by fungal feeders provide nutrients, such as sterols, that are essential for general maintenance and stimulating growth and reproduction, whereas insects lack the ability to synthesize sterols de novo [9,10]. These beetles complete their lifecycles on three fungal genera (*Aspergillus, Eurotium,* and *Penicillium*) that colonize stored grains [6]. Diet impacts the physiology and fitness of *T. stercorea* [6]; for example, beetles had a shorter larval development period and females laid more viable eggs when fed on *Aspergillus* followed by *Eurotium* and *Penicillum* fungal strains [6]. In addition, these beetles are able to complete their lifecycles when reared on high levels of mycotoxins produced by *A. flavus*, which are toxic to humans and other animals [6].

The insect body serves as an inclusive reservoir for microbial communities including bacteria, archaea, and fungi [11]. Diverse groups of microbial organisms can be found in most insects' alimentary canals [12,13]. As a result, microbiota account for up to 10% of the insect's biomass [11]. Insects and their associated microorganisms perform important functions that contribute to the degradation of organic matter, including fungi [14–16]. For example, two endosymbiotic bacteria, *Bacillus* and *Serratia*, secrete chitinolytic enzymes that degrade the chitin present in the cell walls of fungi, which supply nitrogen and carbon as a source of nutrients to their insect host [16,17]. In addition, several species of *Bacillus* have the capacity to express cyclic lipopeptides that inhibit the growth of certain species of *Aspergillus* [18,19]. These microbes affect the fitness of their insect hosts in different ways, such as nutrition uptake, immune defenses, and protection from pathogens [20]. Therefore, gut microbial communities tend to mediate functions that are essential to the fitness of their insect host.

In the past, it was difficult to identify microbiota and their roles within the gut microbiome. However, the advent of Next-Gen Sequencing (NGS) has allowed researchers to investigate the microbiome of insects by using amplicon sequencing. This technique is used to produce a taxonomic profile, allowing for the identification of underrepresented communities of microbes within an environment. Furthermore, this approach has been used to identify correlations between microbial communities and insect fitness [21]. The first step in understanding the interaction between gut microbes and their host is to characterize the microbiota. Measuring the diversity of the gut microbiome will serve as a starting point for understanding biological mechanisms.

The broad goal of this study was to understand the components and roles of gut microbial communities that are present in the alimentary canal of *T. stercorea.* The objectives of this study were to (1) characterize the microbial community residing in the alimentary canal of *T. stercorea* and (2) compare the microbial compositions of field-collected and laboratory-reared populations. Our hypotheses were (1) there will be a decreased microbiota composition from larval to adult life stages and (2) the gut microbiota of laboratory-reared adults and larvae will be less diverse when compared to field-collected adults. Using 16s rRNA amplicon sequencing, we examined the microbiota across all life stages, and compared diversity patterns in microbial communities of *T. stercorea* under laboratory and field conditions.

2. Materials and Methods

2.1. *Typhaea stercorea* Laboratory and Field Strains

Typhaea stercorea laboratory colonies were obtained from The Ohio State University where insect colonies were collected from spilled, moldy grain along the rail spur of a food manufacturer west of Columbus, OH, during the summer of 1986. Stock colonies were maintained on an oat/yeast/agar diet (50 g of rolled oats, 5 g of brewers' yeast, 2 g of agar, and 15 mL of water) in glass jars (800 mL) sealed with a double layer of filter paper as a lid. Insect cultures were incubated at 30 ± 0.5 °C and 72% r.h.

Field strains of *T. stercorea* were collected from Throckmorton Purdue Agricultural Center (TPAC) using pitfall traps during the months of June and July 2018 (Figure S1). The piftall traps included a Mason jar, polyvinyl chloride (PVC) sheet, and wire rods. The Mason jars contained maize, which was infested with *A. flavus* to attract *T. stercorea*. The presence of *T. stercorea* was monitored weekly and adult *T. stercorea* were removed from the pitfall traps using an aspirator.

2.2. Insect Preparation and Dissection

Newly emerged late instar larvae (Figure 1A) and adults (Figure 1B) of *T. stercorea* obtained from the lab colony were collected and placed in a soufflé cup (5.5 oz) containing a cotton ball saturated with 200 µL of water, then re-treated daily with 100 µL of water for three days. After the three-day holding period, insects were randomly selected for dissection. Field-collected insects were dissected within three hours of trapping. Insect specimens were prepared for gut dissection by surface sterilization in 70% ethanol. Late instar larva and adult stages were dissected in ethanol, and the whole alimentary canal was separated from the body with the help of fine-tip forceps under microscope. Fifteen whole alimentary canals were pooled in sterile 1.5 mL centrifuge tubes containing 800 µL of ethanol and stored at −20 °C until use. A total of 5 tubes were prepared for each treatment group: laboratory-reared newly emerged late instar larva, laboratory-reared newly emerged adult and field-collected adult *T. stercorea*.

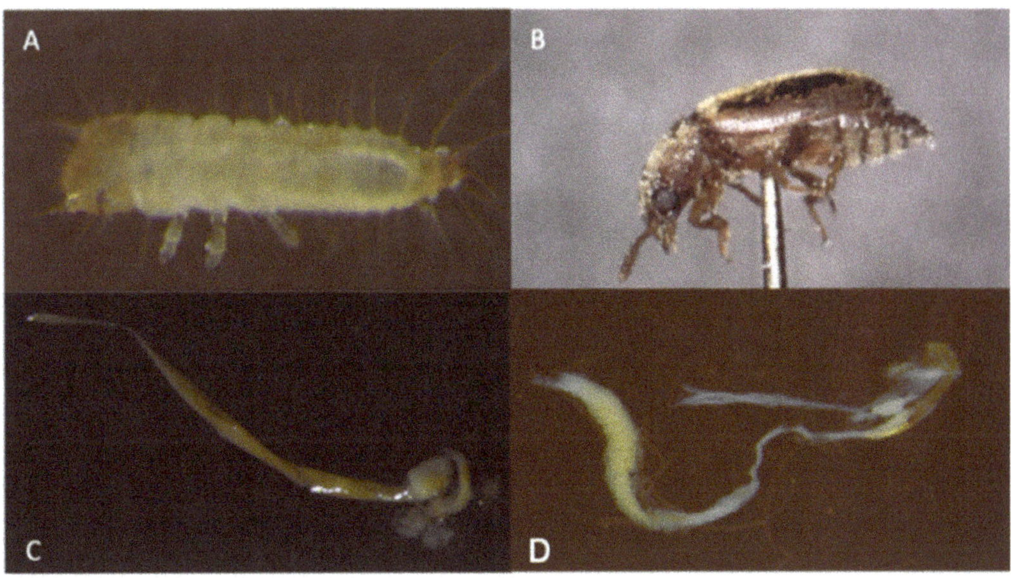

Figure 1. Different life stages of *T. stercorea*, including their alimentary canal. (**A**) Late instar larva. (**B**) Adult. (**C**) Alimentary canal of late instar larva. (**D**) Alimentary canal of adult.

2.3. DNA Extraction and Sequencing of Bacterial 16S rRNA Gene

Five replicates of each group (late instar larva laboratory, adult laboratory, and adult field) contained 15 dissected tissue samples, which were used for DNA extraction (Qiagen DNeasy Blood and Tissue Kit: Valencia, CA, USA). Prior to DNA extraction, samples were centrifuged for 20 min at 15000 rpm to separate the tissues from the 70% ethanol, and then the ethanol was removed. Afterwards, the tissue was pulverized using 0.5 mL pellet pestle for 1 min. All subsequent steps were completed using the standard manufacturer's protocol, including 4 h proteinase K digestion. Elutions were carried out with 75 µL buffer AE. The total DNA concentrations of all samples were determined on a NanoDrop 2000c Spectrophotometer (Thermo Scientific: Wilmington, DE, USA).

Samples were sent to the University of Minnesota Genomic Center (UMGC) for sequencing by high-throughput, paired-end (2 × 300 bp) sequencing, MiSeq technology (Illumina). The UMGC measured the amount of bacterial DNA present in the extracts with quantitative PCRs (qPCRs) of the bacterial 16S rRNA gene for quality control. Samples that were above 500 copy number (molecules/µL) were processed for library construction and sequencing. We used the V3F_Nextera_375F (5′-TCGTCGGCAGCGTCAGATGTGTATAAGAG ACAGCCTACGGGAGGCAGCAG) and Meta_V4_806R (5′-GTCTCGTGGGCTCGGAGAT GTGTATAAGAGACAGGGACTACHVGGGTWTCTAAT) universal bacterial primers to target the V3V4 region of the 16S rRNA gene of all bacteria and archaea present.

2.4. Sequence Curation

Demultiplexed raw sequences were extracted from the Illumina MiSeq system in FASTQ format. After removing low-quality sequences, paired-end reads were merged and curated using the Qiime2 software package v2 [22]. We imported the raw reads into Qiime2 using the format PairedEndFasqManifestPhred33, which allowed us to determine how many sequences per sample. For quality control, we used the DADA2 method to filter sequences, denoise, merge, and remove chimeras [23]. Quality filtering allowed us to trim off bases of each sequence, which removed low-quality regions of the sequences. Based on the demultiplex summary stat, we kept all 300 bp of the forward reads, but we removed 30 bp from the tail end of the reverse reads. After identifying the unique sequences and their frequency in each sample, sequences were aligned to the rRNA database project Silva_v132 and split into groups corresponding to their taxonomy at the level of genus and then assigned to operational taxonomic units (OTUs) at a 1% dissimilarity level. The analysis of composition of microbes (ANCOM) was applied to identify features that are differentially abundant across sample groups assuming that less than 25% of the features are changing between treatment groups [24].

2.5. Statistical Analyses

2.5.1. Statistical Analyses of *T. stercorea* Life Stage Microbiota

For *T. stercorea*, life stage analysis was carried out in Qiime2. The sequence data were rarefied to a sequencing depth of 7848 sequence count (larva—lab strain (N = 4) and adult—lab strain (N = 4)). The results were plotted on an alpha–rarefaction curve using a max depth of 24,000 (Figure S2) and Permanova test was performed to distinguish the differences between life stages. Permanova is a robust non-parametric test of the general multivariate hypothesis of differences in the composition and/or relative abundances of organisms of different species in samples from different groups or treatments [25]. To characterize microbial alpha-diversity (species richness, choa 1 index, evenness, and Shannon's diversity index), between life stages were statistically tested by Kruskal–Wallis test using the observed OTU table generated in Qiime2. The Kruskal–Wallis test is a non-parametric test used to observe the mean differences between treatments [26]. Observed OTU was used for a qualitative measure of community richness, and evenness (Pielou's Evenness) was used to measure the community evenness [27]. Beta diversity was analyzed using the Jaccard, Bray–Curtis, unweighted UniFrac and weighted UniFrac distances to compare the different groups and plotted in a principle coordinate analysis (PCoA). The Jaccard distance is a qualitative measure of the community dissimilarity, whereas the unweighted UniFrac is also qualitative measure that incorporates phylogenetic relationships between the features (OTUs) [27]. The Bray–Curtis distance is a quantitative measure of community dissimilarity, whereas weighted UniFrac is also a quantitative measure of community dissimilarity that incorporates phylogenetic relationships between the features [27]. A p-value of ≤ 0.05 was used to indicate significant differences between groups.

2.5.2. Statistical Analysis Comparing Field-Collected Adults to Laboratory-Reared Larvae and Adults

To evaluate if the microbiota vary between laboratory-reared and field-collected *T. stercorea*, we compared larva (N = 4) and adult (N = 4) that were reared on artificial diet in laboratory conditions to field-collected adults (N = 2). Sequences were rarefied to a sequencing depth of 66 sequences counts. The results were plotted on an alpha–rarefaction curve to max depth of 200 (Figure S3). The statistical analysis followed the procedures described above for *T. sterocera* life stage microbiota.

3. Results

3.1. Analysis of Bacterial 16S rRNA Gene Sequences

The V3V4 region of the 16S rRNA was amplified and sequenced from the alimentary canal of the larvae and adult *T. stercorea*. A MiSeq (Illumina: Minneapolis, MN, USA) instrument was used to obtain a mean of 20,773 sequences per sample. The sequences were processed and filtered through the QIIME2 pipeline [22], and a total of 112 unique operational taxonomic units (OTUs) were obtained among the samples.

3.2. Alpha and Beta Diversity of T. stercorea

Observed OTUs and Chao1 were used to measure richness within a sample, while evenness measured the relative abundance of species richness (Figure 2). The Shannon Index was used to measure alpha diversity by measuring the number of different species and their relative abundance within a sample. The analyses of *T. stercorea* microbiota diversity showed that species richness using both observed OTUs and Chao 1 was not significantly different across life stages (Figure 2A,C). The adults showed larger variation in species richness, which had higher and more constant value when compared to larvae. In addition, evenness exhibited no significant difference between life stages (Figure 2B). The alpha diversity using the Shannon Index revealed that there were no significant differences across life stages (Figure 2D). Although larva samples showed more variation, adult samples had higher and more constant values. The beta diversities (Jaccard, Bray–Curtis, unweighted UniFrac and weighted UniFrac) were used to measure the gut microbiota composition. These dissimilarity matrices were observed in a principle coordinates analysis plot, where samples were separated according to their life stage (Figure 3A,D). PERMANOVA was calculated using distance matrices (Jaccard and Bray–Curtis), which indicated that the gut microbiota composition was significantly different between life stages (Table 1, $p < 0.05$). However, unweighted UniFrac and weighted UniFrac analyses exhibited no significant differences in species composition between larva and adult *T. stercorea*, suggesting that the phylogenetic distances between observed OTUs were weakening the beta diversity (Table 1, $p > 0.05$).

Table 1. PERMANOVA analysis based on distance matrices between laboratory-reared *T. stercorea* larvae and adults.

Beta Diversity Measure (PERMANOVA)	Pseudo-F	p-Value
Jaccard	1.91	0.02
Bray–Curtis	8.22	0.03
Unweighted UniFrac	1.33	0.26
Weighted UniFrac	8.05	0.06

A p-value of < 0.05 was used to indicate significant differences between groups.

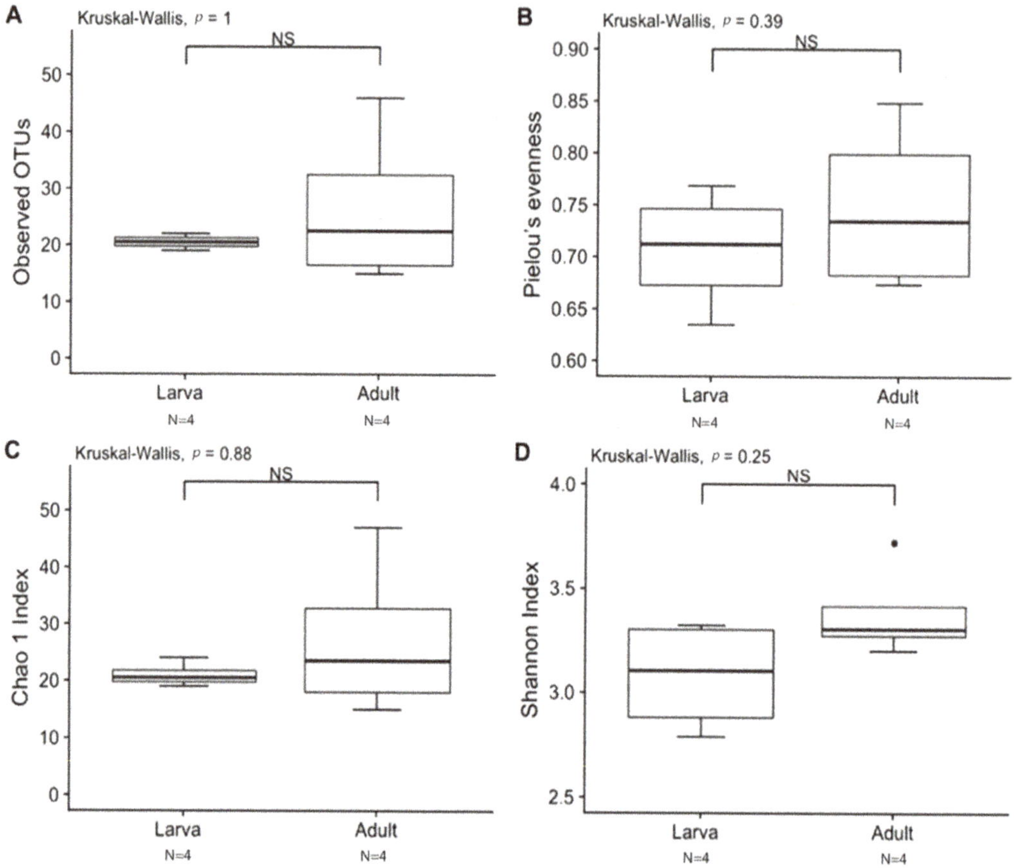

Figure 2. Alpha diversity between laboratory-reared larva and adult *T. stercorea* gut microbiota. A *p*-value of ≤ 0.05 was used to indicate significant differences between groups. NS denotes non-significant differences between the groups. N denotes the sample size. (**A**) Species richness boxplot. (**B**) Pielou's evenness boxplot. (**C**) Chao 1 Index boxplot. (**D**) Shannon Index boxplot. For each group, the bars delineate the means, the hinges represent the lower and upper quartiles, the whiskers extend to the most extreme values, and black dots (•) denote outliers are plotted if present.

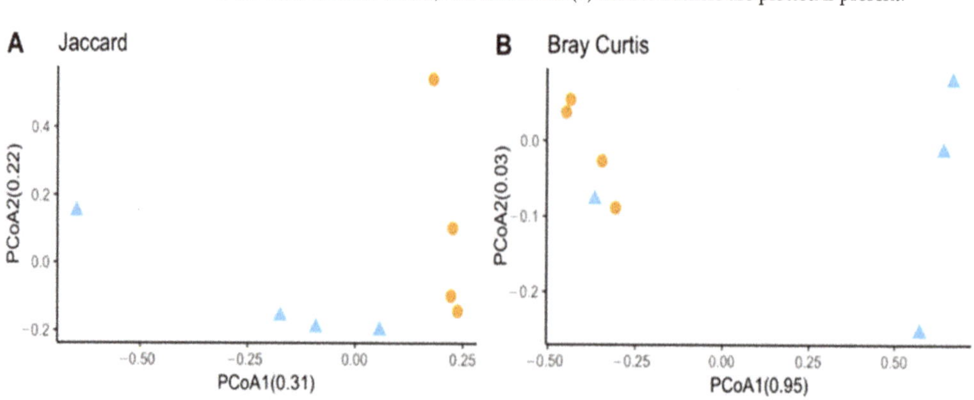

Figure 3. *Cont.*

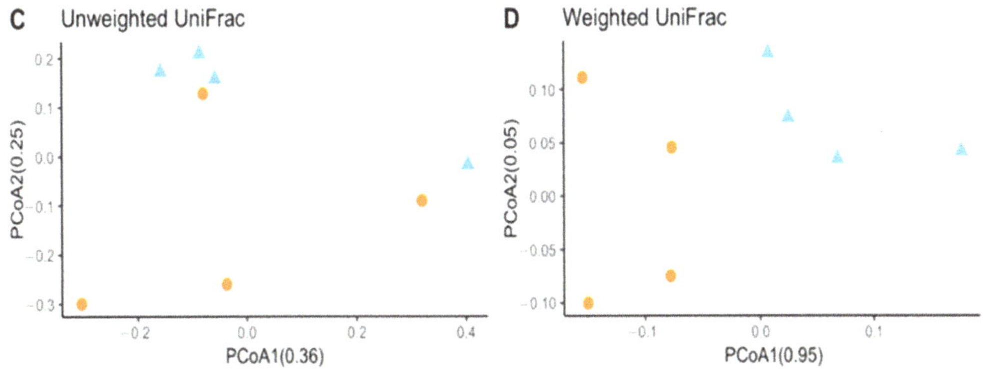

Figure 3. Beta diversity between laboratory-reared *T. stercorea* larva and adult gut microbiota composition. (**A**) Jaccard PCoA graph showing PCoA1 (0.31 variation) and PCoA2 (0.22 variation). (**B**) Bray–Curtis PCoA graph showing PCoA1 (0.95 variation) and PCoA2 (0.02 variation). (**C**) Unweighted UniFrac PCoA graph showing PCoA1 (0.36 variation) and PCoA2 (0.25 variation). (**D**) Weighted UniFrac PCoA graph showing PCoA1 (0.95 variation) and PCoA2 (0.05 variation). Blue triangles (▲) denote larvae and orange circles (•) denote adults.

3.3. *Microbial Taxonomic Composition in the Alimentary Canal Shows Differences between Life Stages of Typhaea stercorea*

According to the Silva v.132 database, there were 112 OTUs aggregated into 38 genera and 8 phyla. Of these phylotypes, 97.81% belong to the phylum Proteobacteria (data not shown). The two most abundant genera in the larvae and adults were an unspecified genus of the Enterobacteriaceae family (46.60% and 90.97%, respectively) and *Enterobacter* (3.89% and 5.75%, respectively) (Figure 4A). The analysis of composition of microbiomes (ANCOM) was conducted to identify the taxa that was differentially abundant across life stages of *T. stercorea*. With the use of ANCOM, we observed that the genus *Pseudomonas* was significantly different in the gut composition between larvae and adults (47.67% and 0.21%, respectively) (Figure 4A). Larvae had the most genera present at greater than 1%, with three genera identified, and adults had two genera (Figure 4A).

In order to describe which OTUs were present in both larva and adult life stages, an OTU was assumed to be present if it was observed in at least two of replications in each life stage. The data show that adults have higher numbers of OTUs present in their alimentary canal compared to larvae (Figure 4B). The intersection between the larva and adult stages of *T. stercorea* shares four OTUs throughout their lifecycle (Figure 4B). The shared OTUs belong to the genera *Enterobacter*, *Streptomyces*, *Staphylococcus*, and an unspecified genus belonging to the family Enterobacteriaceae.

3.4. *Variation in Diversity and Microbial Composition between Laboratory-Reared and Field-Collected Populations of T. stercorea*

The species richness of laboratory-reared (larva and adult) and field-collected populations of *T. stercorea* was not significantly different (Chao 1 Index: Kruskal–Wallis $p = 0.11$) (Figure 5A), although the laboratory-reared insects exhibited a trend of higher species richness than field-collected specimens. Field-collected adults tended to have a lower alpha diversity using the Shannon Index compared to laboratory-reared larvae and adults (Kruskal–Wallis (all groups): $H = 5.72$; $p = 0.057$) (Figure 5B). The species diversity of laboratory-reared larvae was not significantly different when compared to laboratory-reared adults (Kruskal–Wallis (pairwise: $H = 2.08$; $p = 0.148$), which it corroborated the previous objective. The Jaccard, Bray–Curtis, and weighted UniFrac analyses showed three distinct clusters, suggesting that the microbial compositions in the alimentary canal between laboratory-reared and field-collected populations are different (Figure 6 and

Table 2). The unspecified genus of the Enterobacteriaceae family was still the dominant taxa, representing 27.4% and 90.5% of the microbiota in laboratory-reared larvae and adults, respectively, and 50.0% in field-collected adults (Figure 7). The field-collected adults had the most genera present at greater than 5%, with seven genera identified (Figure 7). Of these seven genera, *Apibacter*, *Alcaligenes*, and *Enterobacter* were identified at the genus level, while the other phylotypes were classified as unspecified genera in the Bacillaceae, Bacillales, Enterobacteriaceae, and Weeksellaceae families.

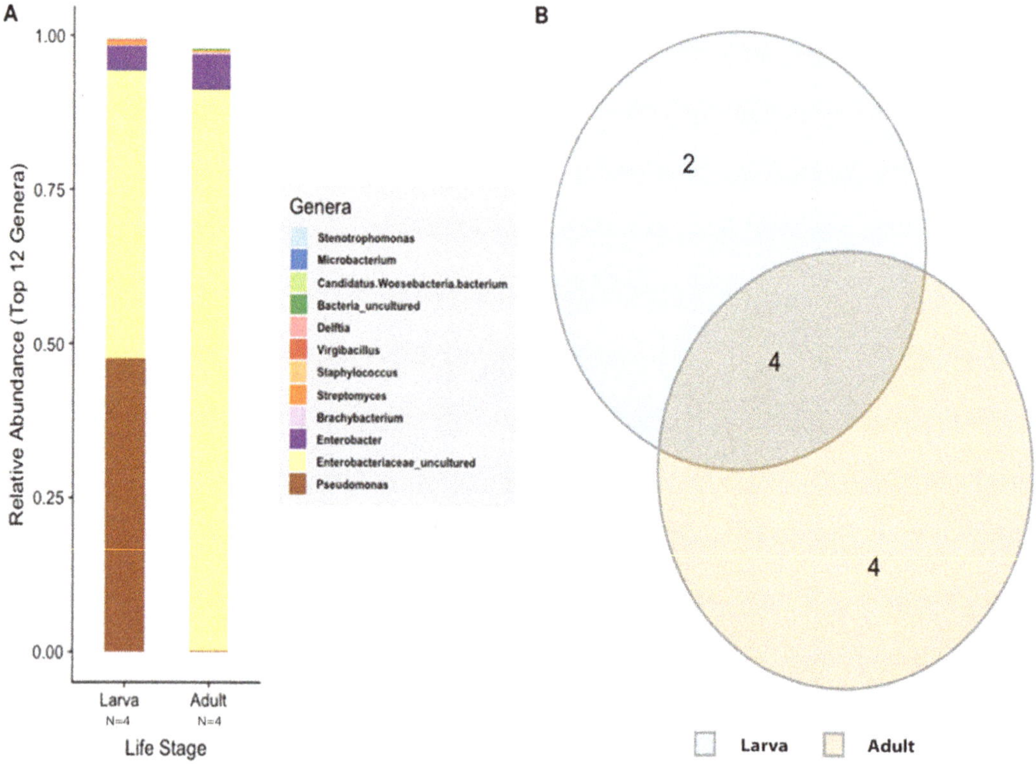

Figure 4. Gut microbiota composition of *T. stercorea*. (**A**) Taxonomy graph comparing the relative abundances of genera present between larva and adult *T. stercorea*. The 12 most abundant genera are shown. (**B**) Two-part Venn diagram comparison between laboratory-reared larva and adult of *T. stercorea* gut microbiota, showing the OTUs shared among life stages: larva (blue) and adult (orange). The numbers in the diagrams represent how many OTUs were unique in life stages or shared between life stages as their areas intersect.

Table 2. PERMANOVA analysis based on distance matrices between laboratory-reared larvae and adults of *T. stercorea* gut microbiota vs. field-collected adults.

Beta Diversity Measure (PERMANOVA)	Pseudo-F	*p*-Value
Jaccard	4.23	0.002
Bray–Curtis	4.41	0.006
Unweighted UniFrac	1.05	0.41
Weighted UniFrac	3.70	0.01

A *p*-value of < 0.05 was used to indicate significant differences between groups.

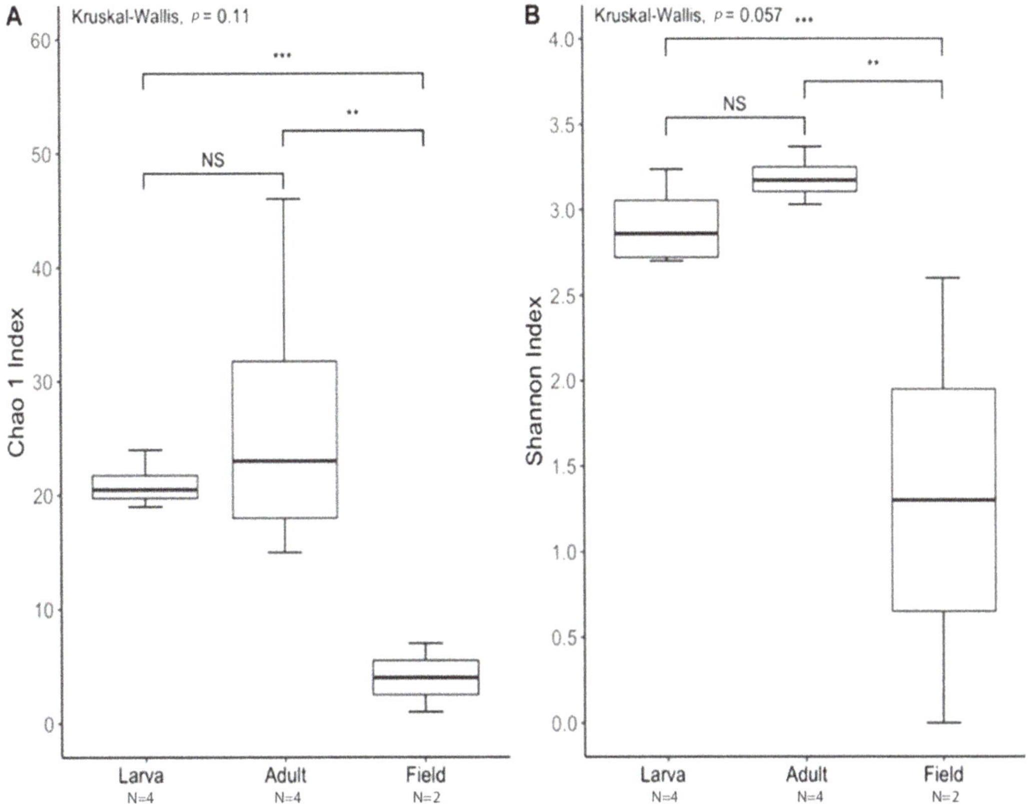

Figure 5. Alpha diversity between laboratory-reared larvae and adults vs. field-collected *T. stercorea* adults. A *p*-value of ≤ 0.05 was used to indicate significant differences between groups. ** denotes significant differences between laboratory-reared adult and field-collected adult. *** denotes significant differences between laboratory-reared larva and field-collected adult. NS denotes non-significant differences between groups. N denotes the sample size. (**A**) Chao 1 Index. (**B**) Shannon Index. For each group, the bars delineate the means, the hinges represent the lower and upper quartiles, the whiskers extend to the most extreme values, and outliers are plotted if present.

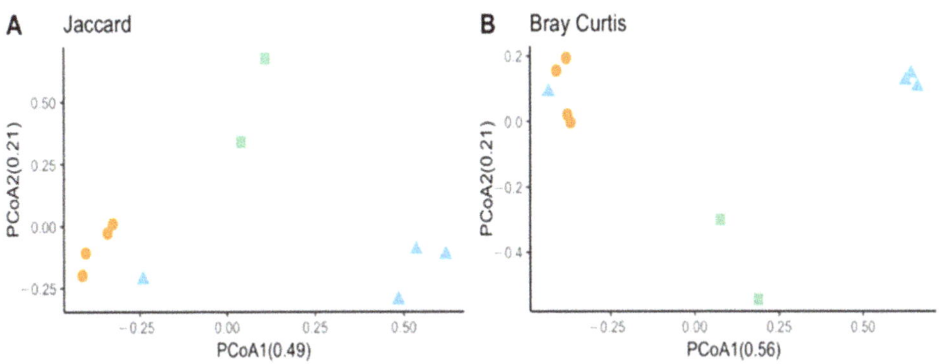

Figure 6. *Cont.*

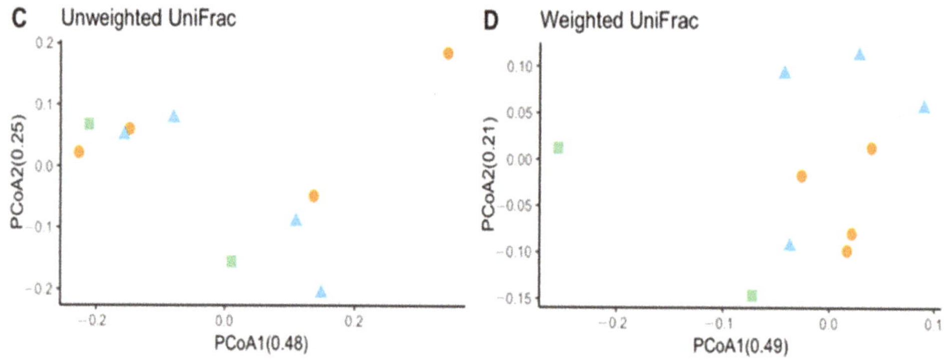

Figure 6. Beta diversity between laboratory-reared *T. stercorea* larva and adult gut microbiota composition vs. field-collected adult. (**A**) Jaccard PCoA graph showing PCoA1 (0.48 variation) and PCoA2 (0.25 variation). (**B**) Bray–Curtis PCoA graph showing PCoA1 (0.56 variation) and PCoA2 (0.21 variation). (**C**) Unweighted UniFrac PCoA graph showing PCoA1 (0.48 variation) and PCoA2 (0.25 variation). (**D**) Weighted UniFrac PCoA graph showing PCoA1 (0.49 variation) and PCoA2 (0.21 variation). Blue triangles (▲) denote laboratory-reared larvae, orange circles (•) denote laboratory-reared adults, and green squares (■) denote field-collected adults.

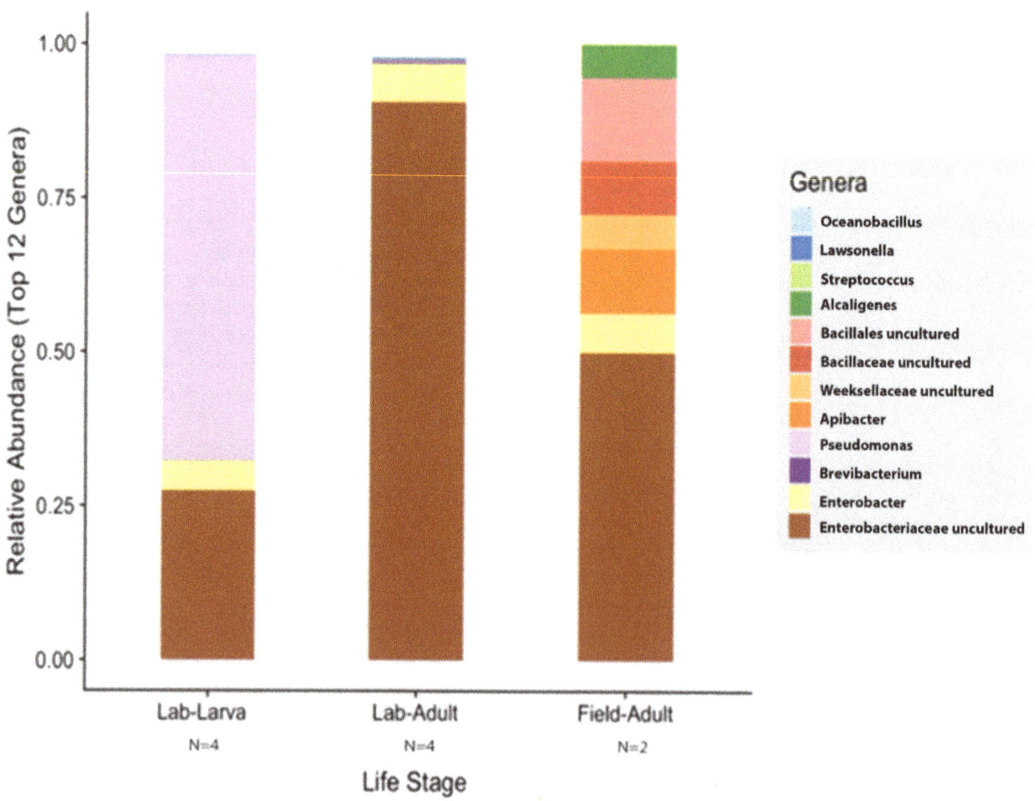

Figure 7. Comparison of laboratory-reared larvae and adults vs. field-collected adult gut microbiota composition of *T. stercorea*. Taxonomy graph comparing the relative abundance of genera present in *T. stercorea*. The 12 most abundance genera are shown.

4. Discussion

Numerous studies have focused on identifying and characterizing microbial communities from the alimentary canal of insects that include, but are not limited to, aphids, bees, cockroaches, termites, and thrips [28]. While many microbial communities in the alimentary canals of insects have been described, many still await characterization, especially with respect to stored product insects. Therefore, we describe and have analyzed a new, important aspect of this species of economically important stored product insects. Previous work has identified microorganisms that exhibit a wide diversity of specialized interactions with their hosts, relating to nutrients, growth, development and other physiological processes.

Typhaea stercorea (L.) is an important pest of stored products that serves as an indicator of increases in fungal biomass during storage, leading to grain quality loss. Many pathogenic fungi found on stored grain products can produce mycotoxins which cause a variety of adverse health threats to both humans and livestock. *Typhaea stercorea* actively feeds on these pathogenic fungi and mycotoxins with no apparent adverse effects. This suggests that the gut microbiota metabolizes fungal diets and, in return, provides sterols and other nutrients that facilitate the growth and development of *T. stercorea*. Many insect-associated microorganisms promote an insect's capacity to utilize diets of low or unbalanced nutritional content by providing specific nutrients that the insect cannot synthesize, including amino acids, vitamins, and sterols [11].

Here, we surveyed the microbial composition in the alimentary canal of *T. stercorea* as a first step towards understanding how the gut microbiota may play a role in growth and development. The initial hypothesis was that there will be a decrease in bacterial diversity in the gut composition from the larval to adult life stages. We showed that there is low bacterial diversity found in the alimentary canal of *T. stercorea* across life stages. From the larval to adult stages, the gut bacterial community is potentially purged from the onset of the radical remodeling of the alimentary canal during metamorphosis (Figure 1C,D), which has been observed in other holometabolous insects [29]. For alpha diversity, the observed OTUs and Shannon Index showed no differences across life stage, indicating that the distribution of OTUs from *T. stercorea's* environment remains constant. In contrast to the lack of alpha diversity, we found differences between the beta diversity of the gut microbiota and life stages of *T. stercorea*, which suggests that diet or radical remodeling of the gut morphology/physiology are two of the most important factors that influence the assemblage of the gut microbiota [13,30–34]. Throughout *T. stercoreae's* development, larvae possess two unique genera, *Pseudomonas* and *Virgibacillus*, whereas adults had four rare genera *Brachybacterium*, *Brevibacterium*, *Lawsonella*, and *Oceanobacillus*. Furthermore, four phylotypes (*Streptomyces*, *Staphylococcus*, *Enterobacter* and an unidentified genus belonging to the family Enterobacteriaceae) were shared across life stages. Of these diverse genera found in the alimentary canal of *T. stercorea*, *Brevibacterium*, *Enterobacter*, *Pseudomonas*, *Staphlococcus*, and *Streptomyces* are known to exhibit antagonistic effects against pathogenic fungi (e.g., *Aspergillus* spp, *Fusarium* spp, and *Penicillum* spp) [19].

For all developmental stages of *T. stercorea*, the most dominant phylum was Proteobacteria (%), followed by Actinobacteria and Firmicutes. A microbiome study of the fungivore *Hoplothirps carpathicus* (Thysanoptera: Phlaeothripidae) found similar results, i.e., that the microbiome of *H. carpathicus* was also dominated by the phylum of Proteobacteria (57.49%). Kaczmarcyzk et al. [35] also noted that there was an increase (>2-fold) in the phylum Proteobacteria during developmental stages from pupa to adult. Other studies have indicated that the phylum Proteobacteria shows enriched diversity in adults compared to larvae [35–37]. However, our findings did not agree with these previous studies, since the *T. stercorea* larvae and adults had similar relative abundances of the phylum Proteobacteria throughout their lifecycles.

There is now persuasive evidence that insects that predominately feed on fungal diets are host to unique bacterial communities with mycolytic properties [38,39]. Mycolytic bacterial communities belong to groups of the phyla Actinobacteria, Bacteroidetes, Firmicutes, and Proteobacteria [28,40–42]. These phyla produce key enzymes which exhibit chitinase activity that supports a shift to fungal diets [38,43,44]. These enzymes (e.g., α-mannanases, β-1,3/1,6-glucanases, and chitnases) target the cell walls of fungi composed of complex and dynamic structures of mannan, glucan, and chitin [42,45]. Moreover, these mycolytic enzymes, such as β-1,3-glucananase, serve addition roles by protecting their host from fungal infections [46]. Their relative abundance in the alimentary canal may be caused by the digestion of a protein-rich fungal diet [44,47]. Similarities in host diet have been shown to drive convergence in the functional potential of gut microbes in other insects [38]. The degradation of chitin in nature is primarily carried out by bacterial taxa, such as pseudomonads, enteric bacteria, gliding bacteria, actinomycetes, and members of the genera *Bacillus*, *Vibrio*, and *Clostridium* [48,49].

Our study has allowed us to track marked changes in microbiota profiles associated with development. The gut microbiota composition was different between life stages. We found that the genus *Pseudomonas* was enriched in *T. stercorea* larvae as opposed to adults. *Pseudomonas* is classified as a Gram-negative bacterium that is commonly found in a variety of environments (e.g., insects, soil, and water). Previous literature states that the presence of *Pseudomonas melophtora* is necessary for larval survival and development in various stages of the apple maggot, *Rhagoletis pomonella* (Diptera: Tephritidae) [50]. *Pseudomonas* has been shown to synthesize amino acids, such as methionine and cystine, which are required for the development and growth of their insect host [51]. Other studies have shown that *Pseudomonas savastanoi* produces methionine and threonine, which are required for the olive fly, *Dacus oleae* (Diptera: Tephritidae), to complete its lifecycle from larva to adult [52]. Therefore, we suggest that the genus *Pseudomonas* is a facultative secondary endosymbiont that resides in the alimentary canal and breaks down the chitin walls of fungi, providing nutrients to its host. In this case, *Pseudomonas* synthesizes amino acids that may be rare in the fungal diet of their host, *T. stercorea*. In addition, other strains of *Pseudomonas* can metabolize insecticides [53,54], such as neonicotinoids, which suggests that *Pseudomonas* species perform protective functions during the developmental stages of *T. stercorea*.

Lastly, we compared the microbiota from the alimentary canal of laboratory strain specimens (larvae and adults) to field-collected *T. stercorea* adults. We hypothesized that the gut microbiota of reared larvae and adults will lack diversity compared to field-collected adults. Our data indicate that field-collected *T. stercorea* tend to have lower species richness than laboratory-reared beetles. In contrast, other studies have demonstrated that the gut microbiota of field-collected insects typically possess more diverse bacterial species than laboratory-reared insects [32,55,56]. There was no clear explanation as to why our study showed that field-collected *T. stercorea* tended to have lower species richness than laboratory-reared members. However, it has been suggested that ecological exogenous factors can influence microbiota diversity, which indicates that the host's diet and habitat may affect the insect gut microbiota's species richness and composition [55]. As expected, the microbial gut composition of adult field populations was different to those of laboratory-reared larvae and adults. Field populations possessed several known chitinolytic bacteria belonging to the families Bacillaceae and Enterobacteriacea. Furthermore, the family Enterobacteriacea, which belongs to the phylum Proteobacteria, showed the highest relative abundance in laboratory-reared adults and field-collected adults. Our findings thus suggest that insects associated with fungal communities depend on the family Enterobacteriaceae in the adult stage. A previous study indicated that the family Enterobacteriaceae was one of the dominant bacterial families in the gut microbiota of the bark beetles *Dendroctonus* spp. Interestingly, the genus *Enterobacter* contributes to host nutrition by fixing atmospheric nitrogen [57–59]. Based on previous reports, several chitinolytic bacterial strains of Bacillaceae, Enterobacteriaceae, and Pseudomonadaceae families were classified as antagonistic microbes towards pathogenic fungi. Thus, it is imperative to continue to study these chitinolytic microbes and their properties for potential use as a biological control of pathogenic fungi associated with stored grain products.

In conclusion, this report presents data from a profile analysis of gut bacterial communities of *T. stercorea* through the use of NGS 16S rRNA amplicon sequence data. The current study adds to our understanding of how important mutualistic prokaryotes found within the gut microbiota may provide essential nutrients (e.g., sterols, vitamins, carbohydrates, and amino acid synthesis/metabolism) during the growth and development of *T. stercorea*. For future investigations, manipulating bacterial communities through the use of antibiotics will allow for the testing of emerging hypotheses on the role of gut microbes in their host's lifecycle and fecundity.

Supplementary Materials: The following supporting information can be downloaded at: https://www.mdpi.com/article/10.3390/insects13080685/s1, Figure S1: Diagram of the pitfall trap used to attract field strains of *T. stercorea*. Figure S2: Alpha–rarefaction curve of the averaged observed OTUs detected in the *T. stercorea*'s alimentary canal. Figure S3: Alpha–rarefaction curve of the averaged observed OTUs detected in *T. stercorea*'s alimentary canal.

Author Contributions: Data curation, J.E.; Formal analysis, J.E.; Investigation, J.E.; Methodology, J.E.; Project administration, L.M.; Visualization, J.E.; Writing—original draft, J.E. and L.M.; Writing—review and editing, J.E. and L.M. All authors have read and agreed to the published version of the manuscript.

Funding: This research received no external funding.

Institutional Review Board Statement: Not applicable.

Informed Consent Statement: Not applicable.

Data Availability Statement: Data are contained within the article or Supplementary Files. Raw data are available upon request from the corresponding author.

Acknowledgments: We thank Ohio State University for supplying a starter colony of the *T. stercorea*. We are also grateful to Hannah E. Quellhorst, Bradley W. Smith, and Aaron Rodriques and Peter E. Dunn for providing useful discussions and reviews of the manuscript. Furthermore, we appreciate the help and services of our committee members: Laramy S. Enders, Michael E. Scharf, and Charles P. Woloshuk.

Conflicts of Interest: The authors declare no conflict of interest.

References

1. Pitt, J.I. Toxigenic fungi and mycotoxins. *Br. Med. Bull.* **2000**, *56*, 184–192. [CrossRef] [PubMed]
2. Hussein, H.S.; Brasel, J.M. Toxicity, metabolism, and impact of mycotoxins on humans and animals. *Toxicology* **2001**, *167*, 101–134. [CrossRef]
3. Christensen, C.M.; Kaufmann, H.H. *Grain Storage: The Role of Fungi in Quality Loss*; University of Minnesota Press: Minneapolis, MN, USA, 1969; pp. 3–35.
4. FAO. FAO Cereal Supply and Demand Brief: World Food Situation. Available online: http://www.fao.org/worldfoodsituation/csdb/en/ (accessed on 17 July 2016).
5. Dunkel, F.V. The relationship of insects to the deterioration of stored grain by fungi. *Int. J. Food Microbiol.* **1988**, *7*, 227–244. [CrossRef]
6. Tsai, W.-T.; Mason, L.J.; Woloshuk, C.P. Effect of three stored-grain fungi on the development of *Typhaea stercorea*. *J. Stored Prod. Res.* **2007**, *43*, 129–133. [CrossRef]
7. Calvin, D. Hairy Fungus Beetle Fact Sheet. 1988. Available online: https://ento.psu.edu/extension/factsheets/pdf/hairyfungusbtl.pdf (accessed on 3 February 2018).
8. Mason, L.J.; McDonough, M. Biology, behavior, and ecology of stored grain and legume insects. In *Stored Product Protection*; Hagstrum, D.W., Phillips, T.W., Cuperus, G., Eds.; Kansas State University: Manhattan, KS, USA, 2012; pp. 7–20.
9. Capinera, J.L. Native American culture and insects. In *Encyclopedia of Entomology*, 2nd ed.; Capinera, J.L., Ed.; Springer: Boston, MA, USA, 2008; pp. 2546–2550.
10. Janson, E.M.; Grebenok, R.J.; Behmer, S.T.; Abbot, P. Same Host-Plant, Different Sterols: Variation in Sterol Metabolism in an Insect Herbivore Community. *J. Chem. Ecol.* **2009**, *35*, 1309–1319. [CrossRef]
11. Douglas, A.E. Multiorganismal Insects: Diversity and Function of Resident Microorganisms. *Annu. Rev. Èntomol.* **2015**, *60*, 17–34. [CrossRef]
12. Steinhaus, E.A. The Microbiology of Insects: With Special Reference to the Biologic Relationships between Bacteria and Insects. *Bacteriol. Rev.* **1940**, *4*, 17–57. [CrossRef]
13. Engel, P.; Moran, N.A. The gut microbiota of insects–Diversity in structure and function. *FEMS Microbiol. Rev.* **2013**, *37*, 699–735. [CrossRef]

14. Egert, M.; Wagner, B.; Lemke, T.; Brune, A.; Friedrich, M.W. Microbial Community Structure in Midgut and Hindgut of the Humus-Feeding Larva of *Pachnoda ephippiata* (Coleoptera: Scarabaeidae). *Appl. Environ. Microbiol.* **2003**, *69*, 6659–6668. [CrossRef]
15. Hanski, I.; Cambefort, Y. *Dung Beetle Ecology*; Princeton University Press: Princeton, NJ, USA, 2014; pp. 22–35.
16. Rojas-Jiménez, K.; Hernández, M. Isolation of Fungi and Bacteria Associated with the Guts of Tropical Wood-Feeding Coleoptera and Determination of Their Lignocellulolytic Activities. *Int. J. Microbiol.* **2015**, *2015*, 285018. [CrossRef]
17. Rathore, A.S.; Gupta, R.D. Chitinases from Bacteria to Human: Properties, Applications, and Future Perspectives. *Enzym. Res.* **2015**, *2015*, 791907. [CrossRef]
18. Kugler, M.; Loeffler, W.; Rapp, C.; Kern, A. Rhizocticin A, an antifungal phosphono-oligopeptide of Bacillus subtilis ATCC 6633: Biological properties. *Arch. Microbiol.* **1990**, *153*, 276–281. [CrossRef]
19. Kerr, J.R. Bacterial inhibition of fungal growth and pathogenicity. *Microb. Ecol. Health Dis.* **1999**, *11*, 129–142.
20. Douglas, A.E. Symbiosis as a General Principle in Eukaryotic Evolution. *Cold Spring Harb. Perspect. Biol.* **2014**, *6*, a016113. [CrossRef]
21. Huang, H.; Li, H.; Ren, L.; Cheng, D. Microbial Communities in Different Developmental Stages of the Oriental Fruit Fly, Bactrocera dorsalis, Are Associated with Differentially Expressed Peptidoglycan Recognition Protein-Encoding Genes. *Appl. Environ. Microbiol.* **2019**, *85*, e00803-19. [CrossRef]
22. Bolyen, E.; Rideout, J.R.; Dillon, M.R.; Bokulich, N.A.; Abnet, C.; Al-Ghalith, G.A.; Alexander, H.; Alm, E.J.; Arumugam, M.; Asnicar, F. QIIME 2: Reproducible, interactive, scalable, and extensible microbiome data science. *PeerJ Prepr.* **2018**, *6*, e27295v2. [CrossRef]
23. Callahan, B.J.; McMurdie, P.J.; Rosen, M.J.; Han, A.W.; Johnson, A.J.A.; Holmes, S.P. DADA2: High-resoultion sample inference from Illumina amplicon data. *Nat. Methods* **2016**, *13*, 581–583. [CrossRef]
24. Mandal, S.; Van Treuren, W.; White, R.A.; Eggesbø, M.Å.; Knight, R.T.; Peddada, S.D. Analysis of composition of microbiomes: A novel method for studying microbial composition. *Microb. Ecol. Health Dis.* **2015**, *26*, 27663. [CrossRef]
25. Anderson, M.J.; Gorley, R.N.; Clarke, K.R. *Permanova+ for Primer: Guide to Software and Statistical Methods*; Primer-E: Plymouth, UK, 2008.
26. Kruskal, W.H.; Wallis, W.A. Use of ranks in one-criterion variance analysis. *J. Am. Stat. Assoc.* **1952**, *47*, 583–621. [CrossRef]
27. Birmingham, A. Microbiome Analysis with QIIME2: A Hands-on Tutorial. 2018. Available online: http://compbio.ucsd.edu/wp-content/uploads/2018/07/20180621_oslo_university_microbiome_analysis_with_qiime2_tutorial.pdf (accessed on 8 July 2022).
28. Yun, J.-H.; Roh, S.W.; Whon, T.W.; Jung, M.-J.; Kim, M.-S.; Park, D.-S.; Yoon, C.; Nam, Y.-D.; Kim, Y.-J.; Choi, J.-H.; et al. Insect Gut Bacterial Diversity Determined by Environmental Habitat, Diet, Developmental Stage, and Phylogeny of Host. *Appl. Environ. Microbiol.* **2014**, *80*, 5254–5264. [CrossRef]
29. Johnston, P.R.; Rolff, J. Host and Symbiont Jointly Control Gut Microbiota during Complete Metamorphosis. *PLOS Pathog.* **2015**, *11*, e1005246. [CrossRef] [PubMed]
30. Turnbaugh, P.J.; Bäckhed, F.; Fulton, L.; Gordon, J.I. Diet-Induced Obesity Is Linked to Marked but Reversible Alterations in the Mouse Distal Gut Microbiome. *Cell Host Microbe* **2008**, *3*, 213–223. [CrossRef] [PubMed]
31. Muegge, B.D.; Kuczynski, J.; Knights, D.; Clemente, J.C.; González, A.; Fontana, L.; Henrissat, B.; Knight, R.; Gordon, J.I. Diet Drives Convergence in Gut Microbiome Functions Across Mammalian Phylogeny and within Humans. *Science* **2011**, *332*, 970–974. [CrossRef] [PubMed]
32. Pérez-Cobas, A.E.; Maiques, E.; Angelova, A.; Carrasco, P.; Moya, A.; Latorre, A. Diet shapes the gut microbiota of the om-nivorous cockroach *Blattella germanica*. *FEMS Microbiol. Ecol.* **2015**, *91*, fiv022. [CrossRef]
33. Carmody, R.N.; Gerber, G.K.; Luevano, J.M., Jr.; Gatti, D.M.; Somes, L.; Svenson, K.L.; Turnbaugh, P.J. Diet dominates host genotype in shaping the murine gut microbiota. *Cell Host Microbe* **2015**, *17*, 72–84. [CrossRef]
34. Li, H.; Li, T.; Beasley, D.; Hedenec, P.; Xiao, Z.; Zhang, S.; Li, J.; Lin, Q.; Li, X. Diet Diversity Is Associated with Beta but not Alpha Diversity of Pika Gut Microbiota. *Front. Microbiol.* **2016**, *7*, 1169. [CrossRef]
35. Kaczmarczyk, A.; Kucharczyk, H.; Kucharczyk, M.; Kapusta, P.; Sell, J.; Zielińska, S. First insight into microbiome profile of fungivorous thrips *Hoplothrips carpathicus* (Insecta: Thysanoptera) at different developmental stages: Molecular evidence of Wolbachia endosymbiosis. *Sci. Rep.* **2018**, *8*, 14376. [CrossRef]
36. Wang, H.; Jin, L.; Zhang, H. Comparison of the diversity of the bacterial communities in the intestinal tract of adult *Bactrocera dorsalis* from three different populations. *J. Appl. Microbiol.* **2011**, *110*, 1390–1401. [CrossRef]
37. Chen, B.; Teh, B.-S.; Sun, C.; Hu, S.; Lu, X.; Boland, W.; Shao, Y. Biodiversity and Activity of the Gut Microbiota across the Life History of the Insect Herbivore *Spodoptera littoralis*. *Sci. Rep.* **2016**, *6*, 29505. [CrossRef]
38. Hu, H.; da Costa, R.R.; Pilgaard, B.; Schiøtt, M.; Lange, L.; Poulsen, M. The origin of fungi-culture in termites was associated with a shift to a mycolytic gut bacteria community. *BioRxiv* **2018**. [CrossRef]
39. Bost, A.; Martinson, V.G.; Franzenburg, S.; Adair, K.L.; Albasi, A.; Wells, M.T.; Douglas, A.E. Functional variation in the gut microbiome of wild *Drosophila* populations. *Mol. Ecol.* **2018**, *27*, 2834–2845. [CrossRef]
40. Colman, D.R.; Toolson, E.C.; Takacs-Vesbach, C. Do diet and taxonomy influence insect gut bacterial communities? *Mol. Ecol.* **2012**, *21*, 5124–5137. [CrossRef]
41. Chen, B.; Du, K.; Sun, C.; Vimalanathan, A.; Liang, X.; Li, Y.; Wang, B.; Lu, X.; Li, L.; Shao, Y. Gut bacterial and fungal com-munities of the domesticated silkworm (*Bombyx mori*) and wild mulberry-feeding relatives. *ISME J.* **2018**, *12*, 2252–2262. [CrossRef]

42. Hu, H.; da Costa, R.R.; Pilgaard, B.; Schiøtt, M.; Lange, L.; Poulsen, M. Fungiculture in Termites Is Associated with a Mycolytic Gut Bacterial Community. *mSphere* **2019**, *4*, e00165-19. [CrossRef]
43. Mitchell, R.; Alexander, M. The Mycolytic Phenomenon and Biological Control of Fusarium in Soil. *Nature* **1961**, *190*, 109–110. [CrossRef]
44. Dietrich, C.; Köhler, T.; Brune, A. The Cockroach Origin of the Termite Gut Microbiota: Patterns in Bacterial Community Structure Reflect Major Evolutionary Events. *Appl. Environ. Microbiol.* **2014**, *80*, 2261–2269. [CrossRef]
45. Reilly, M.C.; Doering, T.L. Biosynthesis of fungal and yeast glycans. In *Microbial Glycobiology*; Holst, O., Brennan, P.J., von Itzstein, M., Eds.; Academic Press: Cambridge, MA, USA, 2010; pp. 393–412.
46. Rosengaus, R.B.; Schultheis, K.F.; Yalonetskaya, A.; Bulmer, M.S.; DuComb, W.S.; Benson, R.W.; Thottam, J.P.; Godoy-Carter, V. Symbiont-derived Î²-1,3-glucanases in a social insect: Mutualism beyond nutrition. *Front. Microbiol.* **2014**, *5*, 607. [CrossRef]
47. Hyodo, F.; Tayasu, I.; Inoue, T.; Azuma, J.-I.; Kudo, T.; Abe, T. Differential role of symbiotic fungi in lignin degradation and food provision for fungus-growing termites (Macrotermitinae: Isoptera). *Funct. Ecol.* **2003**, *17*, 186–193. [CrossRef]
48. Gooday, G.W. The ecology of chitin degradation. In *Advances in Microbial Ecology*; Marshall, K.C., Ed.; Springer: Boston, MA, USA, 1990; pp. 387–430.
49. Delsuc, F.; Metcalf, J.L.; Wegener Parfrey, L.; Song, S.J.; González, A.; Knight, R. Convergence of gut microbiomes in myrmecophagous mammals. *Mol. Ecol.* **2014**, *23*, 1301–1317. [CrossRef]
50. Allen, T.C.; Pinckard, J.A.; Riker, A.J. Frequent association of *Phytomonas melophthora* with various stages in the lifecycle of the apple maggot, *Rhagoletis pomonella*. *Phytopathology* **1934**, *22*, 228–238.
51. Miyazaki, S.; Boush, G.M.; Baerwald, R.J. Amino acid synthesis by *Pseudomonas melophthora*, bacterial symbiote of *Rhagoletis pomonella* (Diptera). *J. Insect Physiol.* **1968**, *14*, 513–518. [CrossRef]
52. Hagen, K.S. Dependence of the Olive Fly, Dacusoleae, Larvae on Symbiosis with Pseudomonas savastanoi for the Utilization of Olive. *Nature* **1966**, *209*, 423–424. [CrossRef]
53. Xie, W.; Meng, Q.-S.; Wu, Q.-J.; Wang, S.-L.; Yang, X.; Yang, N.-N.; Li, R.-M.; Jiao, X.-G.; Pan, H.-P.; Liu, B.-M.; et al. Pyrosequencing the *Bemisia tabaci* Transcriptome Reveals a Highly Diverse Bacterial Community and a Robust System for Insecticide Resistance. *PLoS ONE* **2012**, *7*, e35181. [CrossRef] [PubMed]
54. Hussain, S.; Hartley, C.J.; Shettigar, M.; Pandey, G. Bacterial biodegradation of neonicotinoid pesticides in soil and water systems. *FEMS Microbiol. Lett.* **2016**, *363*, fnw252. [CrossRef]
55. Ng, S.H.; Stat, M.; Bunce, M.; Simmons, L.W. The influence of diet and environment on the gut microbial community of field crickets. *Ecol. Evol.* **2018**, *8*, 4704–4720. [CrossRef]
56. Belda, E.; Pedrola, L.; Pereto, J.; Martínez-Blanch, J.F.; Montagud, A.; Navarro, E.; Urchueguía, J.; Ramón, D.; Moya, A.; Porcar, M. Microbial diversity in the midguts of field and lab-reared populations of the European corn borer *Ostrinia nubilalis*. *PLoS ONE* **2011**, *6*, e21751. [CrossRef]
57. Durand, A.A.; Bergeron, A.; Constant, P.; Buffet, J.P.; Déziel, E.; Guertin, C. Surveying the endomicrobiome and ectomicro-biome of bark beetles: The case of *Dendroctonus simplex*. *Sci. Rep.* **2015**, *5*, 17190. [CrossRef]
58. Gauthier, J.-P.; Outreman, Y.; Mieuzet, L.; Simon, J.-C. Bacterial Communities Associated with Host-Adapted Populations of Pea Aphids Revealed by Deep Sequencing of 16S Ribosomal DNA. *PLoS ONE* **2015**, *10*, e0120664. [CrossRef]
59. Gurung, K.; Wertheim, B.; Salles, J.F. The microbiome of pest insects: It is not just bacteria. *Èntomol. Exp. Appl.* **2019**, *167*, 156–170. [CrossRef]

insects

MDPI

Article

Investigation of Gut Bacterial Communities of Asian Citrus Psyllid (*Diaphorina citri*) Reared on Different Host Plants

Lixue Meng [1,2], Changxiu Xia [3], Zhixiong Jin [4] and Hongyu Zhang [2,*]

[1] School of Basic Medical Sciences, Hubei University of Medicine, Shiyan 442000, China; ningxueyt@126.com
[2] State Key Laboratory of Agricultural Microbiology, Key Laboratory of Horticultural Plant Biology (MOE), Institute of Urban and Horticultural Entomology, College of Plant Science and Technology, Huazhong Agricultural University, Wuhan 430070, China
[3] Ganzhou Citrus Science Institute, Ganzhou 341000, China; xia8295739@126.com
[4] The Department of Clinical Laboratory, Sinopharm Dongfeng Hospital, Hubei University of Medicine, Shiyan 442000, China; 20090569@hbmu.edu.cn
* Correspondence: hongyu.zhang@mail.hzau.edu.cn; Tel.: +86-027-87286962

Simple Summary: *Diaphorina citri* is a crucial natural vector of the Huanglongbing pathogen, which has devastated the citrus industry. The host plant is a critical factor that affects insect biology and its symbiont abundance. However, little is known about how host plants affect the bacterial community located in *D. citri*. In this work, the guts of five different host-plant-feeding populations (i.e., *Citrus reticulata* cv. Shatangju, *Citrus poonensis* cv. Ponkan, *Murraya paniculata* (orange jasmine), *Citrus limon* (lemon), and *Citrus sinensis* (navel orange)) were analyzed for bacterial communities by next-generation sequencing. The dominant phylum was Proteobacteria. The most common and abundant bacterial genera in *D. citri* were *Wolbachia*, *Escherichia-Shigella*, and *Candidatus Profftella*, but their relative abundance varied among the different host plant groups. There were obvious differences in the gut microbiota among the different hosts, and the gut microbe diversity was the highest in the ponkan-feeding population, while the lowest was in the Shatangju-feeding population. Overall, our findings indicate that the host plant can significantly affect the gut microbial community of *D. citri*. This result can provide new insights into the co-adaptation of *D. citri* and its symbionts.

Abstract: *Diaphorina citri* Kuwayama (Hemiptera: Liviidae) can cause severe damage to citrus plants, as it transmits *Candidatus* Liberibacter spp., a causative agent of Huanglongbing disease. Symbiotic bacteria play vital roles in the ecology and biology of herbivore hosts, thereby affecting host growth and adaptation. In our research, the effects of Rutaceous plants (i.e., *Citrus reticulata* cv. Shatangju, *Citrus poonensis* cv. Ponkan, *Murraya paniculata* (orange jasmine), *Citrus limon* (lemon), and *Citrus sinensis* (navel orange)) on the gut microbiota (GM) and microbial diversity of *D. citri* adults were investigated by 16S rRNA high-throughput sequencing. It was found that Proteobacteria dominated the GM communities. The gut microbe diversity was the highest in the ponkan-feeding population, and the lowest in the Shatangju-feeding population. The NMDS analysis revealed that there were obvious differences in the GM communities among the different hosts. PICRUSt function prediction indicated significant differences in host function, and those pathways were crucial for maintaining population reproduction, growth, development, and adaptation to environmental stress in *D. citri*. Our study sheds new light on the interactions between symbionts, herbivores, and host plants and expands our knowledge on host adaptation related to GM in *D. citri*.

Keywords: *Diaphorina citri*; host species; 16S rRNA sequencing; gut microbiota

Citation: Meng, L.; Xia, C.; Jin, Z.; Zhang, H. Investigation of Gut Bacterial Communities of Asian Citrus Psyllid (*Diaphorina citri*) Reared on Different Host Plants. *Insects* **2022**, *13*, 694. https://doi.org/10.3390/insects13080694

Academic Editors: Maurizio Francesco Brivio and Angela Marie Smilanich

Received: 12 May 2022
Accepted: 18 July 2022
Published: 2 August 2022

Publisher's Note: MDPI stays neutral with regard to jurisdictional claims in published maps and institutional affiliations.

Copyright: © 2022 by the authors. Licensee MDPI, Basel, Switzerland. This article is an open access article distributed under the terms and conditions of the Creative Commons Attribution (CC BY) license (https://creativecommons.org/licenses/by/4.0/).

1. Introduction

The adaptation of insects to new foods and environments can be facilitated by microbial symbionts [1,2] and is critical for nutritional supplementation in insects [3,4]. Similarly, the insect gut microbiota (GM) is closely associated with plasticity, which can quickly adapt to different diets, even with alterations in the GM population structures [5]. This plasticity is crucial for insects to exploit various food sources, thus contributing to the development of host-associated differentiation, which represents an adaptive ecological strategy that reflects high species diversity in insects [6–8]. The characterization of GM community is crucial for revealing the ecology and biology of host insects and developing a new pest management strategy. As reviewed by Crotti et al. [9], microbes can be manipulated to enhance an SIT program, control pathogens transmitted by insects, and protect beneficial insects.

Diaphorina citri Kuwayama (Hemiptera: Liviidae) can cause severe damage to citrus plants, as it acts a vector of "*Candidatus* Liberibacter" (CLas), the causative agent related to Huanglongbing disease [10]. Excessive use of pesticides can lead to residual citrus pesticides, affect the population dynamics of natural enemies and beneficial insects, and increase the risk of environmental pollution [11]. Thus, new green control measures are of great importance for enacting control and prevention mechanisms for *D. citri*, which feeds on the phloem sap of plants in the Rutaceae family. Because of its specialized diet, *D. citri* is highly required to overcome imbalances (e.g., limited vitamins and essential amino acids) in their diet, which are commonly supplemented by microbial symbionts [12–14]. Thus, how does the GM communities react when *D. citri* exploits various plants as a food source? By answering this question, new environmentally friendly pest management strategy can be designed.

Insect–microbial interactions have a severe impact on natural and agricultural ecosystems. The research on insect host-related GM will provide a basic framework for further functional experiments. Despite the economic importance of *D. citri*, little is known regarding how its gut bacterial communities are influenced by host plant feeding. In this work, we used bacterial 16S-rRNA sequencing to characterize gut communities from different populations (i.e., *Citrus reticulata* cv. Shatangju, *Citrus poonensis* cv. Ponkan, *Murraya paniculata* (orange jasmine), *Citrus limon* (lemon), and *Citrus sinensis* (navel orange)). This study aimed to assess the effects of host plants on the GM communities of *D. citri* and to provide a basis for the development of efficient and green control measures.

2. Materials and Methods

2.1. Insect Sampling and Storage

Adult *D. citri* psyllids were collected from the *Citrus sinensis* (navel orange) at the Citrus Scientific Research Institute, Ganzhou, Jiangxi, and were raised on five host plant species (i.e., *Citrus reticulata* cv. Shatangju, *Citrus poonensis* cv. Ponkan, *Murraya paniculata* (orange jasmine), *Citrus limon* (lemon), and *Citrus sinensis* (navel orange)) in the institute's net room. Regular pruning of host plant branches, fertilization, and watering ensured high numbers of tender shoots in the net. All cages and experiments were kept in a climate-controlled chamber at 27 ± 1 °C, 70–75% RH, and at 14:10 h of light:dark illumination. Adult *D. citri* of the third generation were selected as the test specimen source after 3 generations of continuous feeding to allow for the development of an adapted gut microbiota.

Each treatment (host plant) was repeated 5 times, and 200 live adult specimens were collected from each treatment. Immediately after collection, the adults were frozen at −20 °C for 5 min before dissection. The dead insects were then sterilized superficially with 70% alcohol twice for 60 s and washed twice in sterile distilled water [15]. The specimens were then placed in a phosphate-buffered solution to excise the complete gut with sterile forceps. The samples were then stored at −20 °C until DNA extraction.

2.2. DNA Isolation and Sequencing

The E.Z.N.ATM Soil DNA kit (Omega, Alpharetta, GA, USA) was used to isolate DNA from adult guts. The yield and quality of DNA samples were assessed by NanoDrop ND-1000 spectrophotometry (Thermo Fisher Scientific, Waltham, MA, USA) and agarose gel electrophoresis, respectively. Amplification of the 16S-rRNA V3–V4 region was conducted with 338F/806R (5′-ACT CCT ACG GGA GGC AGC A-3′ and 5′-GGA CTA CHV GGG TWT CTA AT-3′) [16]. The PCR conditions were 3 min at 95 °C, followed by

25 cycles for 30 s at 95 °C, 30 s at 55 °C, 30 s at 72 °C, and 5 min at 72 °C. All assays were performed in triplicate and then pooled to reduce PCR bias. Sequencing was conducted using an MISEQ REAGENT KIT (v2; Illumina, Inc., San Diego, CA, USA) with an Illumina MiSeq platform [17]. All data were deposited in the NCBI's Short Read Archive (accession number: PRJNA515577).

2.3. Sequence Analysis and Diversity Measures

Quantitative insights into microbial ecology (QIIME, v1.9.1) was used to process the sequencing data [18]. Paired-end sequences were aligned by Trimmomatic and FLASH, and those with >97% pairwise identity were mapped to the operational taxonomic unit (OTU) [19,20]. An open reference OTU picking strategy was used for taxonomic assignment via the Greengenes taxonomic database [21].

After rarefying the sequencing data to a depth of 20,000 reads, the microbial diversity was assessed to uncover the bacterial diversity (Shannon and Simpson) and species richness (ACE and Chao1). Beta diversity analysis was employed to assess the structural variations of microbial communities among samples based on UniFrac distance metrics [22,23] visualized via the unweighted pair-group method with arithmetic means (UPGMA), hierarchical clustering, and nonmetric multidimensional scaling (NMDS) [24]. To identify the taxa with differential abundances and indicative in each treatment, linear discriminant analysis effect size (LEfSe) analysis was conducted [25]. A logarithmic LDA score > 2 and $p < 0.05$ were deemed statistically significant. To estimate microbial functions, phylogenetic analysis of communities by reconstruction of unobserved states (PICRUSt) was conducted based on the high-quality sequences [26].

2.4. Statistical Analyses

The Student's t-test and one-way ANOVA followed by Tukey's multiple comparison test were employed to compare the differences among the different treatments, and $p < 0.05$ was considered statistically significant [27]. Species abundances were determined using MetaPhlAn2 [28]. GraphPad Prism v8.0 (GraphPad Software, San Diego, CA, USA) and the R package were applied to obtain the diagrams in this study.

3. Results

3.1. High-Throughput Sequencing Data and the Diversity of GM in D. citri Populations from Various Hosts

A total of 1,111,356 sequences were derived from 25 specimens. After cleaning and trimming, 969,472 were subjected to further analysis. The length of the sequences ranged from 200 to 500 bp, with 99.93% of them having 400 to 450 bp. These sequences were further clustered into 120 bacterial OTUs. The rarefaction curves of different samples became flat, implying effective sampling and successful recovery of OTUs (Figure S1).

The alpha diversity indices were calculated to assess the bacterial diversity (Shannon and Simpson indices) and species richness (OTUs and Chao1). Analysis of species richness calculated by OTUs and Chao1 demonstrated obvious differences among the different treatments. The average species richness of the ponkan- and orange jasmine-feeding populations was significantly higher than that of the navel orange-, lemon- and Shatangju-feeding populations ($p < 0.01$) (Figure 1a,b). There were remarkable differences in the bacterial diversity calculated by the Shannon and Simpson indices. The bacterial diversities of the ponkan-feeding population were the highest as shown by the Shannon index (1.812) and Simpson index (0.288), and the lowest was found in the population feeding on Shatangju ($p < 0.01$) as shown by the Shannon index (0.554) and Simpson index (0.796) (Figure 1c,d).

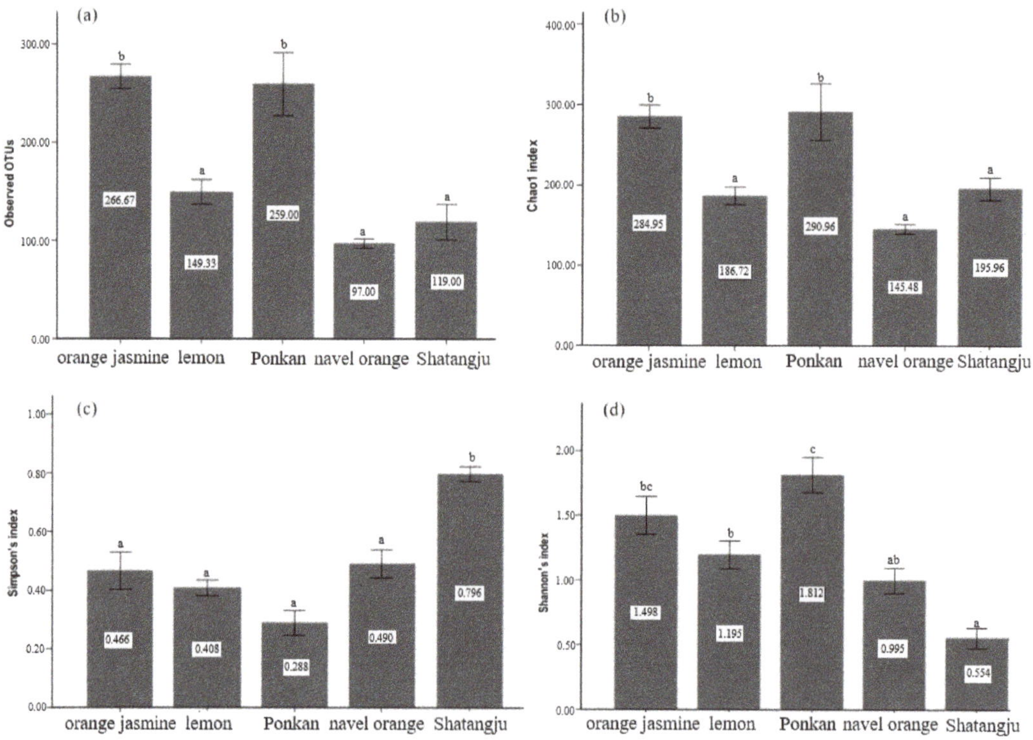

Figure 1. Measures of the α-diversity for each treatment: (**a**) observed OTUs; (**b**) Chao richness estimator; (**c**) Simpson's index; (**d**) Shannon's index. Different letters indicate that the values are significantly different ($p < 0.01$).

3.2. Comparison of GM in D. citri Populations from Various Hosts

The bacterial communities of all samples were dominated by Proteobacteria, but several groups contained high abundances of Actinobacteria and Firmicutes (Table S1). At the family level (Figure 2), Enterobacteriaceae was prevalent in most samples (44.69 ± 33.37%). The relative abundance of Anaplasmataceae was much higher in the orange jasmine-feeding population (65.01 ± 0.92%) than the others. A much higher relative abundance of Oxalobacteraceae in the orange jasmine- (16.73 ± 6.59%) and ponkan-feeding populations (22.69 ± 9.70%) was also observed. The most dominant bacterial genera in *D. citri* were *Wolbachia*, *Escherichia-Shigella*, and *Candidatus Profftella*, but their relative abundances varied among the different host plant groups (Figure 3, Table S2). The LSD multiple-range test showed a significantly higher relative abundance of *Escherichia-Shigella* in the Shatangju- and navel orange-feeding populations than in other host samples, especially in the orange jasmine- and ponkan-feeding populations ($p < 0.01$) (Figure 4a). The relative abundance of *Wolbachia* was significantly higher in the orange jasmine-feeding population ($p < 0.01$) (Figure 4b). *Candidatus Profftella* was present in a lower abundance in the Shatangju-feeding population but higher in the ponkan-feeding population ($p < 0.01$) (Figure 4c). Less abundant but prevalent bacteria, including *Pantoea*, *Stenotrophomonas*, *Lactobacillus*, *Microbacterium*, *Sphingomonas*, *Streptomyces*, and *Methylobacterium*, were also detected in our study (Figure 4d–f). Among them, *Microbacterium* was detected in all treatments, except in the Shatangju-feeding population. Other bacteria appeared in all treatments.

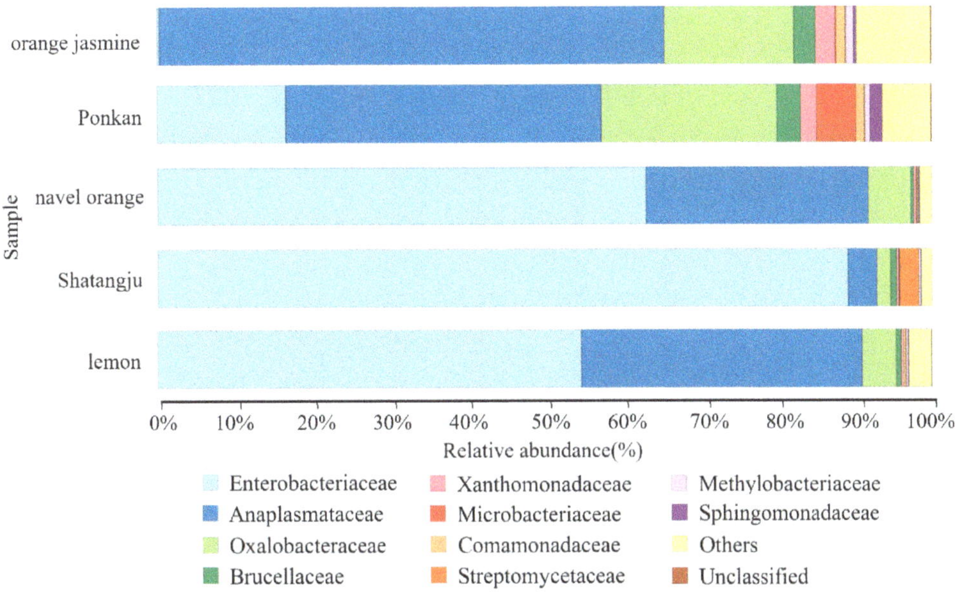

Figure 2. Relative abundance of families in *D. citri* gut. Only the taxa within the 10 most abundant were considered.

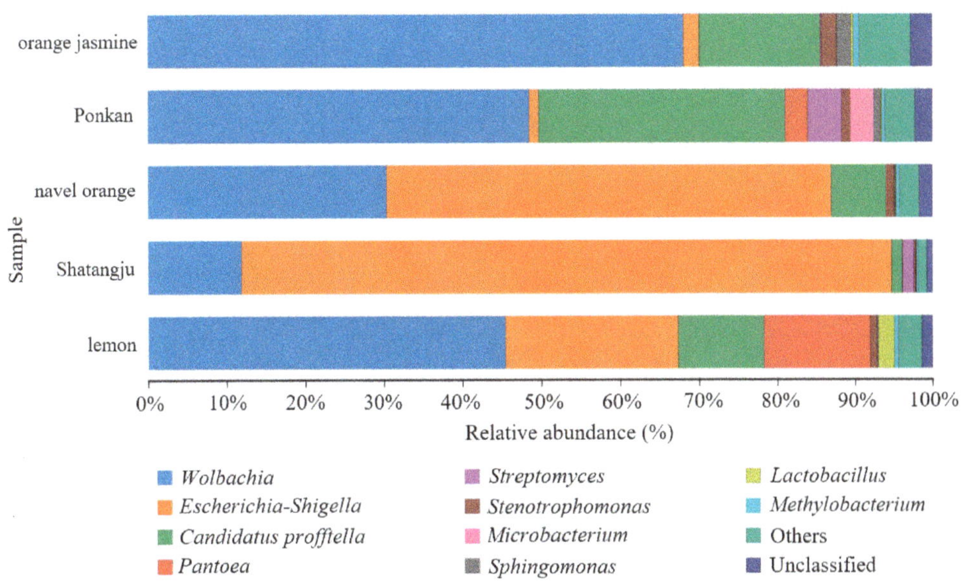

Figure 3. Relative abundance of genera in the gut of *D. citri*. Only the taxa within the 10 most abundant were considered.

The UPGMA analysis indicated that host plants affected the sample groupings. The Shatangju-feeding population samples were more characteristic of the lemon- and navel orange-feeding populations (Figure 5). The NMDS analysis of species diversity also demonstrated obvious differences among the five treatments (Figure 6). The Shatangju-feeding population, which had similar bacterial communities to the navel orange-feeding population, was separated from the orange jasmine-feeding population on NMDS1 and from the ponkan-feeding population on NMDS2 (Figure 6). While the lemon-feeding population was represented on two coordinates. The stress was 0.067, indicating that NMDS most likely reflected the degree of difference in the various samples.

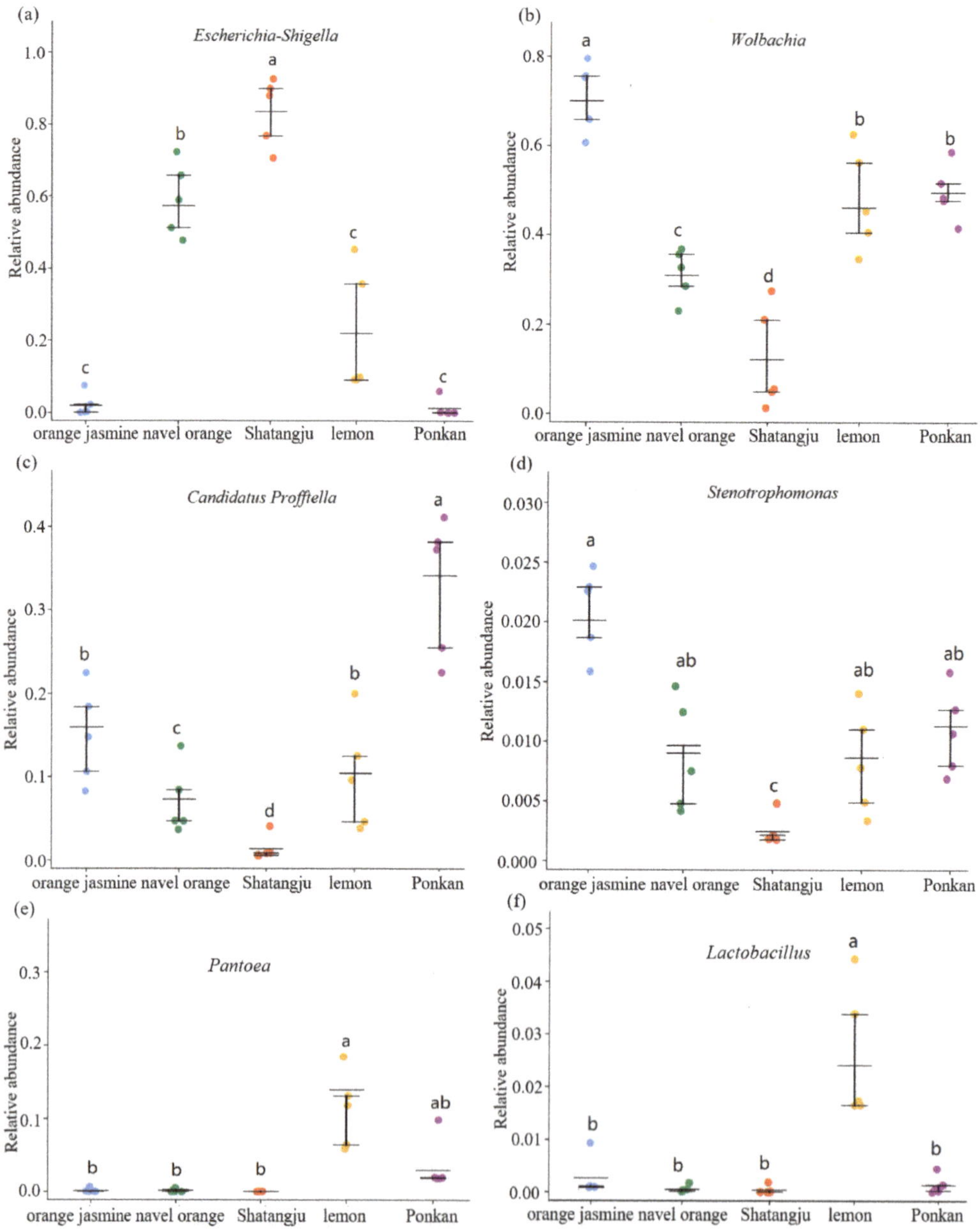

Figure 4. The distribution of six genera (*Escherichia-Shigella* (**a**), *Wolbachia* (**b**), *Candidatus Profftella* (**c**), *Stenotrophomonas* (**d**), *Pantoea* (**e**) and *Lactobacillus* (**f**)) differed significantly in *D. citri* gut microbes feeding on different host plants. Different letters indicate that the values are significantly different ($p < 0.01$).

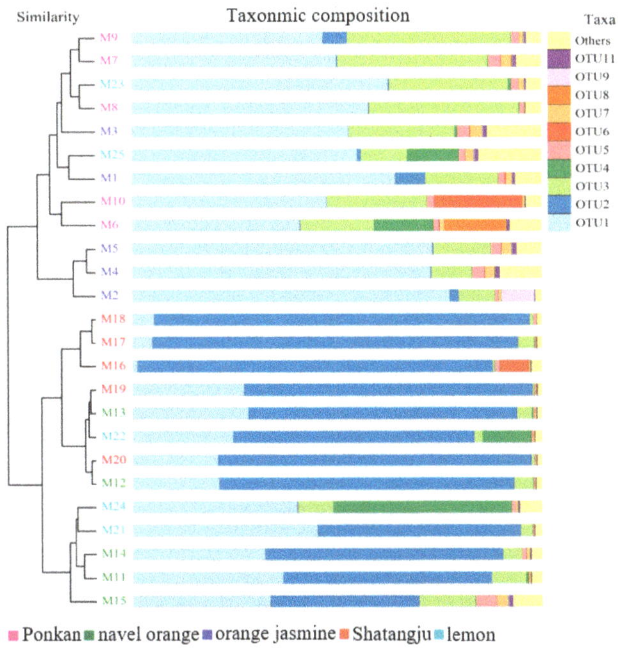

Figure 5. UPGMA clustering analysis of the microbiota based on weighted UniFrac distances.

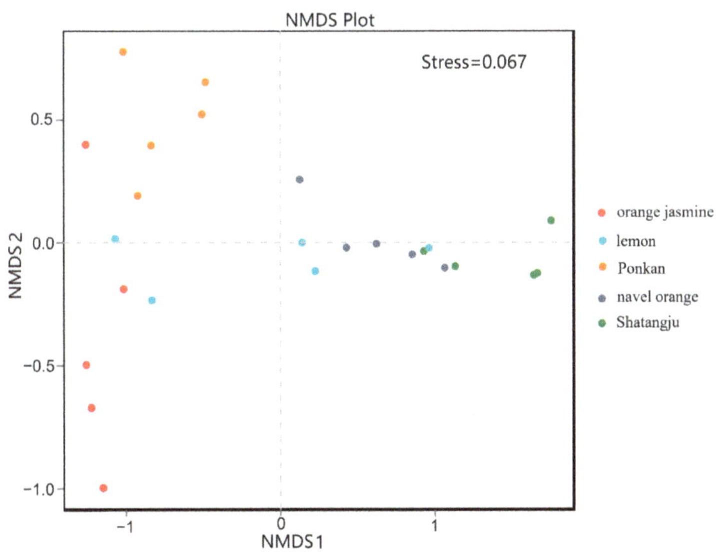

Figure 6. NMDS analysis of the microbial communities based on weighted UniFrac distances. Each signal represents one sample; the distance between samples demonstrates the degree of difference. Stress less than 0.2 indicates the reliability of the NMDS analysis.

LEfSe was conducted to identify specific taxa that consistently varied in abundance among the five treatments and could thus be used as biomarkers. Based on a logarithmic LDA score of 2.0 as the cutoff, a total of 14 taxa were significantly differentially represented in the five populations (Figure S2).

3.3. *PICRUSt Analysis and Functional Prediction*

To determine the effects of host plants on the GM and metabolism, PICRUSt analysis was conducted to predict the functional gene profiles of bacterial communities [26]. The KEGG pathway database was used to enrich the predicted genes. KEGG pathway analysis showed that different feeding hosts had varying "metabolism", "genetic information processing", and "environmental information processing" (Figure 7). The enrichment ratio of "metabolism", which involved amino acid metabolism, energy metabolism, lipid metabolism, metabolism of polyketides and terpenoids, and xenobiotics biodegradation and metabolism, was significantly lower in the Shatangju-feeding population ($p < 0.01$). The enrichment rate for "carbohydrate metabolism" was higher in the Shatangju-feeding population (Figure 7a). Bacterial genes potentially involved in "genetic information processing" (e.g., translation, replicate and repair, folding, sorting, and degradation) were estimated to be significantly enriched in the orange jasmine-feeding population ($p < 0.01$) (Figure 7b). In "environmental information processing", which involved three different functional groups, five populations showed differences, but these differences did not reach a significant level (Figure 7c). It can be observed that feeding on different host plants can cause varying "metabolism", "genetic information processing", and "environmental information processing". These pathways are the most crucial for maintaining the population reproduction, growth, development, and adaptation to environmental stress of *D. citri*.

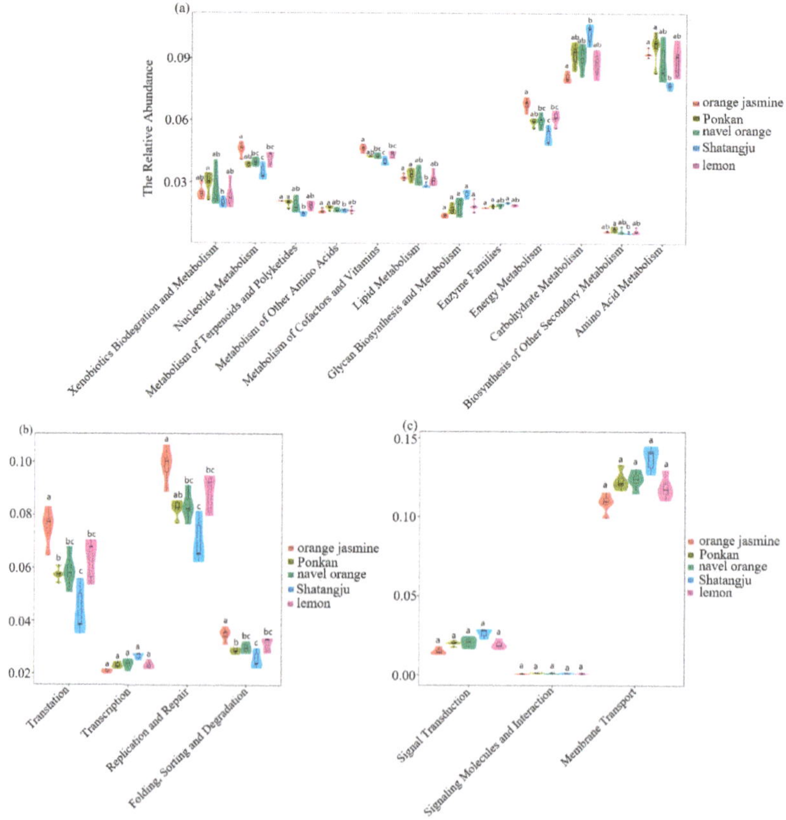

Figure 7. Comparison of predicted KEGG pathways of *D. citri* gut microbes feeding on different host plants. The inferred metabolic pathways are shown with Metabolism (**a**), Genetic Information Processing (**b**), Envionmental Information Processing (**c**). The bars represent the relative abundance predicted for a psyllid sample. Different letters indicate that the values are significantly different ($p < 0.01$).

4. Discussion

Proteobacteria, Actinobacteria, Bacteroidetes, Firmicutes, and Cyanobacteria were common in all *D. citri* populations. Similar to *Grapholita molesta* [29], *Bombyx mori* [30], *Nilaparvata lugens* [31,32], mosquitoes [33,34], and *Triatoma sordida* [35], Proteobacteria had absolute dominance in *D. citri* (relative abundance > 90%). A previous study on the compositional shifts in *D. citri* microbiota through all of the life stages (i.e., egg, nymph 1–5 stages, and adult) also reported that Proteobacteria were dominant in all of the life stages of *D. citri* reared on navel orange [36]. The differences in the relative abundance of phyla provided us with a comprehensive evaluation of the differences in bacterial composition, since each phyla is usually functionally different.

Enterobacteriaceae within Proteobacteria was observed in all adult samples, and this family has been reported to play a role in sugar metabolism. Researchers have suspected that Enterobacteriaceae contributes to digestion, protection, courtship, and reproduction [37,38]. The reason this family is present in all insects may be because of its metabolic diversity, which helps insects adapt to different environments. Enterococcaceae dominated in the Shatangju-, navel orange-, and lemon-feeding populations and was significantly higher in the Shatangju-feeding population. Interestingly, our previous studies on the effects of host plants on *D. citri* development and reproduction confirmed that Shatangju was the most appropriate host for *D. citri* [39]. We can hypothesize that Enterobacteriaceae play an important role in the fitness of *D. citri*. However, additional experiments are needed to examine the roles of *D. citri* biology and its gut bacteria in the future. A higher presence of Anaplasmataceae in the ponkan- and orange jasmine-feeding populations was identified. The Anaplasmataceae was widespread in various arthropods, mainly because their versatility in helping hosts adapt to different environments.

A total of six genera that differed significantly in abundance were found in the five test populations, with *Wolbachia*, *Escherichia-Shigella*, and *Candidatus Profftella* having the most abundance. *Escherichia-Shigella*, a type of enteropathogenic genus, can cause human bacillary dysentery or Shigellosis by regulating gut tissue invasion and epithelial physiology [40]. The significance of large amounts of *Escherichia-Shigella* in the adult *D. citri* gut calls for further study. *Candidatus Profftella*, a Betaproteobacterium, is capable of producing a defensive polyketide (diaphorin) [13] and can provide essential vitamins to *D. citri* to ensure its nutritional balance. *Profftella* was reported to be localized in the bacteriome and has currently only been found in *D. citri* [41]. In the present study, *Profftella* had its highest abundance in the ponkan-feeding population, and our study is the first to report that symbiotic bacteria exist in the *D. citri* gut. *Wolbachia*, the highest relative abundance genus in the Anaplasmataceae, is systemic in *D. citri* [42,43], colocalizes in the gut [44,45], and interacts with CLas [45], but its functions remain unclear. In this study, *Wolbachia* had the highest abundance in the orange jasmine-feeding population, and the lowest abundance in the Shatangju-feeding population. As two dominant genera in the *D. citri* gut, it is unclear why *Profftella* and *Wolbachia* changed significantly with host plants. In addition to host plants, temperature [46], gender [46], and CLas-infected or not [47] could significantly affect the change of these two symbionts. *Pantoea*, *Stenotrophomonas*, *Sphingomonas*, *Methylobacterium*, *Microbacterium*, and *Lactobacillus* were classified as environmental or plant-associated microbes, which can be found during *D. citri* landing/evaluation/feeding steps on the host plants. These observed changes in the microbiome composition may serve as indicators of the ecological processes that form the host-associated microbial communities [48].

There were significant differences in GM community structures between different hosts. In the present study, NMDS revealed a distinct difference in the compositions of GM communities in *D. citri* (stress: 0.067). However, there was no significant difference between the navel orange-feeding population and the Shatangju-feeding population. One possible explanation is that Shatangju and navel orange have similar nutritional components needed for *D. citri* growth. Plant secondary metabolites and host nutrient requirements affect the composition of GM [49]. The interactions between GM and insect hosts are complex, involving morphology, behavior, physiology, and biochemistry. The microbial diversity in the Shatangju-feeding population was the lowest, suggesting Shatangju provides different nutrients for the growth and development of *D. citri*. Our previous research has shown that *D. citri* feeding on Shatangju produces more eggs than when they feed on the other four species [39]. On Ponkan, the high diversity in the GM community may contribute to the absorption of specific nutrients from unbalanced feeding and to the adaptation of *D. citri* to the feeding environment [29,50]. These results strongly suggest that the diversity and structure of the GM in *D. citri* are markedly influenced by the diet (host plants), which is in line with previous reports demonstrating that diet influences the insect GM [5,50]. Research on longhorn beetles and higher termites indicated that diet could affect the insect GM [51,52]. In mammals, diet patterns were also shown to affect the microbial community structure [53,54].

The microbial communities within the gut of a *D. citri* adult can perform many key functions, including "metabolism", "genetic information processing", and "environmental information processing", that are essential to the survival of *D. citri*. Statistically significant differences among the five different feeding populations were reported for the metabolism of carbohydrate and amino acid, membrane transport, and replication and repair. We hypothesize that the differences in the function prediction could mainly be caused by the sugar, amino acid, and toxic contents as well as the secondary metabolites in the host plants. Because of the differences in nutrients and secondary metabolites, the function of the dominant GM is also different [49]. However, our results may serve as a preliminary indication of bacterial community function. Metagomeric and meta-transcriptome analysis are needed to elucidate the host–microbiome interaction and the important functions of GM so as to find new targets for controlling *D. citri*. In addition, it is necessary to compare the differences between the insect GM and host–plant microbiome. For instance, some bacteria that can be only obtained from a host plant by the insect are required to be identified, which can extend our knowledge on how certain environmental microbes are able to establish recurrent associations with hemipteran insects. In our study, the results were entirely in silico, which will require some validation work in the future due to the potential inaccuracy of high-throughput sequencing.

5. Conclusions

This present study conducted a detailed investigation of the GM communities present in five different host-plant-feeding populations (i.e., *Citrus reticulata* cv. Shatangju, *Citrus poonensis* cv. Ponkan, *Murraya paniculata* (orange jasmine), *Citrus limon* (lemon), and *Citrus sinensis* (navel orange)) using high-throughput sequencing technology. It was observed that host diet had a considerable influence on the formation of insect bacterial communities. Our study showed that the highest bacterial richness and diversity were in the ponkan-feeding population, and the lowest bacterial diversity were in the Shatangju-feeding population. The PICRUSt analysis showed that most functional prediction categories were related to "metabolism", "genetic information processing", and "environmental information processing", which are essential for the survival of *D. citri*. Research on the GM of *D. citri* is of great significance for the development of biological control technology based on the complex relationship between vector insects and their gut bacterial communities.

Supplementary Materials: The following supporting information can be downloaded at: https://www.mdpi.com/article/10.3390/insects13080694/s1, Figure S1: Rarefaction curve analysis of gut samples.; Figure S2: Linear Discriminant Analysis Effect Size (LEfSe) results for *D. citri* gut microbes feeding on different host plants. Table S1: The contents of main phyla in *D. citri* (%); Table S2: Mean relative abundance of the 10 most abundant genera in gut samples of *D. citri* from different host plants (%).

Author Contributions: Conceptualization, L.M. and H.Z.; methodology, L.M., C.X. and Z.J.; software, L.M., C.X. and Z.J.; validation, L.M., C.X. and Z.J.; formal analysis, L.M.; investigation, L.M. and Z.J.; resources, L.M. and C.X.; data curation, L.M., C.X. and Z.J.; writing—original draft preparation, L.M.; writing—review and editing, L.M., C.X., Z.J. and H.Z.; visualization, L.M.; supervision, H.Z.; funding acquisition, L.M. and H.Z. All authors have read and agreed to the published version of the manuscript.

Funding: This work was supported by the National Key R&D Program of China (2021YFD1400802) to Hongyu Zhang, the Faculty Development Grants from Hubei University of Medicine (2020QDJZR004) to Lixue Meng, and the China Agriculture Research System of MOF and MARA (No. CARS-26) to Hongyu Zhang.

Institutional Review Board Statement: Not applicable.

Informed Consent Statement: Not applicable.

Data Availability Statement: The data presented in this study are available upon request from the corresponding author.

Conflicts of Interest: The authors declare no conflict of interest.

References

1. Chen, B.; Xie, S.; Zhang, X.; Zhang, N.; Feng, H.; Sun, C.; Lu, X.; Shao, Y. Gut microbiota metabolic potential correlates with body size between mulberry-feeding lepidopteran pest species. *Pest Manag. Sci.* **2020**, *76*, 1313–1323. [CrossRef]
2. Van Moll, L.; De Smet, J.; Cos, P.; Van Campenhout, L. Microbial symbionts of insects as a source of new antimicrobials: A review. *Crit. Rev. Microbiol.* **2021**, *47*, 562–579. [CrossRef]
3. Giron, D.; Dedeine, F.; Dubreuil, G.; Huguet, E.; Simon, J.C. Influence of microbial symbionts on plant–insect interactions. In *Advances in Botanical Research*, 2nd ed.; Nicolas, S., Denis, T., Paul-André, C., Eds.; Academic Press: Le Rheu, France, 2017; Volume 81, pp. 225–257. [CrossRef]
4. Douglas, A.E. Editorial overview: Insect microbial symbionts. *Curr. Opin. Insect Sci.* **2014**, *4*, v–vii. [CrossRef]
5. Colman, D.R.; Toolson, E.C.; Takacs-Vesbach, C.D. Do diet and taxonomy influence insect gut bacterial communities? *Mol. Ecol.* **2012**, *21*, 5124–5137. [CrossRef]
6. Abrahamson, W.G.; Weis, A.E. *Evolutionary Ecology across Three Trophic Levels: Goldenrods, Gallmakers, and Natural Enemies (MPB-29)*; Princeton University Press: New Haven, CT, USA, 2020. [CrossRef]
7. Berlocher, S.H.; Feder, J.L. Sympatric speciation in phytophagous insects: Moving beyond controversy? *Annu. Rev. Entomol.* **2002**, *47*, 773–815. [CrossRef]
8. Jaenike, J. Criteria for ascertaining the existence of host races. *Am. Nat.* **1981**, *117*, 830–834. [CrossRef]
9. Crotti, E.; Balloi, A.; Hamdi, C.; Sansonno, L.; Marzorati, M.; Gonella, E.; Favia, G.; Cherif, A.; Bandi, C.; Alma, A.; et al. Microbial symbionts: A resource for the management of insect-related problems. *Microb. Biotechnol.* **2012**, *5*, 307–317. [CrossRef]
10. Bové, J.M. Huanglongbing: A destructive, newly-emerging, century-old disease of citrus. *J. Plant Pathol.* **2006**, *88*, 7–37.
11. Grafton-Cardwell, E.E.; Stelinski, L.L.; Stansly, P.A. Biology and management of Asian citrus psyllid, vector of the huanglongbing pathogens. *Annu. Rev. Entomol.* **2013**, *58*, 413–432. [CrossRef]
12. Sétamou, M.; Simpson, C.R.; Alabi, O.J.; Nelson, S.D.; Telagamsetty, S.; Jifon, J.L. Quality Matters: Influences of Citrus Flush Physicochemical Characteristics on Population Dynamics of the Asian Citrus Psyllid (Hemiptera: Liviidae). *PLoS ONE* **2016**, *11*, e0168997. [CrossRef]
13. Nakabachi, A.; Piel, J.; Malenovský, I.; Hirose, Y. Comparative Genomics Underlines Multiple Roles of *Profftella*, an Obligate Symbiont of Psyllids: Providing Toxins, Vitamins, and Carotenoids. *Genome Biol. Evol.* **2020**, *12*, 1975–1987. [CrossRef] [PubMed]
14. Ashraf, H.J.; Ramos Aguila, L.C.; Akutse, K.S.; Ilyas, M.; Abbasi, A.; Li, X.; Wang, L. Comparative microbiome analysis of *Diaphorina citri* and its associated parasitoids *Tamarixia radiata* and *Diaphorencyrtus aligarhensis* reveals *Wolbachia* as a dominant endosymbiont. *Environ. Microbiol.* **2022**, *24*, 1638–1652. [CrossRef] [PubMed]
15. Fawole, M.O.; Oso, B.A. *Laboratory Manual of Microbiology*; Spectrum Books Ltd. Sunshine House: Ibadan, Nigeria, 1988; p. 257.
16. Caporaso, J.G.; Lauber, C.L.; Walters, W.A.; Berglyons, D.; Lozupone, C.A.; Turnbaugh, P.J.; Fierer, N.; Knight, R. Global patterns of 16S rRNA diversity at a depth of millions of sequences per sample. *Proc. Natl. Acad. Sci. USA* **2011**, *108*, 4516–4522. [CrossRef] [PubMed]
17. Zhang, G.; Wei, L.; Chang, C.C.; Zhang, Y.; Wei, D. Molecular Biological Methods in Environmental Engineering. *Water Environ. Res.* **2016**, *88*, 930–953. [CrossRef]
18. Caporaso, J.G.; Kuczynski, J.; Stombaugh, J.; Bittinger, K.; Bushman, F.D.; Costello, E.K.; Fierer, N.; Peña, A.G.; Goodrich, J.K.; Gordon, J.I.; et al. QIIME allows analysis of high-throughput community sequencing data. *Nat. Methods* **2010**, *7*, 335–336. [CrossRef] [PubMed]
19. Magoč, T.; Salzberg, S.L. FLASH: Fast length adjustment of short reads to improve genome assemblies. *Bioinformatics* **2011**, *27*, 2957–2963. [CrossRef] [PubMed]
20. Bolger, A.M.; Lohse, M.; Usadel, B. Trimmomatic: A flexible trimmer for Illumina sequence data. *Bioinformatics* **2014**, *30*, 2114–2120. [CrossRef]
21. McDonald, D.; Price, M.N.; Goodrich, J.; Nawrocki, E.P.; DeSantis, T.Z.; Probst, A.; Andersen, G.L.; Knight, R.; Hugenholtz, P. An improved Greengenes taxonomy with explicit ranks for ecological and evolutionary analyses of bacteria and archaea. *ISME J.* **2012**, *6*, 610–618. [CrossRef]
22. Lozupone, C.; Knight, R. UniFrac: A new phylogenetic method for comparing microbial communities. *Appl. Environ. Microbiol.* **2005**, *71*, 8228–8235. [CrossRef]

23. Lozupone, C.A.; Hamady, M.; Kelley, S.T.; Knight, R. Quantitative and qualitative beta diversity measures lead to different insights into factors that structure microbial communities. *Appl. Environ. Microbiol.* **2007**, *73*, 1576–1585. [CrossRef]
24. Ramette, A. Multivariate analyses in microbial ecology. *FEMS Microbiol. Ecol.* **2007**, *62*, 142–160. [CrossRef] [PubMed]
25. Segata, N.; Izard, J.; Waldron, L.; Gevers, D.; Miropolsky, L.; Garrett, W.S.; Huttenhower, C. Metagenomic biomarker discovery and explanation. *Genome Biol.* **2011**, *12*, R60. [CrossRef] [PubMed]
26. Langille, M.G.; Zaneveld, J.; Caporaso, J.G.; McDonald, D.; Knights, D.; Reyes, J.A.; Clemente, J.C.; Burkepile, D.E.; Vega Thurber, R.L.; Knight, R.; et al. Predictive functional profiling of microbial communities using 16S rRNA marker gene sequences. *Nat. Biotechnol.* **2013**, *31*, 814–821. [CrossRef] [PubMed]
27. Hazra, A.; Gogtay, N. Biostatistics Series Module 3: Comparing Groups: Numerical Variables. *Indian J. Dermatol.* **2016**, *61*, 251–260. [CrossRef]
28. Truong, D.T.; Franzosa, E.A.; Tickle, T.L.; Scholz, M.; Weingart, G.; Pasolli, E.; Tett, A.; Huttenhower, C.; Segata, N. MetaPhlAn2 for enhanced metagenomic taxonomic profiling. *Nat. Methods* **2015**, *12*, 902–903. [CrossRef]
29. Wang, X.L.; Sun, S.J.; Yang, X.L.; Cheng, J.; Wei, H.S.; Li, Z.; Michaud, J.P.; Liu, X.X. Variability of gut microbiota across the life cycle of *Grapholita molesta* (Lepidoptera: Tortricidae). *Front. Microbiol.* **2020**, *11*, 1366. [CrossRef]
30. Chen, B.S.; Du, K.Q.; Sun, C.; Vimalanathan, A.; Liang, X.L.; Li, Y.; Wang, B.H.; Lu, X.M.; Li, L.J.; Shao, Y.Q. Gut bacterial and fungal communities of the domesticated silkworm (*Bombyx mori*) and wild mulberry-feeding relatives. *ISME J.* **2018**, *12*, 2252–2262. [CrossRef]
31. Yun, J.H.; Roh, S.W.; Whon, T.W.; Jung, M.J.; Kim, M.S.; Park, D.S.; Yoon, C.M.; Nam, Y.D.; Kim, Y.J.; Choi, J.H.; et al. Insect gut bacterial diversity determined by environmental habitat, diet, developmental stage, and phylogeny of host. *Appl. Environ. Microbiol.* **2014**, *80*, 5254–5264. [CrossRef]
32. Wang, Z.L.; Wang, T.Z.; Zhu, H.F.; Pan, H.B.; Yu, X.P. Diversity and dynamics of microbial communities in brown planthopper at different developmental stages revealed by high-throughput amplicon sequencing. *Insect Sci.* **2020**, *27*, 883–894. [CrossRef]
33. Muturi, E.J.; Ramirez, J.L.; Rooney, A.P.; Kim, C.H. Comparative analysis of gut microbiota of mosquito communities in central Illinois. *PLoS Negl. Trop. Dis.* **2017**, *11*, e0005377. [CrossRef]
34. Kang, X.; Wang, Y.H.; Li, S.P.; Sun, X.M.; Lu, X.Y.; Rajaofera, M.J.N.; Lu, Y.J.; Kang, L.; Zheng, A.H.; Zou, Z.; et al. Comparative analysis of the gut microbiota of adult mosquitoes from eight locations in Hainan, China. *Front. Cell. Infect. Microbiol.* **2020**, *10*, 596750. [CrossRef] [PubMed]
35. Oliveira, J.L.; Cury, J.C.; Gurgel-Goncalves, R.; Bahia, A.C.; Monteiro, F.A. Field-collected *Triatoma sordida* from central Brazil display high microbiota diversity that varies with regard to developmental stage and intestinal segmentation. *PLoS Negl. Trop. Dis.* **2018**, *12*, e0006709. [CrossRef] [PubMed]
36. Meng, L.X.; Li, X.Y.; Cheng, X.Q.; Zhang, H.Y. 16S rRNA Gene Sequencing Reveals a Shift in the Microbiota of *Diaphorina citri* During the Psyllid Life Cycle. *Front. Microbiol.* **2019**, *10*, 1948. [CrossRef] [PubMed]
37. Zhao, X.; Zhang, X.; Chen, Z.; Wang, Z.; Lu, Y.; Cheng, D. The divergence in bacterial components associated with *Bactrocera dorsalis* across developmental stages. *Front. Microbiol.* **2018**, *9*, 114. [CrossRef]
38. Ben Ami, E.; Yuval, B.; Jurkevitch, E. Manipulation of the microbiota of mass-reared Mediterranean fruit flies *Ceratitis capitata* (Diptera: Tephritidae) improves sterile male sexual performance. *ISME J.* **2010**, *4*, 28–37.
39. Meng, L.X.; Cheng, X.Q.; Xia, C.X.; Zhang, H.Y. Effect of host plants on development and reproduction of *Diaphorina citri* and their host preference. *Entomol. Exp. Appl.* **2022**, *170*, 700–707. [CrossRef]
40. Ferrari, M.L.; Malardé, V.; Grassart, A.; Salavessa, L.; Nigro, G.; Descorps-Declere, S.; Rohde, J.R.; Schnupf, P.; Masson, V.; Arras, G.; et al. Shigella promotes major alteration of gut epithelial physiology and tissue invasion by shutting off host intracellular transport. *Proc. Natl. Acad. Sci. USA* **2019**, *116*, 13582–13591. [CrossRef]
41. Nakabachi, A.; Nikoh, N.; Oshima, K.; Inoue, H.; Ohkuma, M.; Hongoh, Y.; Miyagishima, S.Y.; Hattori, M.; Fukatsu, T. Horizontal gene acquisition of Liberibacter plant pathogens from a bacteriome-confined endosymbiont of their psyllid vector. *PLoS ONE* **2013**, *8*, e82612. [CrossRef]
42. Ren, S.L.; Li, Y.H.; Ou, D.; Guo, Y.J.; Qureshi, J.A.; Stansly, P.A.; Qiu, B.L. Localization and dynamics of Wolbachia infection in Asian citrus psyllid *Diaphorina citri*, the insect vector of the causal pathogens of Huanglongbing. *MicrobiologyOpen* **2018**, *7*, e00561. [CrossRef]
43. Hosseinzadeh, S.; Shams-Bakhsh, M.; Mann, M.; Fattah-Hosseini, S.; Bagheri, A.; Mehrabadi, M.; Heck, M. Distribution and Variation of Bacterial Endosymbiont and "*Candidatus* Liberibacter asiaticus" Titer in the Huanglongbing Insect Vector, *Diaphorina citri* Kuwayama. *Microb. Ecol.* **2019**, *78*, 206–222. [CrossRef]
44. Kruse, A.; Fattah-Hosseini, S.; Saha, S.; Johnson, R.; Warwick, E.; Sturgeon, K.; Mueller, L.; MacCoss, M.J.; Shatters, R.G., Jr.; Cilia Heck, M. Combining 'omics and microscopy to visualize interactions between the Asian citrus psyllid vector and the Huanglongbing pathogen *Candidatus* Liberibacter asiaticus in the insect gut. *PLoS ONE* **2017**, *12*, e0179531. [CrossRef] [PubMed]
45. Mann, M.; Fattah-Hosseini, S.; Ammar, E.D.; Stange, R.; Warrick, E.; Sturgeon, K.; Shatters, R.; Heck, M. *Diaphorina citri* Nymphs Are Resistant to Morphological Changes Induced by "*Candidatus* Liberibacter asiaticus" in Midgut Epithelial Cells. *Infect. Immun.* **2018**, *86*, e00889-17. [CrossRef] [PubMed]
46. Jiang, R.X.; Shang, F.; Jiang, H.B.; Dou, W.; Cernava, T.; Wang, J.J. The Influence of Temperature and Host Gender on Bacterial Communities in the Asian Citrus Psyllid. *Insects* **2021**, *12*, 1054. [CrossRef] [PubMed]

47. Liu, K.; Pang, R.; Guan, Z.Y.; Zhong, M.Z.; He, J.W.; Han, Q.X. Comparative microbiome analysis reveals bacterial communities associated with *Candidatus* Liberibacter asiaticus infection in the Huanglongbing insect vector *Diaphorina citri*. *J. Asia Pac. Entomol.* **2022**, *25*, 101884. [CrossRef]
48. Paniagua Voirol, L.R.; Frago, E.; Kaltenpoth, M.; Hilker, M.; Fatouros, N.E. Bacterial Symbionts in Lepidoptera: Their Diversity, Transmission, and Impact on the Host. *Front. Microbiol.* **2018**, *9*, 556. [CrossRef]
49. Yang, F.Y.; Saqib, H.; Chen, J.H.; Ruan, Q.Q.; Vasseur, L.; He, W.Y.; You, M.S. Differential Profiles of Gut Microbiota and Metabolites Associated with Host Shift of *Plutella xylostella*. *Int. J. Mol. Sci.* **2020**, *21*, 6283. [CrossRef]
50. Liu, Y.; Shen, Z.; Yu, J.; Li, Z.; Liu, X.; Xu, H. Comparison of gut bacterial communities and their associations with host diets in four fruit borers. *Pest Manag. Sci.* **2020**, *76*, 1353–1362. [CrossRef]
51. Mikaelyan, A.; Dietrich, C.; Köhler, T.; Poulsen, M.; Sillam-Dussès, D.; Brune, A. Diet is the primary determinant of bacterial community structure in the guts of higher termites. *Mol. Ecol.* **2015**, *24*, 5284–5295. [CrossRef]
52. Kim, J.M.; Choi, M.Y.; Kim, J.W.; Lee, S.A.; Ahn, J.H.; Song, J.; Kim, S.H.; Weon, H.Y. Effects of diet type, developmental stage, and gut compartment in the gut bacterial communities of two Cerambycidae species (Coleoptera). *J. Microbiol.* **2017**, *55*, 21–30. [CrossRef]
53. Muegge, B.D.; Kuczynski, J.; Knights, D.; Clemente, J.C.; González, A.; Fontana, L.; Henrissat, B.; Knight, R.; Gordon, J.I. Diet drives convergence in gut microbiome functions across mammalian phylogeny and within humans. *Science* **2011**, *332*, 970–974. [CrossRef]
54. Delsuc, F.; Metcalf, J.; Wegener Parfrey, L.; Song, S.; González, A.; Knight, R. Convergence of gut microbiomes in myrmecophagous mammals. *Mol. Ecol.* **2014**, *23*, 1301–1317. [CrossRef] [PubMed]

insects

MDPI

Article

The Role of Feeding Characteristics in Shaping Gut Microbiota Composition and Function of Ensifera (Orthoptera)

Xiang Zheng [1,2], Qidi Zhu [1], Meng Qin [2,3], Zhijun Zhou [1], Chunmao Liu [2], Liyuan Wang [2] and Fuming Shi [1,*]

1 Key Laboratory of Zoological Systematics and Application of Hebei Province, College of Life Sciences, Hebei University, Baoding 071002, China
2 Laboratory of Enzyme Preparation, Hebei Research Institute of Microbiology Co., Ltd., Baoding 071051, China
3 College of Life Sciences, Hebei Agricultural University, Baoding 071001, China
* Correspondence: shif_m@126.com

Simple Summary: Feeding habits were the main factor affecting the gut microbial community structure of Ensifera (Insecta: Orthoptera). The gut microbial communities of Ensifera with different feeding habits were significantly different, as insects with more diverse feeding habits had gut microorganisms with less specific functions. However, feeding habits are not the only factors that affect the gut microbial community structure of Ensifera. Factors related to energy and nutrition acquisition also affect them, such as the abundance of some microbial functional genes unrelated to feeding habits but related to survival.

Abstract: Feeding habits were the primary factor affecting the gut bacterial communities in Ensifera. However, the interaction mechanism between the gut microbiota and feeding characteristics is not precisely understood. Here, the gut microbiota of Ensifera with diverse feeding habits was analyzed by shotgun metagenomic sequencing to further clarify the composition and function of the gut microbiota and its relationship with feeding characteristics. Our results indicate that under the influence of feeding habits, the gut microbial communities of Ensifera showed specific characteristics. Firstly, the gut microbial communities of the Ensifera with different feeding habits differed significantly, among which the gut microbial diversity of the herbivorous *Mecopoda niponensis* was the highest. Secondly, the functional genes related to feeding habits were in high abundance. Thirdly, the specific function of the gut microbial species in the omnivorous *Gryllotalpa orientalis* showed that the more diverse the feeding behavior of Ensifera, the worse the functional specificity related to the feeding characteristics of its gut microbiota. However, feeding habits were not the only factors affecting the gut microbiota of Ensifera. Some microorganisms' genes, whose functions were unrelated to feeding characteristics but were relevant to energy acquisition and nutrient absorption, were detected in high abundance. Our results were the first to report on the composition and function of the gut microbiota of Ensifera based on shotgun metagenomic sequencing and to explore the potential mechanism of the gut microbiota's association with diverse feeding habits.

Keywords: metagenomic; gut microbiota; feeding habits; KEGG; CAZymes; Ensifera

Citation: Zheng, X.; Zhu, Q.; Qin, M.; Zhou, Z.; Liu, C.; Wang, L.; Shi, F. The Role of Feeding Characteristics in Shaping Gut Microbiota Composition and Function of Ensifera (Orthoptera). *Insects* **2022**, *13*, 719. https://doi.org/10.3390/insects13080719

Academic Editors: Hongyu Zhang, Yin Wang and Xiaoxue Li

Received: 20 June 2022
Accepted: 8 August 2022
Published: 10 August 2022

Publisher's Note: MDPI stays neutral with regard to jurisdictional claims in published maps and institutional affiliations.

Copyright: © 2022 by the authors. Licensee MDPI, Basel, Switzerland. This article is an open access article distributed under the terms and conditions of the Creative Commons Attribution (CC BY) license (https://creativecommons.org/licenses/by/4.0/).

1. Introduction

Insects are the main group of arthropods, as well as one of the most diverse groups of animals on earth [1,2]. The diversification and successful evolution of insects were closely related to the symbiotic relationship between them and gut microorganisms in the long-term coevolution process [3,4]. In particular, many symbiotic microorganisms have explicitly adapted to insects as hosts and may participate in numerous metabolic activities. Microorganisms play crucial roles in acquiring and absorbing nutrients [5], secreting digestive enzymes [6], secreting immune-related compounds [7], enhancing pathogen resistance [8,9], and influencing social interactions [10], among other roles. However, identifying these microorganisms and determining their function remains challenging [11].

Microbiota is studied using both culture-dependent [12] and culture-independent [13] methods. However, due to the limitations associated with culture-dependent techniques, most gut microorganisms remain uncultured, limiting the possibility of describing the gut microbial community characteristics through culture techniques [14]. With the development of high-throughput sequencing (HTS) technology, significant progress was made in studying the gut microbiota through molecular biotechnology [15–17]. Furthermore, it allowed us to better understand the microbiota's structure, function, and diversity without culturing [18]. The 16S rRNA gene and shotgun metagenomic sequencing methods, with the characteristic of being microbial culture-independent, are the two main HTS tools that provide insights into microbial community composition and function [18,19]. However, 16S rRNA gene sequencing has limited genomic scope and amplification biases toward particular taxonomic groups. Therefore, no data are provided regarding the functional capacity of a microbial community [20]. More recently, shotgun metagenomic sequencing was applied to describe the viruses, bacteria, archaea, and eukaryotes that compose a given microbiota and explore their implications in metabolism [21–23].

Many factors can influence insect gut microbial community structure, such as feeding behavior, host taxonomy, life stage, and host environment [2,24–27]. A previous study found that the feeding characteristics affect the structure of the insect gut microbial community, which in turn affects the growth and development of the insect [28]. Studies also found that changes in the ecological environment will affect the type of food consumed, leading to changes in the gut microbial community of insects, which may harm their survival [29]. Meanwhile, gut microorganisms play an essential role in promoting the digestion and absorption of the host. Some insects evolved to use lignocellulose substrates as energy by using microbial metabolites [30]. Some symbiotic microorganisms secrete gut enzymes through the hydrolysis of ingested plant cell wall polysaccharides [31]. In other words, the host's feeding behavior and gut microbial community interact. However, the mechanism by which the host feeding behavior modifies the gut microbial community is unclear.

The suborder Ensifera (Insecta: Orthoptera) records about 7971 species and 19 subfamilies, many of which are endemic to China. Chinese Ensifera includes species of katydids, crickets, mole crickets, and wetas, with high species diversity and diversified feeding habits [32–34]. These species are major agricultural and forestry pests and potential resources for biological control. Ensifera is thoroughly studied taxonomically and phylogenetically, providing resources for studying the gut microbiota [35]. In our previous study, the 16S rRNA amplicon sequencing technology was used for sequence analysis of the gut bacterial communities of 12 species of Ensifera from 5 families. It was found that feeding habits were the primary factors affecting the gut bacterial communities, and samples from different taxa with the same feeding habit showed similar gut bacterial community structures [36]. However, the microbial composition and functional diversity of the gut microbiota in Ensifera, especially the potential mechanism of the relationship between the function of the gut microbiota and feeding habits, remains undetermined. Here, we performed shotgun metagenomic sequencing of three selected Ensifera species from our previous study mentioned above. Then, we directed our attention to the composition of the gut microbial community to better describe the relationship between specific microbiota functions and feeding characteristics.

2. Materials and Methods

2.1. Sample Collection and DNA Extraction

The adult Ensifera species were collected from two national nature reserve sites and one farmland site in China in 2019 (Table 1), and were immediately submerged in 99% (v/v) ethanol after capture until identification and dissection [37]. Based on morphological characteristics, these samples were identified by the Katydids Laboratory of Hebei University as *Mecopoda niponensis* (Mec) belonging to the family Tettigoniidae, *Ocellarnaca emeiensis* (Oce) belonging to the family Gryllacrididae, and *Gryllotalpa orientalis* (Gry) belonging to the family Gryllotalpidae, which were characterized as herbivore, carnivore, and omnivore feeding habits, respectively. The Department of Forestry of Guangxi Zhuang Autonomous Region approved our entry to the Daming Mountains National Nature Reserve to collect insect samples. The Zhejiang Tianmu Mountains National Nature Reserve Administration approved our entry to the Tianmu Mountains National Nature Reserve to collect insect samples. Samples from Hebei Province were collected from farmland. No endangered or protected species were used in this study.

Table 1. Overview of sample information.

Taxonomy		Feeding Habits	Location		Abbreviations
Family	Species		County/Mountain, Province	Geographic Coordinates	
Tettigoniidae	*Mecopoda niponensis*	Herbivore	Tianmu Mountains National Nature Reserve, Zhejiang	30°35′ N 119°43′ E	Mec
Gryllacrididae	*Ocellarnaca emeiensis*	Carnivore	Daming Mountains National Nature Reserve, Guangxi	23°52′ N 108°34′ E	Oce
Gryllotalpidae	*Gryllotalpa orientalis*	Omnivore	Quyang County, Hebei	38°59′ N 114°78′ E	Gry

Gut dissection was performed as follows: Each insect species (n = 15) was gently dissected by collecting the midgut and hindgut with gut contents under sterilized conditions [38–40], and 5 guts were randomly pooled together as one biological replicate sample. Then each sample was processed to extract DNA individually, using the TIANamp Stool DNA Kit (TIANGEN, Beijing, China) according to the manufacturer's protocols [13,41]. Sample blanks consisted of unused swabs processed through DNA extraction and were tested to contain no DNA amplicons. Following the extraction, the total DNA in each gut sample was measured using a NanoDrop 2000 spectrophotometer (Thermo Fisher Scientific, Waltham, MA, USA) and stored at −20 °C until sequencing by LC-Bio Technologies Co., Ltd., Hangzhou, China.

2.2. DNA Library Construction

After the DNA library was constructed and passed the quality test, Novaseq 6000 was used for high-throughput sequencing. The sequencing mode was PE150. The sequencing kit was the TruSeq Nano DNA LT Library Preparation Kit (FC-121-4001, Illumina, San Diego, CA, USA) and was fragmented by dsDNA Fragmentase (NEB, M0348S, Ipswich, MA, USA) at 37 °C for 30 min. The construction began with fragmented cDNA generated using a combination of fill-in reactions and exonuclease activity, and size selection was performed with the provided sample purification beads. The A-base was added to each strand's blunt ends to prepare them for ligation to the indexed adapters. Each adapter contained a T-base overhang for ligating the adapter to the A-tailed fragmented DNA and the full complement of the sequencing primer hybridization sites for single, paired-end, and indexed reads. Single- or dual-index adapters were ligated to the fragments [42]. Then, they were amplified with PCR using the following conditions: initial denaturation at 95 °C for 3 min; 8 cycles of denaturation at 98 °C for 15 s, annealing at 60 °C for 15 s, and extension at 72 °C for 30 s; and then final extension at 72 °C for 5 min.

2.3. Metagenomics Data Assembly and Analysis

Raw sequencing data were removed from the connector sequences to obtain valid reads. Firstly, sequencing adapters were removed using cutadapt (v 1.9, https://github.com/marcelm/cutadapt, accessed on 8 November 2020). Secondly, low-quality reads were trimmed by fqtrim (v 0.94, http://ccb.jhu.edu/software/fqtrim/, accessed on 8 November 2020) using a sliding window algorithm. Thirdly, reads were aligned to the host genome using bowtie2 (v2.2.0, http://bowtie-bio.sourceforge.net/bowtie2/, accessed on 10 November 2020) to remove host contamination. After data preprocessing, the quality-filtered reads were de novo assembled into contigs using IDBA-UD (v1.1.1, http:

//i.cs.hku.hk/~alse/hkubrg/projects/idba_ud/, accessed on 12 November 2020) [43]. QUAST (v3.2, St. Petersburg Academic University of the Russian Academy of Sciences, St Petersburg, Russian) was used to visualize the mapping of the genome bin contigs against the closest reference genome [44]. MetaGeneMark (v3.26, Georgia Tech, Atlanta, GA, USA) was used to predict the coding region (CDS) of assembled contigs ($\geq$500 bp), and the CDS sequences less than 100 NT were filtered. Then, CD-HIT (v4.6.1, http://www.bioinformatics.org/cd-hit/, accessed on 12 November 2020) was used to remove redundancy, and 95% of identity and 90% of coverage were used to cluster [45]. Then, bowtie2 (v2.2.0, http://bowtie-bio.sourceforge.net/bowtie2/, accessed on 13 November 2020) was used to compare each sample's clean data to the gene sequence, and the number of reads was calculated. The genome sequence with fewer reads ($\leq$2) was filtered out to obtain unigenes for subsequent analysis. The taxonomy was annotated by searching against the NR_ Meta database (blastp, evalue $\leq 1 \times 10^{-5}$) using DIAMOND (v0.9.14, Max Planck Institute for Biology, Tübingen, Germany) [46]. Combined with NCBI's species classification system, species annotation information at different taxonomic levels was obtained [47]. Similarly, the unigenes' functional annotation by the KEGG and CAZymes databases was obtained.

The reference genomes from *Teleogryllus occipitalis* (https://www.ncbi.nlm.nih.gov/genome/?term=GCA_011170035.1, accessed on 10 November 2020) and *Laupala kohalensis* (https://www.ncbi.nlm.nih.gov/genome/?term=GCA_002313205.1, accessed on 10 November 2020), which are closely related to the samples in this study, were retrieved from the NCBI database.

2.4. Statistical Analyses

Statistical analyses were carried out via R software (v4.1.2, http://cran.r-project.org, accessed on 25 May 2022) [48]. The alpha diversity was calculated using species-level annotation information statistics, and differences between the groups were assessed using the Kruskal–Wallis test [47], with $p < 0.05$ as a significant difference. The beta diversity of PCoA was tested with analysis of similarities (ANOSIM) [49] to analyze the differences between samples, with $p < 0.05$ as a significant difference. LEfSe analysis [50] (LDA score > 4) was used to find the biomarkers of the samples, with $p < 0.05$ as a significant difference. UpSet analysis (threshold > 0.1) was used to show each sample's shared and unique microorganism. Unigenes were compared with the KEGG [51] and CAZymes databases [52,53] (blastp, evalue $\leq 1 \times 10^{-5}$) to obtain the annotation enrichment of each database. The statistical analysis was carried out by pairwise comparisons with Welch's *t*-test [54] at KEGG levels 1 and 2, and CAZy level 2, with $p < 0.05$ as a significant difference. In addition, among the KO entry metabolic pathways of different feeding habits, the pathways related to carbohydrate metabolism, lipid metabolism, and amino acid metabolism with significant differences ($p < 0.05$) were compared and analyzed. The genes with 100% identity in the CAZymes database were screened, and their annotated species information, as well as the annotated enzyme information of CAZy level 2, was analyzed for correlation.

3. Results

In a previous study, we investigated the diversity of the gut bacterial communities of Ensifera from twelve species of five families. We found that feeding characteristics were the main factor affecting the structure of the gut bacterial communities. The gut bacterial communities' structure in Ensifera, from different taxa but with the same feeding habit, was similar [36]. Therefore, we selected three Ensifera species (Mec, Oce, and Gry) with high bacterial community diversity and minor intraspecific differences from the above samples to explore the similarities and differences in the gut microbiota's composition and function mediated by feeding characteristics.

After extracting DNA from each gut sample, the collected samples were analyzed by shotgun metagenomic sequencing. Since the raw sequencing data may contain splice sequences and a certain proportion of low-quality data, clean data for subsequent analysis could be obtained after quality trimming and host genome filtering. The preprocessing results are shown in Table S1. Gry obtained the most sequencing reads, with a value of 85,662,373, followed by Oce and Mec, with 82,575,455 and 79,442,693, respectively. After data preprocessing, IDBA-UD was assembled using a single sample, and QUAST evaluated the assembly results. The assembly results are shown in Table S2. A co-assembly of samples of Mec, Oce, and Gry generated 71,476, 129,921, and 417,677 contigs, respectively.

3.1. The Diversity and Composition of the Gut Microbiota

The alpha diversity analysis of the three species showed that the richness and diversity of the gut microbiota in Mec were the highest. In contrast, Gry and Oce's gut microbiota was with high diversity similarity. There was no significant difference in the alpha diversity index among the three species, however, with a significant difference between Mec and Oce (Kruskal–Wallis, $p < 0.05$) (Figure 1A). Based on Bray–Curtis distances, the beta diversity from the principal coordinate analysis (PCoA) showed significant differences in gut microbial structure among the three species (ANOSIM: $R = 1$, $p = 0.005$). Meanwhile, the sample of each feeding habit clustered together indicated that the intraspecific similarity of the species was high (Figure 1B).

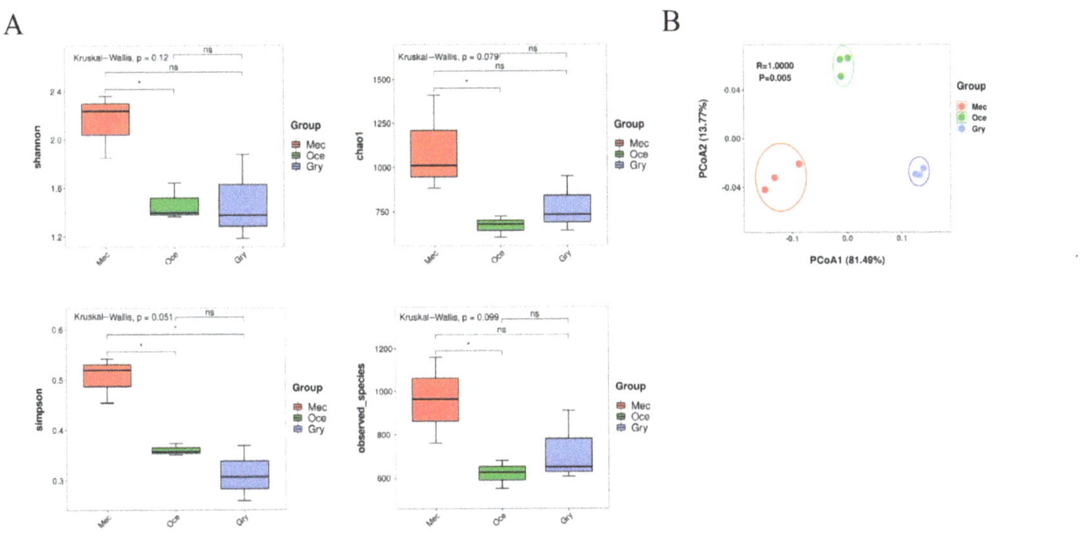

Figure 1. Gut microbiota composition of the three species. Relative abundance of gut microbial composition, (**A**) Alpha diversity of the gut microbial community based on Shannon, Chao1, Simpson, and observed_species (ns, $p > 0.05$; *, $p < 0.05$). (**B**) Beta diversity of PCoA analysis based on Bray–Curtis distances to compare differences between species.

The genes obtained from preprocessing were compared in the NR database (blastp, evalue $\leq 1 \times 10^{-5}$), and then the species annotation at different taxonomic levels was obtained. Subsequently, the species abundance at each taxonomic level was obtained by combining the species classification with the gene abundance. The classified sequences were assigned to bacteria, eukaryotes, viruses, and archaea. Among the annotated classified species, the abundance of bacteria was the highest, and the bacterial abundances of Mec, Oce, and Gry were 81.91%, 88.71%, and 86.03%, respectively. Viruses were the next most abundant in the gut microbiota of Mec and Gry, while eukaryotes were the next most abundant in Oce (Table S3).

At the phylum level, the structure of the gut microbiota of the three species was different. For instance, Firmicutes were the highest abundance of microbiota in Gry, Whereas

Proteobacteria had the highest abundance of microbiota in Mec and Oce. Although the bacterial abundance of Mec and Oce occupied the first place, there were several eukaryotic phyla with high abundances, such as Mucoromycota and Basidiomycota. At the species level, the dominant species of the three species were quite different. In particular, *Intestinimonas massiliensis*, with the highest abundance in Gry, were not detected in Mec and Oce. Meanwhile, both Gry and Oce had a high abundance of microbiota as dominant species (Gry: *Intestinimonas massiliensis*; Oce: *Lactococcus lactis*). However, although Mec had a highly diverse microbiota, no dominant taxonomic group was detected (Table 2).

At the phylum level, the main microbial phyla of bacteria (Figure 2A), eukaryotes (Figure 2C), and viruses (Figure S1A) were the same, except archaea (Figure S1C). However, the relative abundance and the dominant microbial components differed. Furthermore, the microbial composition of viruses and archaea was low diversity, and few viruses were detected in Oce. At the species level, bacteria (Figure 2B) specific to three species were found in the dominant bacterial populations, such as *Intestinimonas massiliensis* in Gry, *Leclercia adecarboxylata* in Mec, and *Lactococcus lactis* in Oce. However, although more specific eukaryotes with high abundance (Figure 2D) were found in Gry (*Metarhizium majus* and *Endogone* sp. *FLAS-F59071*) and Oce (*Synchytrium microbalum* and *Sparassis crispa*), eukaryotes with high abundance (*Rhizophagus irregularis*, *Puccinia striiformis* and *Rhizophagus clarus*) in Mec were also detected in other species. The composition and structure of gut viruses (Figure S1B) were completely different. Archaea (Figure S1D) accounted for less than 0.2% of Ensifera's gut microbiota and showed little difference except that Mec was the only one containing *Thaumarchaeota archaeon*.

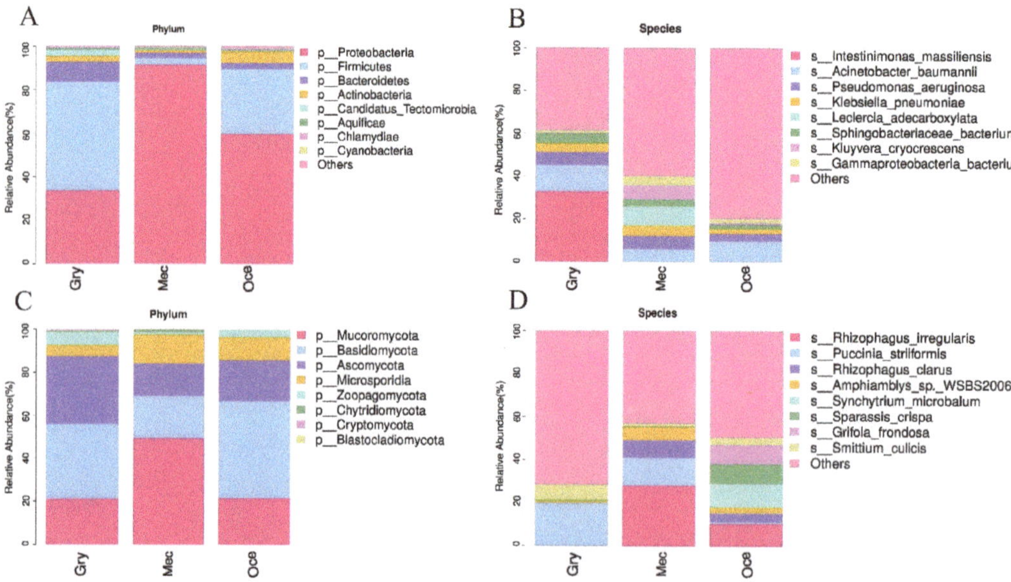

Figure 2. A stacked bar chart revealing the relative abundance of gut bacterial and eukaryotic composition at the (**A**,**C**) phylum and (**B**,**D**) species levels. The results show the phylum and species of the gut microbiota with the highest abundance.

Table 2. Top 4 abundances in the gut microbiota at the phylum and species levels of samples with different feeding habits.

Samples		*Gryllotalpa orientalis*		*Mecopoda niponensis*		*Ocellarnaca emeiensis*	
Microbial Taxonomy		**Phylum/%**	**Species/%**	**Phylum/%**	**Species/%**	**Phylum/%**	**Species/%**
Identification of microbiota	Top1	Firmicutes/47.09	*Intestinimonas massiliensis*/42.12	Proteobacteria/77.65	*Leclercia adecarboxy lata*/12.80	Proteobacteria/40.34	*Lactococcus lactis*/23.82
	Top2	Proteobacteria/31.80	*Acinetobacter baumannii*/15.42	Mucoromycota/7.43	*Kluyvera cryocrescens*/9.51	Firmicutes/20.04	*Acinetobacter baumannii*/20.86
	Top3	Bacteroidetes/8.69	*Cotesia sesamiae bracovirus*/10.82	Basidiomycota/2.94	*Pseudomonas aeruginosa*/9.16	Basidiomycota/14.64	*Rhizophagus irregularis*/8.16
	Top4	Candidatus_Tectomicrobia/2.61	*Pseudomonas aeruginosa*/7.8	Firmicutes/2.51	*Acinetobacter baumannii*/8.47	Mucoromycota/7.01	*Solemya velum gill symbiont*/8.14

3.2. The Characteristics of the Gut Microbiota

Based on the analysis of the three species' gut microbiota structure and diversity, the shared and unique microorganisms at the phylum and species level were displayed by UpSet analysis (threshold > 0.1). At the phylum level (Figure 3A), the number of shared microorganisms (28) of the three species was more than unique microorganisms, and most of them were in high abundance. However, the number of unique microorganisms of Gry was nine, while for Mec and Oce, it was seven and one, respectively. At the species level (Figure 3B), the numbers of unique microorganisms of Mec (483) and Gry (436) were more than the number of shared microorganisms (298), which overlapped in the three species. The number of unique microorganisms in Oce was about 1/4 of that in the other species. From the relationship between the two species, the number of microorganisms that overlapped among Mec and Oce was 527, which was more than the number of unique microorganisms.

To further explore the differences in the gut microbiota, LEfSe analysis (LDA score > 4, $p < 0.05$) was used to find the biomarkers in bacteria (Figure 3C), eukaryotes (Figure S2A), viruses (Figure S2B), and archaea (Figure 3D), with significant abundance differences among the three species. In terms of the biomarker numbers, there was little difference in gut eukaryotes and viruses among the samples. However, with significant differences between gut bacteria and archaea. Among them, no biomarker was found in the gut bacteria of Mec, whereas the number of biomarkers in the gut archaea of Mec was the highest. In terms of the biomarker species, bacteria and viruses were the main biomarkers of Gry, mainly including Firmicutes, Proteobacteria, Bacteroidetes, and *Cotesia sesamiae bracovirus*. Archaea were the main biomarkers of Mec, mainly Thermoplasmata.

3.3. Metabolic Potential Functions of Gut Microbiota According to the KEGG Database

The functional annotation of gut microbiota in the KEGG database was investigated, and these annotated biological functions were divided into six categories. In Mec and Oce, more than 65% of the genes were mapped onto metabolism, followed by genes mapped onto genetic information processing and environmental information processing. However, nearly 85% of the genes were mapped onto human diseases in Gry, followed by genes related to metabolism (Table 3). At KEGG level 1, genetic information processing was the dominant pathway in Oce through pairwise comparison (Welch's t-test, $p < 0.05$), and human diseases were the dominant pathway in Gry (Figure 4A). At KEGG level 2, metabolism of other amino acids and translation was the dominant pathway in Oce through pairwise comparison (Welch's t-test, $p < 0.05$), and drug resistance was the dominant pathway in Gry (Figure 4B). No metabolic pathway significantly higher than the other species was found in Mec.

Table 3. The percent of gut microbiota was assigned to biological functions.

KEGG Pathway	Percent of Genes (%)		
	Mec	Oce	Gry
Organismal systems	0.00	0.06	0.00
Metabolism	72.78	65.35	10.95
Human diseases	1.86	2.03	83.60
Genetic information processing	8.66	17.58	4.00
Environmental information processing	11.22	10.49	0.79
Cellular processes	5.48	4.49	0.66

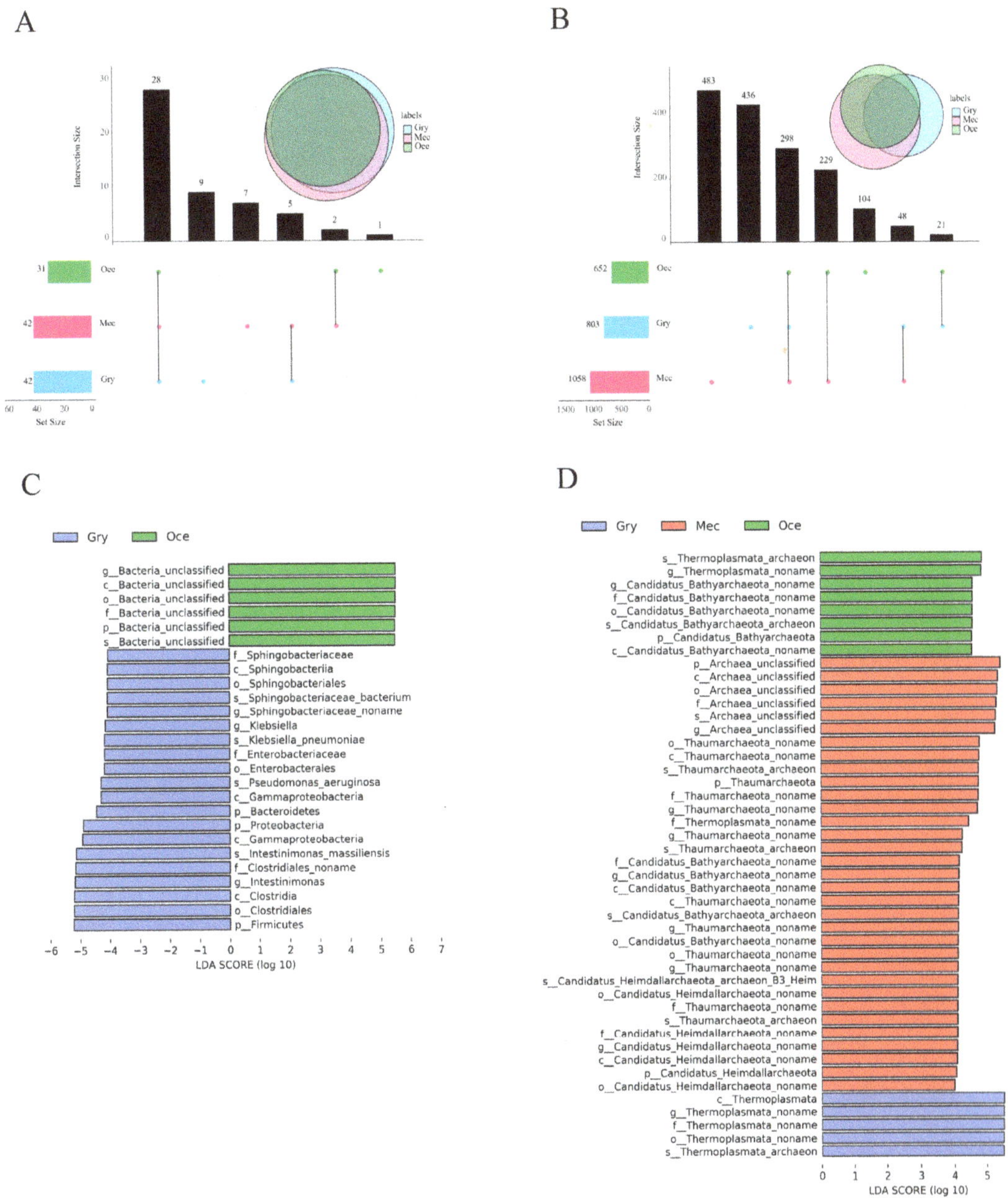

Figure 3. Characteristics of gut microbiota composition. UpSet analysis of the shared and unique microorganisms at the (**A**) phylum and (**B**) species levels between species. Distribution diagram of the LEfSe analysis based on the LDA score of (**C**) bacteria and (**D**) archaea to screen the biomarkers.

In order to deeply explore the relationship between the metabolic function of the gut microbiota and feeding characteristics, we performed a difference analysis on KEGG ORTHOLOGY (KO) database entry. Then, the carbohydrate metabolism, lipid metabolism, and amino acid metabolism pathways, which were significantly different ($p < 0.05$) and related to food digestion and absorption, were further explored. Among them, the gene abundance of Mec in the carbohydrate metabolism pathway was higher than that of the other species, such as fructose and mannose metabolism (ID: map00051) (Figure 4C) and galactose metabolism (ID: map00052) (Figure S3A). However, in starch and sucrose metabolism (ID: map00500) (Figure 4D), Gry showed the most substantial ability to convert cellulose into glucose, and Oce showed the most substantial ability to convert maltose into glucose. Interestingly, there was no high abundance of sequences related to lipid metabolism on Oce. In amino acid metabolism, it was found that Mec and Oce have high gene abundance in the biosynthesis of some amino acids, such as arginine biosynthesis (ID: map00220) (Figure S3B) in Oce and phenylalanine, tyrosine and tryptophan biosynthesis (ID: map00400) (Figure S3C) in Mec. Importantly, we did not find a high abundance of genes related to nutrition in Gry. However, we found a high gene abundance (k02172: bla regulator protein blaR1) involved in beta lactam resistance (ID: map01501) (Figure S3D) in Gry.

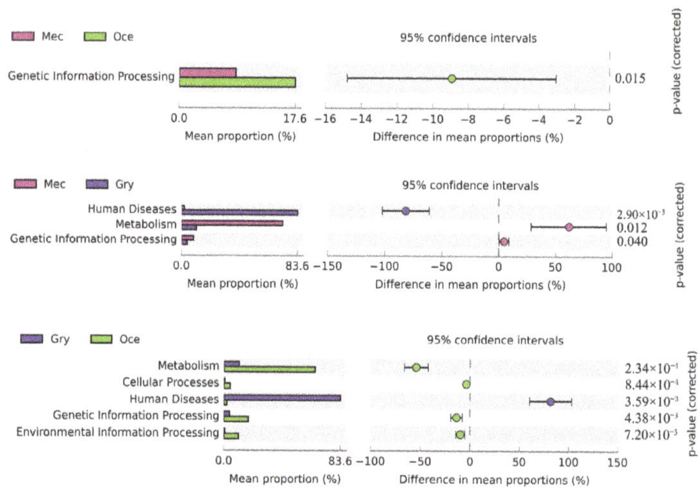

(A)

Figure 4. *Cont.*

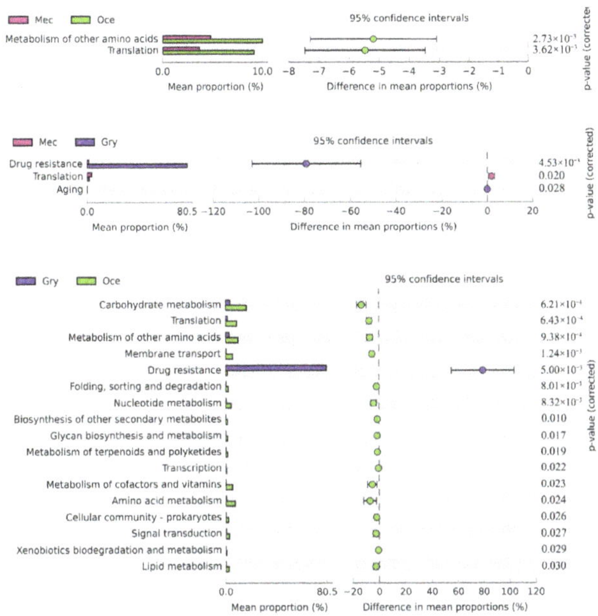

(**B**)

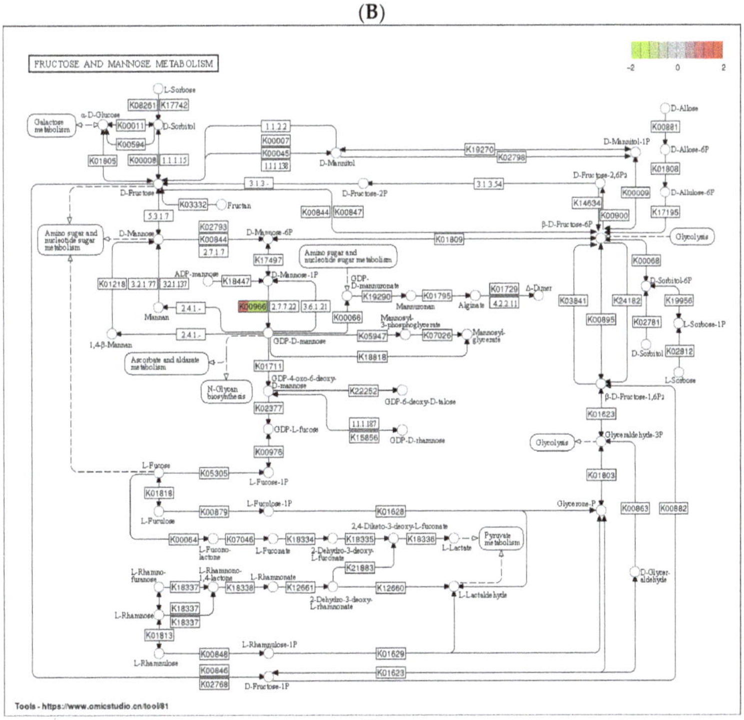

(**C**)

Figure 4. *Cont.*

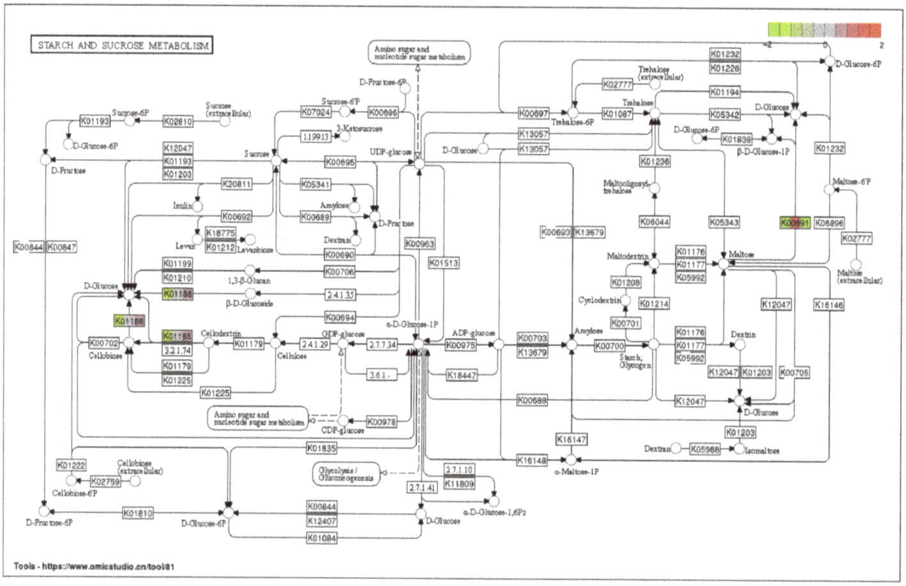

(D)

Figure 4. Functional and pathway annotation comparison of the genes assigned to the KEGG database. (**A**) Function annotation at KEGG level 1. (**B**) Function annotation at KEGG level 2. (**C**) Gene abundance comparison in the fructose and mannose metabolism pathway. (**D**) Gene abundance comparison in the starch and sucrose metabolism pathway. Notes for C and D: The genes belonging to the three samples with significant differences in the metabolic pathway map are marked by a color, in which Mec, Oce, and Gry are from left to right. The red to green color represents the high to low gene abundance. (https://www.omicstudio.cn/tool, accessed on 22 May 2022).

3.4. Metabolic Potential Functions of Gut Microbiota According to CAZymes Database

Based on the analysis of carbohydrate metabolism in Ensifera using the KEGG database, we performed a detailed pathway analysis using the carbohydrate-active enzymes from the CAZymes database. A correlation between the gut microbiota and CAZymes was established. The annotated genes belonged to six classes of CAZymes, mainly glycosyl transferases (GTs), glycoside hydrolases (GHs), and carbohydrate esterases (CEs). GTs involved in catalyzing the transfer of sugar moieties forming glycosidic bonds and GHs involved in the hydrolysis of glycosidic bonds and carbohydrate esters were the most abundant CAZymes, which combined contributed 83.67%, 83.65%, and 94.78% of the abundance of Mec, Oce, and Gry, respectively (Figure 5A). Interestingly, CEs represented a small amount in Mec and were almost absent in Gry, whereas their abundance was 11.77% in Oce.

A deeper analysis of the abundance and diversity of CAZymes (CAZy level 2) revealed that GT1, GH17, GH38, and CE8 were the dominant enzyme families in Oce through pairwise comparison (Welch's *t*-test, $p < 0.01$), while CE3 and CBM10 were the dominant enzyme families in Mec (Welch's *t*-test, $p < 0.05$) (Figure 5B). Notably, the CAZymes family with significant differences in Mec was a unique enzyme family. No enzyme family significantly higher than the other species was found in Gry (Table 4). The enzyme families with the highest abundance among the samples were GT1, GT2, GT4, GT47, GH3, and CE8. GT2, as an enzyme with high abundance, mainly existed in Mec and Gry. It was found in all types of microorganisms, but mainly existed in bacteria. The representative enzyme was cellulose synthase (EC 2.4.1.12), which was involved in cellulose synthesis, and chitin synthase (EC 2.4.1.16), which converted UDP-N-acetyl-α-D-glucosamine into chitin and UDP. CE8, the enzyme family with the second highest abundance in Oce, was almost absent in Gry, as it mainly existed in bacteria and eukaryotes. The representative enzyme was pectin methyl esterase (EC 3.1.1.11), which catalyzed pectin hydrolysis to produce pectinic acid and methanol. Meanwhile, GH3, with β-glucosidase as the representative enzyme, was involved in the hydrolysis β-D-glucosyl residues with the release of β-D-glucose.

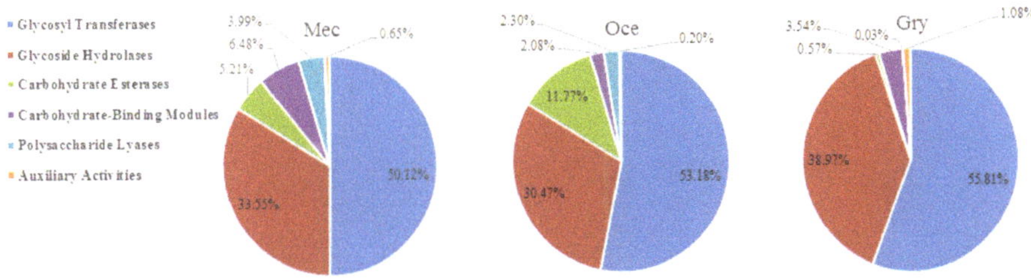

(A)

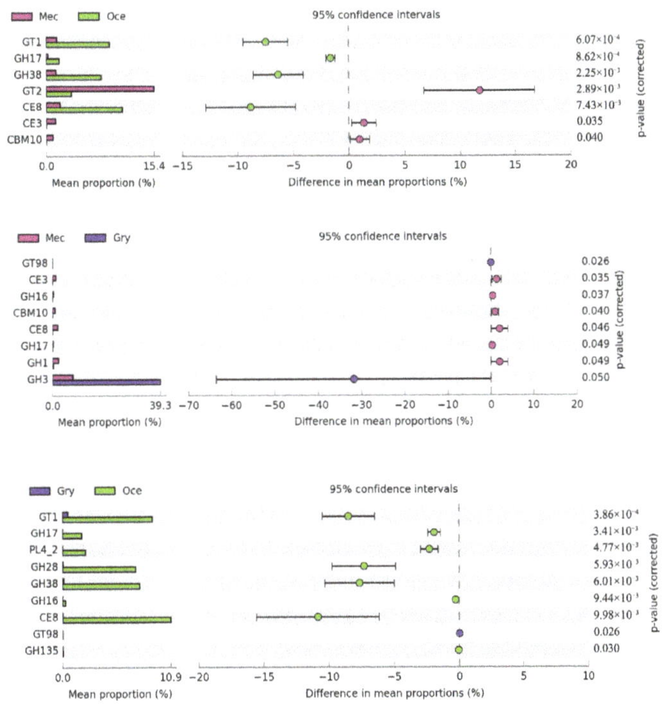

(B)

Figure 5. *Cont.*

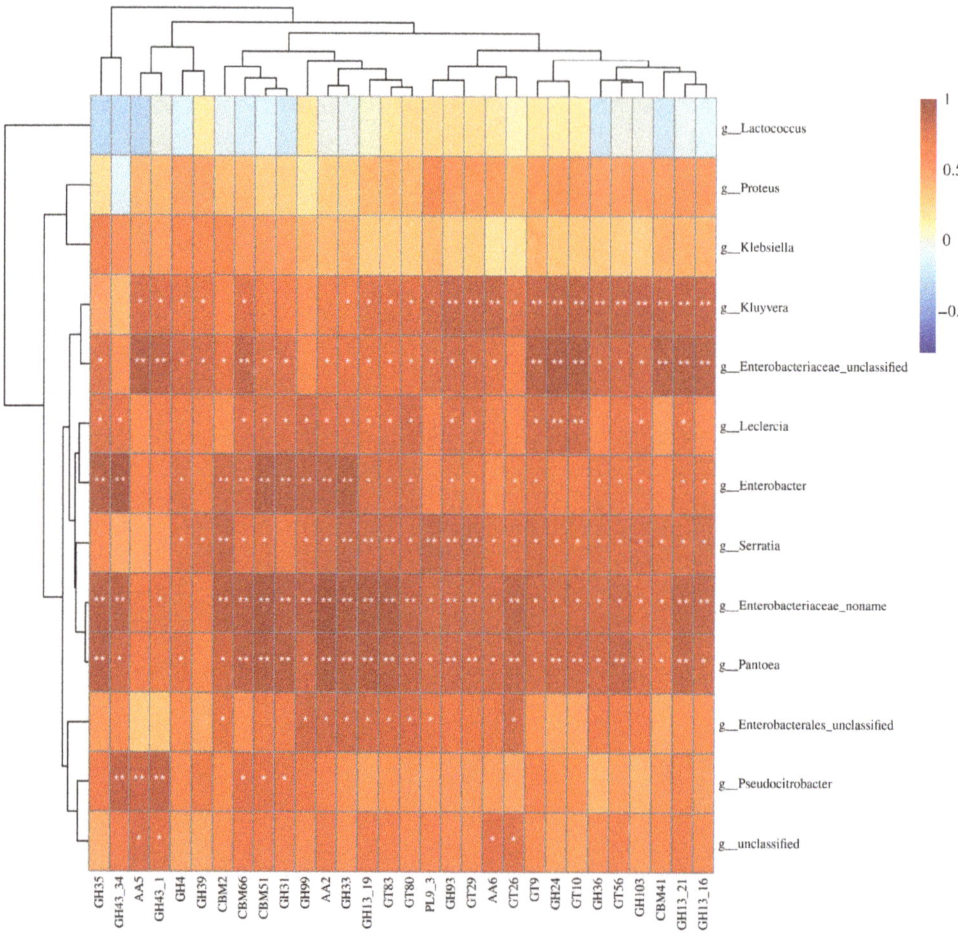

(C)

Figure 5. Comparison of the genes assigned to the CAZymes database. (**A**) Classification of CAZyme families in the samples. (**B**) The different abundant genes assigned to CAZy level 2. (**C**) Correlation analysis between microbiota and CAZymes (*, $p < 0.05$; **, $p < 0.01$).

In order to identify what microbial species play an essential role in the carbohydrate active enzyme families, we screened 109 genes with an identity of 100% in the CAZymes database and analyzed the correlation with the CAZy level 2 (Figure 5C). The results show that *Kluyvera*, Enterobacteriaceae_unclassified, *Enterobacter*, Enterobacteriaceae_noname, and Pantoea were positively correlated microorganisms, and *Lactococcus* was a negatively correlated microorganism. These five microorganisms belonged to a bacterial genus of Mec with high abundance, but their functions associated with the CAZymes database were not highly abundant or unique.

Table 4. Comparison of CAZyme families in the three species.

Groups	High Abundance CAZymes		Significantly Different CAZymes		Unique CAZymes	
	CAZyme Family	Activities in Family	CAZyme Family	Activities in family	CAZyme Family	Activities in Family
Mec	GT47	heparan β-glucuronyltransferase (EC 2.4.1.225); xyloglucan β-galactosyltransferase (EC 2.4.1.-)	CE3	acetyl xylan esterase (EC 3.1.1.72)	CE3	acetyl xylan esterase (EC 3.1.1.72)
	GT2	cellulose synthase (EC 2.4.1.12); chitin synthase (EC 2.4.1.16)	CBM10	cellulose-binding function	CBM10,	cellulose-binding function
	GT4	sucrose synthase (EC 2.4.1.13); sucrose-phosphate synthase (EC 2.4.1.14)			GH5_18	b-mannosidase (EC 3.2.1.25)
					GH5_13	b-D-galactofuranosidase (EC 3.2.1.146);a-L-arabinofuranosidase (EC 3.2.1.55)
					GH100	alkaline and neutral invertase (EC 3.2.1.26)
Oce	GT47	heparan β-glucuronyltransferase (EC 2.4.1.225); xyloglucan β-galactosyltransferase (EC 2.4.1.-)	CE8	pectin methylesterase (EC 3.1.1.11)		
	CE8	pectin methylesterase (EC 3.1.1.11)	GT1	UDP-glucuronosyltransferase (EC 2.4.1.17); zeatin O-β-xylosyltransferase (EC 2.4.2.40)		
	GT1	UDP-glucuronosyltransferase (EC 2.4.1.17); zeatin O-β-xylosyltransferase (EC 2.4.2.40)	GH38	α-mannosidase (EC 3.2.1.24); mannosyl-oligosaccharide α-1,2-mannosidase (EC 3.2.1.113)		
			GH17	glucan endo-1,3-β-glucosidase (EC 3.2.1.39); licheninase (EC 3.2.1.73)		
Gry	GT2	cellulose synthase (EC 2.4.1.12); chitin synthase (EC 2.4.1.16)				
	GH3	β-glucosidase (EC 3.2.1.21); xylan 1,4-β-xylosidase (EC 3.2.1.37)				
	GT47	heparan β-glucuronyltransferase (EC 2.4.1.225); xyloglucan β-galactosyltransferase (EC 2.4.1.-)				

4. Discussion

Feeding characteristics were the primary factor affecting the structure of the gut bacterial communities of Ensifera [36], and they significantly affected the gut microbial composition of other organisms [55–58]. This study explored the microbial community composition and function in three representatives of Ensifera mediated by feeding characteristics to deeply analyze the interaction between feeding habits and the gut microbiota.

Previous studies found that the taxa with complex feeding habits had a higher gut bacterial diversity than those with single feeding habits [25,27]. This was also the result of our previous study on the gut bacterial diversity of Ensifera [36]. However, in this comprehensive study on gut bacteria, eukaryotes, viruses, and archaea by shotgun metagenomic sequencing, it was found that, although the omnivorous Gry had the most significant number of reads and contigs, its microbial diversity was not the highest. On the contrary, the herbivorous Mec had the highest gut microbial diversity. LEfSe analysis found that bacteria were the main biomarkers of the omnivorous Gry, and archaea and eukaryotes played an important role in Mec. Therefore, we speculate that eukaryotes, archaea, and viruses may have high diversity in the herbivorous Mec. Although the gut microbiota of the herbivorous Mec was highly diversified, we did not find a biomarker in the bacterial communities with the highest abundance.

Proteobacteria was involved in degrading cellulose substances in the host gut [59]. Similarly, we also found such characteristics in Ensifera, which showed that Proteobacteria in the herbivorous Mec accounted for about 80% of the total microbiota abundance. Among these species, bacteria were the dominant microbiota, followed by viruses, eukaryotes, and archaea. Notably, the abundance of viruses was higher than that of eukaryotes in the herbivorous Mec and omnivorous Gry. Moreover, it was previously reported that honeybee gut viruses were a microbial population second only to bacteria [60]. Conversely, viruses in the carnivorous Oce were almost absent, and possibly implied a high abundance of gut viruses in plant-feeding Ensifera. In the previous study, Cyanobacteria mainly existed in herbivores compared with other feeding habits [61]; and could even be used as a biomarker for herbivorous insects. Interestingly, with the more accurate shotgun metagenomic sequencing [5,62], the abundance of Cyanobacteria decreased significantly and was almost undetected in the carnivorous Oce and omnivorous Gry.

Diet was the main driving factor for the functional composition of the gut microbial community [27,45]. To explore the functional characteristics of the gut microbiota in Ensifera, we annotated the metagenomic genes using the KEGG and CAZymes databases [61,63,64]. Interestingly, although the herbivorous Mec had a high abundance of genes in the carbohydrate metabolism pathway, we also found that the carnivorous Oce had the most substantial ability to convert maltose into glucose and that the omnivorous Gry had the most substantial ability to convert cellulose into glucose, in the carbohydrate metabolism pathway. This indicated that Ensifera, with different feeding habits, had unique methods of converting polysaccharides into monosaccharides to obtain energy. However, the herbivorous Mec had a more vital ability to metabolize carbohydrates. Moreover, the carnivorous Oce did not show an advantage in the lipid metabolism pathway, which was far from our thoughts. Importantly, we did not find a high abundance of genes related to food digestion and absorption in the omnivorous Gry. It was speculated that the specificity of genes involved in nutrient metabolism in Ensifera with a single feeding habit was higher than that with a broad-spectrum feeding habit. Ensifera, with a single feeding habit, had a more vital ability to obtain nutrition from food. However, we found a high abundance of resistance genes in Gry, suggesting that the omnivorous Gry's defense mechanism was better than that of herbivorous and carnivorous Ensifera species.

In this study, GTs were the most abundant enzymes in all samples, which was inconsistent with some animal reports that GHs were the most highly expressed enzymes [49,65]. GTs played a role in the biosynthesis of disaccharides, oligosaccharides, and polysaccharides, catalyzed the transfer of sugar groups to aglycones, and were very important for synthesizing many natural products [66]. GT2, with the high abundance of chitin synthase, was mainly found in the herbivorous Mec, as chitin mainly existed in the epidermis of insects, which indicated that the herbivorous Mec could digest not only cellulose but also chitin in the insect epidermis. CE8, with the high abundance of pectin methyl esterase, was mainly found in the carnivorous Oce, as pectin primarily existed in the plant cell wall and inner layer [67], which was consistent with the conclusion in the KEGG database that the carnivorous Oce can convert maltose into glucose to obtain nutrition and energy by digesting food derived from plants. Meanwhile, GH3 with the representative enzyme β-glucosidase involved in the hydrolysis release of β-D-glucose, consisting of the star and sucrose metabolic pathways of the KEGG database, was found in Gry, which could efficiently convert glucose. The above results show that in the gut of Ensifera species with a specific feeding habit, the genes of metabolic pathways and enzyme families related to the feeding habit have a high abundance, and those related to energy and nutrient digestion and absorption may also have a high abundance. It was speculated that Ensifera might have to retain or evolve functions in order to adapt to extreme environments [58].

We found that the abundance of genes related to the metabolic pathway of the omnivorous Gry in drug resistance was significantly higher than that of the other species. Gry usually lives in soil; however, we still do not know the role of such a high-abundance metabolic pathway in its life activities. Although we detected the most significant number of contigs and many unique microbial populations at the phylum and species levels, we did not find a good performance in Gry's unique functions of gut microbiota. For example, a low abundance of KEGG metabolic pathways related to food digestion and absorption was found, and no unique CAZyme family was found. This might indicate that the more complex the feeding behavior of Ensifera, the worse the functional specificity related to the feeding behavior of its gut microbiota. Furthermore, microbiota with a high correlation with the CAZymes database had a high abundance, but the enzyme families associated with the microbiota were not highly abundant or unique. This might mean that the functional microbiota in the gut was not necessarily high-abundance microbiota [65,68], but maybe some low-abundance or unique microbiota.

5. Conclusions

Our results show significant differences in the gut microbial community of Ensifera are mediated by feeding behavior and that the main functions of the gut microbiota were consistent with feeding characteristics. Specifically, the microbial community diversity of herbivorous Ensifera species was higher than that of the omnivorous and carnivorous species. At the same time, it was found that the abundance and specificity of the microbial population related to feeding habits in omnivorous Ensifera species was low, indicating that Ensifera species with a single feeding habit had a more vital ability to obtain nutrition from food. We also found that the gut microbiota associated with a higher abundance of metabolic pathways and carbohydrate active enzyme families were highly correlated with feeding characteristics. However, some microorganisms that had nothing to do with feeding characteristics, but were related to energy acquisition and nutrient absorption, also had a high abundance. In addition, gut microbiota with a low abundance may play an essential role in the life activities of Ensifera.

Supplementary Materials: The following supporting information can be downloaded at: https://www.mdpi.com/article/10.3390/insects13080719/s1. Figure S1: A stacked bar chart revealed the relative abundance of gut viruses and archaea composition at (A,C) phylum and (B,D) species levels. The results show the phylum and species of the gut microbiota with the highest abundance; Figure S2: Distribution diagram of the LEfSe analysis based on the LDA score of (A) eukaryote and (B) viruses to screen the biomarkers; Figure S3: Pathway annotation comparisons of the genes assigned to the KEGG database. (A) Gene abundance comparison in galactose metabolism pathway. (B) Gene abundance comparison in arginine biosynthesis pathway. (C) Gene abundance comparison in phenylalanine, tyrosine and tryptophan biosynthesis pathway. (D) Gene abundance comparison in beta lactam resistance pathway. Note: the genes belonging to the three samples with significant differences in the metabolic pathway map are marked by color, in which Mec, Oce, and Gry are from left to right. The color from red to green represents the gene abundance from high to low; Table S1: raw and clean reads from the analyzed samples; Table S2: general assembly and mapping statistics for samples; Table S3: relative abundance of classified gut microbiota in samples.

Author Contributions: Conceptualization: X.Z. and F.S.; methodology: X.Z., Q.Z., M.Q. and Z.Z.; resources: X.Z., Q.Z. and F.S.; formal analysis: X.Z., Q.Z., M.Q. and C.L.; data curation: L.W.; software: L.W.; visualization: X.Z.; writing—original draft preparation: X.Z., M.Q. and C.L.; writing—review and editing: X.Z., Z.Z. and F.S. All authors have read and agreed to the published version of the manuscript.

Funding: This research was funded by the Science and Technology Project of Hebei Academy of Sciences (Grant No. 2022Q2 and 22A05), and the Provincial School Cooperation Development Fund Project of Hebei Province (Grant No. 2020031513-2). The funders had no role in the study design, data collection and analysis, decision to publish, or preparation of the manuscript.

Institutional Review Board Statement: Not applicable.

Informed Consent Statement: Not applicable.

Data Availability Statement: The datasets presented in this study can be found in NCBI under project number PRJNA762197.

Acknowledgments: The authors thank Lixuan Chen and Lidan Zhang of Hebei University for the sample collection. We also thank Haoyu Liu of Hebei University for the sample identification of *Gryllotalpa orientalis*.

Conflicts of Interest: The authors declare that they have no known competing financial interests or personal relationships that could have appeared to influence the work reported in this paper.

References

1. Basset, Y.; Cizek, L.; Cuenoud, P.; Didham, R.K.; Guilhaumon, F.; Missa, O.; Novotny, V.; Odegaard, F.; Roslin, T.; Schmidl, J.; et al. Arthropod diversity in a tropical forest. *Science* **2012**, *338*, 1481–1484. [CrossRef] [PubMed]
2. Yun, J.H.; Roh, S.W.; Whon, T.W.; Jung, M.J.; Kim, M.S.; Park, D.S.; Yoon, C.; Nam, Y.D.; Kim, Y.J.; Choi, J.H.; et al. Insect gut bacterial diversity determined by environmental habitat, diet, developmental stage, and phylogeny of host. *Appl. Environ. Microbiol.* **2014**, *80*, 5254–5264. [CrossRef] [PubMed]
3. Shi, W.; Ding, S.-Y.; Yuan, J.S. Comparison of insect gut cellulase and xylanase activity across different insect species with distinct food sources. *BioEnergy Res.* **2010**, *4*, 1–10. [CrossRef]
4. Engel, P.; Moran, N.A. The gut microbiota of insects—Diversity in structure and function. *FEMS Microbiol. Rev.* **2013**, *37*, 699–735. [CrossRef] [PubMed]
5. Chen, B.; Yu, T.; Xie, S.; Du, K.; Liang, X.; Lan, Y.; Sun, C.; Lu, X.; Shao, Y. Comparative shotgun metagenomic data of the silkworm bombyx mori gut microbiome. *Sci. Data* **2018**, *5*, 180285. [CrossRef] [PubMed]
6. Pennisi, E. How do gut microbes help herbivores? Counting the ways. *Science* **2017**, *355*, 236. [CrossRef]
7. Vogel, H.; Shukla, S.P.; Engl, T.; Weiss, B.; Fischer, R.; Steiger, S.; Heckel, D.G.; Kaltenpoth, M.; Vilcinskas, A. The digestive and defensive basis of carcass utilization by the burying beetle and its microbiota. *Nat. Commun.* **2017**, *8*, 15186. [CrossRef]
8. Dillon, R.J.; Dillon, V.M. The gut bacteria of insects: Nonpathogenic interactions. *Annu. Rev. Entomol.* **2004**, *49*, 71–92. [CrossRef]
9. Nadarasah, G.; Stavrinides, J. Insects as alternative hosts for phytopathogenic bacteria. *FEMS Microbiol. Rev.* **2011**, *35*, 555–575. [CrossRef]
10. Lavy, O.; Gophna, U.; Gefen, E.; Ayali, A. The effect of density-dependent phase on the locust gut bacterial composition. *Front. Microbiol.* **2018**, *9*, 3020. [CrossRef]
11. Brumfield, K.D.; Huq, A.; Colwell, R.R.; Olds, J.L.; Leddy, M.B. Microbial resolution of whole genome shotgun and 16s amplicon metagenomic sequencing using publicly available neon data. *PLoS ONE* **2020**, *15*, e0228899.
12. Huang, S.; Sheng, P.; Zhang, H. Isolation and identification of cellulolytic bacteria from the gut of holotrichia parallela larvae (coleoptera: Scarabaeidae). *Int. J. Mol. Sci.* **2012**, *13*, 2563–2577. [CrossRef] [PubMed]
13. Kwong, W.K.; Medina, L.A.; Koch, H.; Sing, K.W.; Soh, E.J.Y.; Ascher, J.S.; Jaffe, R.; Moran, N.A. Dynamic microbiome evolution in social bees. *Sci. Adv.* **2017**, *3*, e1600513. [CrossRef] [PubMed]
14. Allen-Vercoe, E. Bringing the gut microbiota into focus through microbial culture: Recent progress and future perspective. *Curr. Opin. Microbiol.* **2013**, *16*, 625–629. [CrossRef] [PubMed]
15. Schloss, P.D.; Handelsman, J. Biotechnological prospects from metagenomics. *Curr. Opin. Biotechnol.* **2003**, *14*, 303–310. [CrossRef]
16. Handelsman, J. Metagenomics: Application of genomics to uncultured microorganisms. *Microbiol. Mol. Biol. Rev.* **2004**, *68*, 669–685. [CrossRef]
17. Yen, S.; Johnson, J.S. Metagenomics: A path to understanding the gut microbiome. *Mamm. Genome* **2021**, *32*, 282–296. [CrossRef]
18. Han, D.; Gao, P.; Li, R.; Tan, P.; Xie, J.; Zhang, R.; Li, J. Multicenter assessment of microbial community profiling using 16s rrna gene sequencing and shotgun metagenomic sequencing. *J. Adv. Res.* **2020**, *26*, 111–121. [CrossRef]
19. Beaudry, M.S.; Wang, J.; Kieran, T.J.; Thomas, J.; Bayona-Vasquez, N.J.; Gao, B.; Devault, A.; Brunelle, B.; Lu, K.; Wang, J.S.; et al. Improved microbial community characterization of 16s rrna via metagenome hybridization capture enrichment. *Front. Microbiol.* **2021**, *12*, 644662. [CrossRef]
20. Narayan, N.R.; Weinmaier, T.; Laserna-Mendieta, E.J.; Claesson, M.J.; Shanahan, F.; Dabbagh, K.; Iwai, S.; DeSantis, T.Z. Piphillin predicts metagenomic composition and dynamics from dada2-corrected 16s rdna sequences. *BMC Genom.* **2020**, *21*, 56.
21. Enagbonma, B.J.; Aremu, B.R.; Babalola, O.O. Profiling the functional diversity of termite mound soil bacteria as revealed by shotgun sequencing. *Genes* **2019**, *10*, 637. [CrossRef] [PubMed]
22. Miller, I.J.; Rees, E.R.; Ross, J.; Miller, I.; Baxa, J.; Lopera, J.; Kerby, R.L.; Rey, F.E.; Kwan, J.C. Autometa: Automated extraction of microbial genomes from individual shotgun metagenomes. *Nucleic Acids Res.* **2019**, *47*, e57. [CrossRef] [PubMed]
23. Seol, D.; Jhang, S.Y.; Kim, H.; Kim, S.Y.; Kwak, H.S.; Kim, S.H.; Lee, W.; Park, S.; Kim, H.; Cho, S.; et al. Accurate and strict identification of probiotic species based on coverage of whole-metagenome shotgun sequencing data. *Front. Microbiol.* **2019**, *10*, 1683. [CrossRef]

24. Santo Domingo, J.W.; Kaufman, M.G.; Klug, M.J.; Tiedje, J.M. Characterization of the cricket hindgut microbiota with fluorescently labeled rrna-targeted oligonucleotide probes. *Appl. Environ. Microbiol.* **1998**, *64*, 752–755. [CrossRef]
25. Colman, D.R.; Toolson, E.C.; Takacs-Vesbach, C.D. Do diet and taxonomy influence insect gut bacterial communities? *Mol. Ecol.* **2012**, *21*, 5124–5137. [CrossRef]
26. Otani, S.; Mikaelyan, A.; Nobre, T.; Hansen, L.H.; Kone, N.A.; Sorensen, S.J.; Aanen, D.K.; Boomsma, J.J.; Brune, A.; Poulsen, M. Identifying the core microbial community in the gut of fungus-growing termites. *Mol. Ecol.* **2014**, *23*, 4631–4644. [CrossRef] [PubMed]
27. Perez-Cobas, A.E.; Maiques, E.; Angelova, A.; Carrasco, P.; Moya, A.; Latorre, A. Diet shapes the gut microbiota of the omnivorous cockroach blattella germanica. *FEMS Microbiol. Ecol.* **2015**, *91*, fiv022. [CrossRef]
28. Wang, Y.; Li, Z.; Ma, L.; Li, G.; Han, K.; Liu, Z.; Wang, H.; Xu, B. The native dietary habits of the two sympatric bee species and their effects on shaping midgut microorganisms. *Front. Microbiol.* **2021**, *12*, 738226. [CrossRef]
29. Ptaszynska, A.A.; Latoch, P.; Hurd, P.J.; Polaszek, A.; Michalska-Madej, J.; Grochowalski, L.; Strapagiel, D.; Gnat, S.; Zaluski, D.; Gancarz, M.; et al. Amplicon sequencing of variable 16s rrna from bacteria and its2 regions from fungi and plants, reveals honeybee susceptibility to diseases results from their forage availability under anthropogenic landscapes. *Pathogens* **2021**, *10*, 381. [CrossRef]
30. Hori, C.; Song, R.; Matsumoto, K.; Matsumoto, R.; Minkoff, B.B.; Oita, S.; Hara, H.; Takasuka, T.E. Proteomic characterization of lignocellulolytic enzymes secreted by the insect-associated fungus daldinia decipiens oita, isolated from a forest in northern Japan. *Appl. Environ. Microbiol.* **2020**, *86*, e02350-19. [CrossRef]
31. Kirsch, R.; Wielsch, N.; Vogel, H.; Svatos, A.; Heckel, D.G.; Pauchet, Y. Combining proteomics and transcriptome sequencing to identify active plant-cell-wall-degrading enzymes in a leaf beetle. *BMC Genom.* **2012**, *13*, 587. [CrossRef] [PubMed]
32. Jin, X.B.; Xia, K.L. An index-catalogue of chinese tettigoniodea (orthopteroidea: Grylloptera). *J. Orthoptera Res.* **1994**, *3*, 15–41.
33. Zhou, Z.; Shi, F.; Zhao, L. The first mitochondrial genome for the superfamily hagloidea and implications for its systematic status in ensifera. *PLoS ONE* **2014**, *9*, e86027. [CrossRef]
34. Song, H.; Amédégnato, C.; Ciglianod, M.M.; Grandcolas, L.D.; Headse, S.W.; Huang, Y.; Otteg, D.; Whiting, M.F. 300 million years of diversification: Elucidating the patterns of orthopteran evolution based on comprehensive taxon and gene sampling. *Cladistics* **2015**, *31*, 621–651. [CrossRef] [PubMed]
35. Yuan, H.; Huang, Y.; Mao, Y.; Zhang, N.; Nie, Y.; Zhang, X.; Zhou, Y.; Mao, S. The evolutionary patterns of genome size in ensifera (insecta: Orthoptera). *Front. Genet.* **2021**, *12*, 693541. [CrossRef]
36. Zheng, X.; Zhu, Q.; Zhou, Z.; Wu, F.; Chen, L.; Cao, Q.; Shi, F. Gut bacterial communities across 12 ensifera (orthoptera) at different feeding habits and its prediction for the insect with contrasting feeding habits. *PLoS ONE* **2021**, *16*, e0250675. [CrossRef]
37. Smith, C.C.; Srygley, R.B.; Healy, F.; Swaminath, K.; Mueller, U.G. Spatial structure of the mormon cricket gut microbiome and its predicted contribution to nutrition and immune function. *Front. Microbiol.* **2017**, *8*, 801. [CrossRef]
38. Ng, S.H.; Stat, M.; Bunce, M.; Simmons, L.W. The influence of diet and environment on the gut microbial community of field crickets. *Ecol. Evol.* **2018**, *8*, 4704–4720. [CrossRef]
39. Rubanov, A.; Russell, K.A.; Rothman, J.A.; Nieh, J.C.; McFrederick, Q.S. Intensity of nosema ceranae infection is associated with specific honey bee gut bacteria and weakly associated with gut microbiome structure. *Sci. Rep.* **2019**, *9*, 3820. [CrossRef]
40. Shelomi, M.; Lin, S.S.; Liu, L.Y. Transcriptome and microbiome of coconut rhinoceros beetle (oryctes rhinoceros) larvae. *BMC Genom.* **2019**, *20*, 957. [CrossRef]
41. Kakumanu, M.L.; Maritz, J.M.; Carlton, J.M.; Schal, C. Overlapping community compositions of gut and fecal microbiomes in lab-reared and field-collected german cockroaches. *Appl. Environ. Microbiol.* **2018**, *84*, e01037-18. [CrossRef] [PubMed]
42. Chung, Y.W.; Gwak, H.J.; Moon, S.; Rho, M.; Ryu, J.H. Functional dynamics of bacterial species in the mouse gut microbiome revealed by metagenomic and metatranscriptomic analyses. *PLoS ONE* **2020**, *15*, e0227886. [CrossRef] [PubMed]
43. Bobay, L.M.; Wissel, E.F.; Raymann, K. Strain structure and dynamics revealed by targeted deep sequencing of the honey bee gut microbiome. *Msphere* **2020**, *5*, e00694-20. [CrossRef]
44. Gurevich, A.; Saveliev, V.; Vyahhi, N.; Tesler, G. Quast: Quality assessment tool for genome assemblies. *Bioinformatics* **2013**, *29*, 1072–1075. [CrossRef]
45. Hu, H.; da Costa, R.R.; Pilgaard, B.; Schiott, M.; Lange, L.; Poulsen, M. Fungiculture in termites is associated with a mycolytic gut bacterial community. *Msphere* **2019**, *4*, e00165-19. [CrossRef] [PubMed]
46. Amato, K.R.; Sanders, J.G.; Song, S.J.; Nute, M.; Metcalf, J.L.; Thompson, L.R.; Morton, J.T.; Amir, A.; McKenzie, V.J.; Humphrey, G.; et al. Evolutionary trends in host physiology outweigh dietary niche in structuring primate gut microbiomes. *ISME J.* **2019**, *13*, 576–587. [CrossRef]
47. Rausch, P.; Ruhlemann, M.; Hermes, B.M.; Doms, S.; Dagan, T.; Dierking, K.; Domin, H.; Fraune, S.; von Frieling, J.; Hentschel, U.; et al. Comparative analysis of amplicon and metagenomic sequencing methods reveals key features in the evolution of animal metaorganisms. *Microbiome* **2019**, *7*, 133. [CrossRef]
48. Tap, J.; Storsrud, S.; Le Neve, B.; Cotillard, A.; Pons, N.; Dore, J.; Ohman, L.; Tornblom, H.; Derrien, M.; Simren, M. Diet and gut microbiome interactions of relevance for symptoms in irritable bowel syndrome. *Microbiome* **2021**, *9*, 74. [CrossRef]
49. Marynowska, M.; Goux, X.; Sillam-Dusses, D.; Rouland-Lefevre, C.; Halder, R.; Wilmes, P.; Gawron, P.; Roisin, Y.; Delfosse, P.; Calusinska, M. Compositional and functional characterisation of biomass-degrading microbial communities in guts of plant fibre- and soil-feeding higher termites. *Microbiome* **2020**, *8*, 96. [CrossRef]

50. Fu, X.; Ou, Z.; Zhang, M.; Meng, Y.; Li, Y.; Wen, J.; Hu, Q.; Zhang, X.; Norback, D.; Deng, Y.; et al. Indoor bacterial, fungal and viral species and functional genes in urban and rural schools in shanxi province, china-association with asthma, rhinitis and rhinoconjunctivitis in high school students. *Microbiome* **2021**, *9*, 138. [CrossRef]
51. Slizen, M.V.; Galzitskaya, O.V. Comparative analysis of proteomes of a number of nosocomial pathogens by kegg modules and kegg pathways. *Int. J. Mol. Sci.* **2020**, *21*, 7839. [CrossRef] [PubMed]
52. Maire, J.; Girvan, S.K.; Barkla, S.E.; Perez-Gonzalez, A.; Suggett, D.J.; Blackall, L.L.; van Oppen, M.J.H. Intracellular bacteria are common and taxonomically diverse in cultured and in hospite algal endosymbionts of coral reefs. *ISME J.* **2021**, *15*, 2028–2042. [CrossRef] [PubMed]
53. Liew, K.J.; Liang, C.H.; Lau, Y.T.; Yaakop, A.S.; Chan, K.G.; Shahar, S.; Shamsir, M.S.; Goh, K.M. Thermophiles and carbohydrate-active enzymes (cazymes) in biofilm microbial consortia that decompose lignocellulosic plant litters at high temperatures. *Sci. Rep.* **2022**, *12*, 2850. [CrossRef] [PubMed]
54. He, X.; Sun, J.; Liu, C.; Yu, X.; Li, H.; Zhang, W.; Li, Y.; Geng, Y.; Wang, Z. Compositional alterations of gut microbiota in patients with diabetic kidney disease and type 2 diabetes mellitus. *Diabetes Metab. Syndr. Obes.* **2022**, *15*, 755–765. [CrossRef] [PubMed]
55. Mikaelyan, A.; Dietrich, C.; Kohler, T.; Poulsen, M.; Sillam-Dusses, D.; Brune, A. Diet is the primary determinant of bacterial community structure in the guts of higher termites. *Mol. Ecol.* **2015**, *24*, 5284–5295. [CrossRef]
56. Lyu, T.; Liu, G.; Zhang, H.; Wang, L.; Zhou, S.; Dou, H.; Pang, B.; Sha, W.; Zhang, H. Changes in feeding habits promoted the differentiation of the composition and function of gut microbiotas between domestic dogs (canis lupus familiaris) and gray wolves (canis lupus). *AMB Express* **2018**, *8*, 123. [CrossRef]
57. Rinninella, E.; Cintoni, M.; Raoul, P.; Lopetuso, L.R.; Scaldaferri, F.; Pulcini, G.; Miggiano, G.A.D.; Gasbarrini, A.; Mele, M.C. Food components and dietary habits: Keys for a healthy gut microbiota composition. *Nutrients* **2019**, *11*, 2393. [CrossRef]
58. Greene, L.K.; Williams, C.V.; Junge, R.E.; Mahefarisoa, K.L.; Rajaonarivelo, T.; Rakotondrainibe, H.; O'Connell, T.M.; Drea, C.M. A role for gut microbiota in host niche differentiation. *ISME J.* **2020**, *14*, 1675–1687. [CrossRef]
59. Reid, N.M.; Addison, S.L.; Macdonald, L.J.; Lloyd-Jones, G. Biodiversity of active and inactive bacteria in the gut flora of wood-feeding huhu beetle larvae (prionoplus reticularis). *Appl. Environ. Microbiol.* **2011**, *77*, 7000–7006. [CrossRef]
60. Bovo, S.; Utzeri, V.J.; Ribani, A.; Cabbri, R.; Fontanesi, L. Shotgun sequencing of honey DNA can describe honey bee derived environmental signatures and the honey bee hologenome complexity. *Sci. Rep.* **2020**, *10*, 9279. [CrossRef]
61. Xia, X.; Lan, B.; Tao, X.; Lin, J.; You, M. Characterization of spodoptera litura gut bacteria and their role in feeding and growth of the host. *Front. Microbiol.* **2020**, *11*, 1492. [CrossRef] [PubMed]
62. Ranjan, R.; Rani, A.; Metwally, A.; McGee, H.S.; Perkins, D.L. Analysis of the microbiome: Advantages of whole genome shotgun versus 16s amplicon sequencing. *Biochem. Biophys. Res. Commun.* **2016**, *469*, 967–977. [CrossRef] [PubMed]
63. Brune, A. Symbiotic digestion of lignocellulose in termite guts. *Nat. Rev. Microbiol.* **2014**, *12*, 168–180. [CrossRef] [PubMed]
64. Romero Victorica, M.; Soria, M.A.; Batista-Garcia, R.A.; Ceja-Navarro, J.A.; Vikram, S.; Ortiz, M.; Ontanon, O.; Ghio, S.; Martinez-Avila, L.; Quintero Garcia, O.J.; et al. Neotropical termite microbiomes as sources of novel plant cell wall degrading enzymes. *Sci. Rep.* **2020**, *10*, 3864. [CrossRef] [PubMed]
65. Murphy, R.; Benndorf, R.; de Beer, Z.W.; Vollmers, J.; Kaster, A.K.; Beemelmanns, C.; Poulsen, M. Comparative genomics reveals prophylactic and catabolic capabilities of actinobacteria within the fungus-farming termite symbiosis. *Msphere* **2021**, *6*, e01233-20. [CrossRef]
66. Schmid, J.; Heider, D.; Wendel, N.J.; Sperl, N.; Sieber, V. Bacterial glycosyltransferases: Challenges and opportunities of a highly diverse enzyme class toward tailoring natural products. *Front. Microbiol.* **2016**, *7*, 182. [CrossRef]
67. Liu, L.; Zhou, Y.; Qu, M.; Qiu, Y.; Guo, X.; Zhang, Y.; Liu, T.; Yang, J.; Yang, Q. Structural and biochemical insights into the catalytic mechanisms of two insect chitin deacetylases of the carbohydrate esterase 4 family. *J. Biol. Chem.* **2019**, *294*, 5774–5783. [CrossRef]
68. Benjamino, J.; Lincoln, S.; Srivastava, R.; Graf, J. Low-abundant bacteria drive compositional changes in the gut microbiota after dietary alteration. *Microbiome* **2018**, *6*, 86. [CrossRef]

insects

MDPI

Article

Analysis of Intestinal Microbial Diversity of Four Species of Grasshoppers and Determination of Cellulose Digestibility

Jing Bai †, Yao Ling †, Wen-Jing Li, Li Wang, Xiao-Bao Xue, Yuan-Yi Gao, Fei-Fei Li and Xin-Jiang Li *

The Key Laboratory of Zoological Systematics and Application, School of Life Sciences, Institute of Life Sciences and Green Development, Hebei University, Baoding 071002, China; baijing2566@163.com (J.B.); lingyao0112@163.com (Y.L.); wenjing12022021@163.com (W.-J.L.); hbuwangl@163.com (L.W.); xuexiaobao2021@163.com (X.-B.X.); gaoyy0302@163.com (Y.-Y.G.); lifeiyazhou@163.com (F.-F.L.)

* Correspondence: hbulxj@163.com

† These authors contributed equally to this work.

Simple Summary: Grasshoppers are typical phytophagous pests, which prefer eating monocotyledons with more cellulose and hemicellulose. Due to its large appetite and high utilization rate, the intestinal contents of grasshoppers have the potential to be developed into a bioreactor, which can be applied to improve straw utilization efficiency in the future. The digestive tract of grasshoppers is a complex ecosystem, inhabited by a large number of microorganisms. The existence of these microorganisms enables grasshoppers to have high decomposition and utilization of plant fibers. However, there are few reports on the microflora structure and diversity of the digestive tract of grasshoppers. In this study, the diversity of symbiotic bacteria in the intestinal tract of four species of grasshoppers, namely *Acrida cinerea, Trilophidia annulata, Atractomorpha sinensis* and *Sphingonotus mongolicus*, was studied by using the method of constructing a 16S rRNA gene library and Illumina Miseq sequencing technology. At the same time, the digestibility of cellulose and hemicellulose of the four species of grasshoppers were determined and the relationship between digestibility and intestinal microbial diversity was analyzed. This study provided basic data for the development of the digestible bioreactor of cellulose and hemicellulose, which may provide a new idea for degrading straw.

Citation: Bai, J.; Ling, Y.; Li, W.-J.; Wang, L.; Xue, X.-B.; Gao, Y.-Y.; Li, F.-F.; Li, X.-J. Analysis of Intestinal Microbial Diversity of Four Species of Grasshoppers and Determination of Cellulose Digestibility. *Insects* **2022**, *13*, 432. https://doi.org/10.3390/insects13050432

Academic Editors: Hongyu Zhang, Yin Wang and Xiaoxue Li

Received: 6 April 2022
Accepted: 3 May 2022
Published: 5 May 2022

Publisher's Note: MDPI stays neutral with regard to jurisdictional claims in published maps and institutional affiliations.

Copyright: © 2022 by the authors. Licensee MDPI, Basel, Switzerland. This article is an open access article distributed under the terms and conditions of the Creative Commons Attribution (CC BY) license (https://creativecommons.org/licenses/by/4.0/).

Abstract: Grasshoppers (Insecta, Orthoptera, Acridoidea) are a large group of agricultural and animal husbandry pests. They have a large food intake with high utilization of plants fibers. However, the composition of the grasshopper gut microbial community, especially the relationship between gut microbial community and cellulose digestibility, remains unclear. In this research, 16S rRNA gene sequences were used to determine the intestinal microbial diversity of *Acrida cinerea, Trilophidia annulata, Atractomorpha sinensis* and *Sphingonotus mongolicus*, and Spearman correlation analysis was performed between the intestinal microbes of grasshoppers and the digestibility of cellulose and hemicellulose. The results showed that Proteobacteria was the dominant phylum and *Klebsiella* was the dominant genus in the guts of the four species of grasshoppers; there was no significant difference in the species composition of the gut microbes of the four species of grasshoppers. Spearman correlation analysis showed that *Brevibacterium* and *Stenotrophomonas* were significantly correlated with cellulose digestibility. *Brevibacterium, Clavibacter, Microbacterium* and *Stenotrophomonas* were significantly associated with hemicellulose digestibility. Our results confirmed that the gut microbes of grasshoppers were correlated with the digestibility of cellulose and hemicellulose, and indicated that grasshoppers may have the potential to develop into bioreactors, which can be applied to improve straw utilization efficiency in the future.

Keywords: grasshopper; gut microbiome; microbial diversity; cellulose digestibility; 16S rRNA

1. Introduction

Insects are the largest group of animals, are widely distributed in the world and have a long evolutionary history [1]. The insect gut is the place where various nutrients and metabolic wastes are exchanged with the external environment, and it hosts a large number of microorganisms. There is a co-evolutionary relationship between intestinal microorganisms and the host, and the core microorganisms of the intestinal tract are different among species. Hongoh et al. proved that there is a certain relationship between the phylogeny of different species of termite gut symbiotic microbes and termite species, indicating that termite gut microbes have a co-evolutionary relationship with the host [2]. Different feeding habits and living environments affect the intestinal structure and function of insects, and vice versa [3]. Aziz et al. compared the similarities and differences of intestinal microorganisms of three grasshoppers by biochemical and molecular research methods, purified and isolated the bacteria, and attributed the different reasons of microorganisms to the different geographical environment [4]. Moreover, Lavy et al. summarized the research on the diversity of intestinal microorganisms of desert grasshoppers and migratory grasshoppers, including *Locusta pardalina, Dociostarus marocanus* and *Callipamus Italicus*, and proved that the composition of bacterial colonies in the digestive tract of grasshoppers is largely affected by their specific anatomical structure [5]. In the gut of insects, many types of microorganisms are available, which can be divided into resident microflora and passing microflora. The resident microflora occupies a certain position and performs a specific function in the gut, which exists in the hosts for a long time along with their growth and development. They are closely related to the hosts, and their population is maintained in a dynamic equilibrium mechanism. The passing microflora is transient in the hosts and is excreted as metabolic wastes [6]. Intestinal microbes provide nutrients to the host [7], help digest stubborn food components [8], protect the host from predators [9], parasites and pathogens [10], affect the efficiency of disease vectors [11] and even affect the mating and reproductive system of the host [12]. The microbial community in the insect digestive tract plays an important role, which not only ensures the orderly operation of an insect body, but also has an important impact on human production and life in medicine, agriculture and ecology. Therefore, insect intestinal microorganisms are ideal materials for studying relevant evolutionary mechanisms, as well as a huge microbial resource bank to find key microorganisms with biological functions, studying their functional mechanisms and ultimately applying it to production practice.

A large number of crop stalks cannot be effectively used in the world. Incineration will cause environmental pollution and waste resources. One of the most difficult problems that human beings face is how to solve the problem of crop straw recycling. Grasshoppers are widely distributed, have large appetites, strong reproductive capacity and migrate fast. Swarms of grasshoppers can do great harm to crops or pastures. Termites, grasshoppers and longicorn beetles feed on cellulose and contain a variety of symbiotic bacteria that degrade cellulose, including Enterobacteriaceae, *Bacteroides, Staphylococcus, Streptococcus* and *Bacillus,* and its degradation capacity of lignin model compounds is about 20–100%, which is 30–40% higher than that of large herbivores [13]. Su et al. studied 16 species of grasshopper intestinal symbiotic bacteria through DEEG, and the results showed that cellulolytic enzymes and intestinal microbial communities may reflect the relationship between different species of grasshopper and their feeding patterns [14]. The core gut bacteria of *Cytrotrachelus Buqueti,* a bamboo nose beetle, have carbohydrate-active enzymes that are key to lignocellulosic degradation and are used to break down bamboo cell walls, thereby contributing to the growth of host insects [15]. There are few studies on the relationship between cellulose digestibility and gut microbial community structure in grasshoppers. Grasshoppers maybe have the potential to be developed into bioreactors, which can be applied to improve straw utilization efficiency in the future.

Acrida cinerea Thunberg, 1815 (Ac), *Trilophidia annulata Thunberg,* 1815 (Ta), *Atractomorpha sinensis* Bolivar, 1905 (As) and *Sphingonotus mongolicus* Saussure, 1888 (Sm) are used in the study. The 16S rRNA gene sequences of four species of grasshopper intestinal bacteria were sequenced by the paired-end sequencing method and construction of a small fragment library. The intestinal microbial diversity of these species was further analyzed. The digestibility of cellulose and hemicellulose of four species of grasshoppers were determined by moss black phenol colorimetry and anthrone colorimetry, respectively. Furthermore, the relationship between digestibility and intestinal microbial diversity was analyzed. It provides a new thought for the green utilization of crop straw, which has important theoretical and practical significance.

2. Materials and Methods

2.1. Specimen Collection

Adults of *Acrida cinerea* Thunberg, 1815 (Ac), *Trilophidia annulata* Thunberg, 1815 (Ta), *Atractomorpha sinensis* Bolivar, 1905 (As) and *Sphingonotus mongolicus* Saussure, 1888 (Sm), were collected from Baoding City, Hebei Province, China in July–October 2018 (Table 1). They had the same living environment.

Table 1. Basic information of experimental specimens.

Species	Sample Code	Collection Date	Locality
	Ac1	July 2018	
Acrida cinerea	Ac2	July 2018	Baoding, China
	Ac3	October 2018	
	Ta1	October 2018	
Trilophidia annulate	Ta2	October 2018	Baoding, China
	Ta3	July 2018	
	Sm1	October 2018	
Sphingonotus mongolicus	Sm2	July 2018	Baoding, China
	Sm3	July 2018	
	As1	July 2018	
Atractomorpha sinensis	As2	July 2018	Baoding, China
	As3	July 2018	

2.2. Sample Treatment

Grasshoppers collected from the wild were classified and placed in different cages on an empty stomach for 2 days, so that their intestines were emptied of feces. The grasshoppers were immersed in 70% ethanol solution for 5 min to sterilize the bacteria on the grasshoppers' surface. In the ultra-clean working table, the bodies were placed on the sterilizing glass plate. The legs and wings were cut off by sterilized scissors and the bodies were cut from the anus to the chest along the abdomen. The surface of the bodies was cut open with sterilized dissecting needles, and the guts were removed with sterilized forceps. The guts were put into sterilized 1.5 mL EP tubes. Each tube contained a sample of five female and five male guts of the same species. There were three samples of each species. Intestinal dissection procedures were performed on ice.

2.3. Extraction of Total DNA from the Intestinal Contents

The PowerSoil DNA Isolation Kit was used to extract total DNA from the guts of grasshoppers. The common primers—338F (5′-ACTCCTACGGGAGCAGCA-3′) and 806R (5′-GGACTACHVGGGTWTCTAAT-3′)—in the V3 + V4 region of bacterial 16S rDNA were used as amplification primers [16]. Sequencing adapters were added to the primer ends to perform PCR. Target region PCR was performed in a total reaction volume of 10 μL: KOD FX Neo Buffer, 5.0 μL; DNA template, 50 ± 0 ng; primer1 (10 mmol/L), 0.3 μL; primer2 (10 mmol/L), 0.3 μL; dNTP, 2.0 μL; KOD FX Neo (5 U/mL), 0.2 μL; and constant volume to 10 μL with ddH_2O. After an initial denaturation at 95 °C for 5 min, amplification was performed with 25 cycles of incubations for 30 s at 95 °C, 30 s at 50 °C and 40 s at 72 °C, followed by a final extension at 72 °C for 7 min. Then, Solexa PCR was performed in a total reaction volume of 20 μL: 2 × Q5 HF MM, 10.0 μL; Target region PCR product (100 ng/mL), 5 μL; primer1 (2 mmol/L), 2.5 μL; primer2 (2 mmol/L), 2.5 μL. After an initial denaturation at 98 °C for 30 s, amplification was performed with 10 cycles of incubations for 10 s at 98 °C, 30 s at 65 °C and 30 s at 72 °C, followed by a final extension at 72 °C for 5 min. The amplified products were then purified and recovered using 1.8% agarose gel electrophoresis. The products were purified, quantified and homogenized to form sequencing libraries. The qualified sequencing libraries were sequenced with Illumina HiSeq 2500 (2×250 pairedends) at Biomarker Technologies Corporation, Beijing, China. This process was completed by Beijing Biomarker Technologies Co., Ltd.

2.4. Microbial Diversity Analysis

The original data were paired by FLASH 1.2.7 (overlap > 10 bp, false match rate ≤ 0.2) [17]. The paired raw reads were filtered by Trimmomatic v0.33 [18]. The Clean Reads without primer sequences were obtained by using Cutadapt 1.9.1 to identify and remove primer sequences. After the Clean Reads of each sample were paired, the Usearch v10 was used for length filtering. After removing chimeras by UCHIME v8.1 [19], the tags sequence with a higher quality was finally obtained. With Silva as the reference database, the feature sequences were annotated by using a Naive Bayes classifier. With 0.005% of all sequences as the filtering threshold [20], the sequences were clustered by Operational taxonomic unit (OTU) at the 97% similarity level [21], and OTU was taxonomically annotated. This process was completed by Beijing Biomarker Technologies Co., Ltd.

2.5. Digestibility of Cellulose and Hemicellulose

The grasshoppers collected from the wild were kept in cages and were fed wheat (*Triticum aestivum* Linnaeus, 1753) seedlings. After consecutively feeding for 3 days (no dung was collected during the period), grasshoppers were fasted for 2 days. At the beginning of the formal experiment, the grasshoppers were fed with wheat seedlings regularly and quantitatively, and feces were collected at the same time. The intake and excretion of grasshoppers were recorded in a day. The grasshoppers were fed every day at 9 am and 5 pm, and the feces excreted by the grasshoppers and the remaining wheat seedlings were collected the next morning. The formal experiment lasted for a week. The content of cellulose and hemicellulose in feces and wheat were determined by moss black phenol colorimetry and anthrone colorimetry, respectively. The following formulas were used to calculate the cellulose digestibility and hemicellulose digestibility of grasshoppers. Refer to Wang for specific methods [22].

$$\text{Cellulose (hemicellulose) content (\%)} = \left(\text{c} \times 240 \times 10^{-3}\,\text{L} \times 0.9(0.88)\right)/\text{m} \times \text{dilution mutiple} \times 100\%$$

$$\text{Cellulose (hemicellulose) digestibiliy (\%)} = (\text{a} - \text{b})/\text{a} \times 100\%$$

Note: c is the sugar concentration (g/L) calculated according to the standard curve, 240 is the total volume of sample solution (mL), m is the weighed sample mass (g), 0.9 and 0.88 are coefficients. a is amount of cellulose fed on wheat seedlings (g), b is fecal cellulose content (g).

2.6. Correlation between Digestibility and Intestinal Microbial Diversity

The LefSe analysis and Spearman analysis were performed using R and the Psych, Pheatmap and reshape2 package on the Biomarker Cloud Platform (Biomarker Biotechnology Co., Beijing, China) [23]. The correlation between cellulose digestibility and intestinal microbial diversity of grasshoppers was established.

3. Results

3.1. Analysis of the Gut Microbiota Diversity and Bacterial Composition between Four Species of Grasshoppers

In the guts of *Acrida cinerea* (Ac), *Trilophidia annulata* (Ta), *Atractomorpha sinensis* (As) and *Sphingonotus mongolicus* (Sm), a total of 858,719 pairs of reads were obtained from 12 samples; 758,316 Clean Reads were produced after quality control and splicing of paired-end reads.

A total of 7 phyla, 12 classes, 21 orders, 42 families and 55 genera were annotated for the four species of grasshoppers (Table 2). As shown in Figure 1A, as the sample number increased, the cumulative curve and the common quantity curve tended to be flat, which demonstrates that the new and common species detected in the sample were both approaching saturation, and the sample size was sufficient and could be used for diversity and abundance analysis. The Shannon, Chao1 and ACE indices were used to express the α-diversity of the microorganisms in the samples. As shown in Table 3, the coverage of the 12 samples was relatively high, reaching 99.97~99.99%. The above results showed that the sequencing data were reasonable and that the vast majority of bacteria in the samples were detected.

Table 2. Species statistics table of each grade of the sample.

Sample	Kingdom	Phylum	Class	Order	Family	Genus
Ac1	1	6	10	17	34	36
Ac2	1	6	10	15	28	32
Ac3	1	6	9	13	24	28
As1	1	5	9	14	23	26
As2	1	5	9	15	28	29
As3	1	5	10	14	24	28
Sm1	1	6	10	18	33	41
Sm2	1	6	10	18	32	37
Sm3	1	5	8	15	24	28
Ta1	1	6	10	16	30	35
Ta2	1	5	6	9	18	22
Ta3	1	6	9	16	26	29
Total	1	7	12	21	42	55

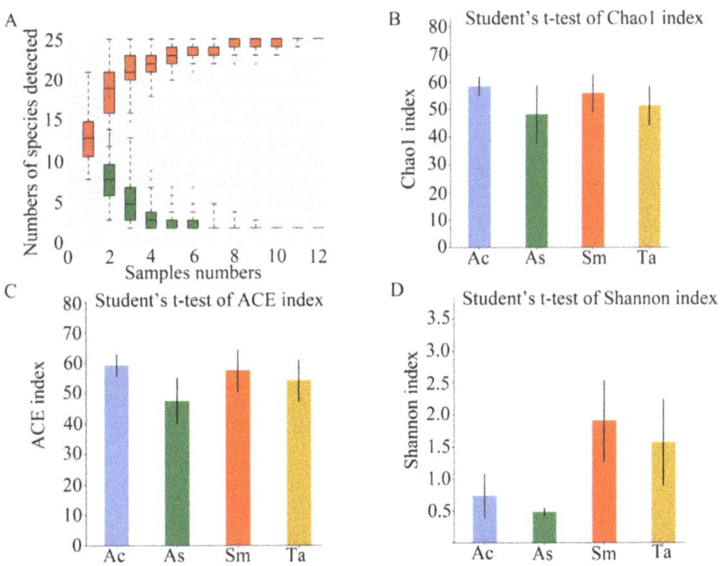

Figure 1. The results of α-diversity analysis. (**A**) Species discovery curve. (A single red box in this figure represents the total number of species detected in randomly selected samples. The cumulative curve is composed of the totality of red boxes, which represents the rate of new species appearing under continuous sampling; a single green box in this figure represents the number of common species detected in a given number of samples. The set of green boxes form the common quantity curve, which represents the rate of common species detected under continuous sampling). (**B**) Chao1 index of the four species of grasshoppers. (**C**) ACE index of the four species of grasshoppers. (**D**) Shannon index of the four species of grasshoppers. Ac, *Acridia chinensis;* As, *Atractomorpha sinensis;* Sm, *Sphingonotus mongolicus;* Ta, *Trilophidia annulate.*

Table 3. The average value of the Alpha diversity index of each sample.

Species	Sample ID	ACE	Chao1	Shannon	Coverage
Acrida cinerea	Ac	59.2633	58.3929	0.7407	0.9997
Atractomorpha sinensis	As	47.4831	48.1513	0.4829	0.9998
Sphingonotus mongolicus	Sm	57.3990	55.7500	1.9040	0.9999
Trilophidia annulate	Ta	54.1572	51.2593	1.5683	0.9998

The average value of each index of the three samples from the same species was calculated (Table 3) and then used to compare and analyze the α-diversity among the different species. In a community, species diversity was affected by the richness and evenness of the species. The ACE index and Chao1 index reflected the species richness. The Chao1 index (Figure 1B) and ACE index (Figure 1C) of *Acrida cinerea* (Ac) was the highest, and *Atractomorpha sinensis* (As) was the lowest. The Shannon index reflected the species diversity. The Shannon index (Figure 1D) of *Sphingonotus mongolicus* (Sm) was the highest, and *Atractomorpha sinensis* (As) was the lowest. In the guts of the four species of grasshoppers, the species richness in descending order was *Acrida cinerea* (Ac), *Sphingonotus mongolicus* (Sm), *Trilophidia annulate* (Ta) and *Atractomorpha sinensis* (As). The species diversity in descending order was *Sphingonotus mongolicus* (Sm), *Trilophidia annulate* (Ta), *Acrida cinerea* (Ac) and *Atractomorpha sinensis* (As).

PCoA was principal coordinate analysis, which can further display the differences in species diversity between samples. In Figure 2, the closer the distance of the graphic indicates that the samples are more similar. Except for one sample of *Acrida cinerea* (Ac), the other samples of *Acrida cinerea* (Ac) and *Atractomorpha sinensis* (As) were close to each other, indicating that the samples of *Acrida cinerea* (Ac) and *Atractomorpha sinensis* (As) were similar. The three samples of *Sphingonotus mongolicus* (Sm) were close in distance and similar, and there was no obvious difference between individuals, but the three samples of *Trilophidia annulate* (Ta) were scattered, and the inter-individual differences of *Trilophidia annulate* (Ta), were larger than those of the other three species of grasshoppers. This result was only affected by the presence or absence of species, not by species abundance. The gut microbial composition of the four grasshopper species was different, but the difference was not significant ($p > 0.05$). This may be related to the same living environment of the four species of grasshoppers.

At the phylum level (Figure 3A), a total of 7 phyla were obtained from 12 samples. According to the abscissa in Figure 3A, from left to right: *Acrida cinerea* (Ac), *Atractomorpha sinensis* (As), *Sphingonotus mongolicus* (Sm), *Trilophidia annulate* (Ta). Proteobacteria accounted for the highest proportion at 96.62, 99.23, 87.66 and 64.25%, respectively. The rest were Firmicutes (2.36, 0.11, 7.43, 25.29%), Cyanobacteria (0.02, 0.58, 0.08, 10.19%), Actinomycetes (0.14, 0.03, 4.87, 0.05%), Bacteroides (0.78, 0.06, 0.94, 0.76%), Tenericutes (0.08%, 0, 0, 0.14%) and Fusobacteria (0, 0, 0.10%, 0). Among them, Proteobacteria was the absolute dominant phylum. Further, Firmicutes had a larger proportion in the *Sphingonotus mongolicus* (Sm) and *Trilophidia annulate* (Ta) guts than the other two species of grasshoppers. At the family level (Figure 3B), Enterobacteriaceae was ubiquitous in most samples. *Trilophidia annulate* (Ta) had the lowest relative abundance of Enterobacteriaceae. However, the relative abundance of Streptoceccaceae in the *Trilophidia annulate* (Ta) group was significantly higher than that in the other three groups. The absolute dominant bacteria of the four species of grasshoppers at the phylum level and the family level were similar.

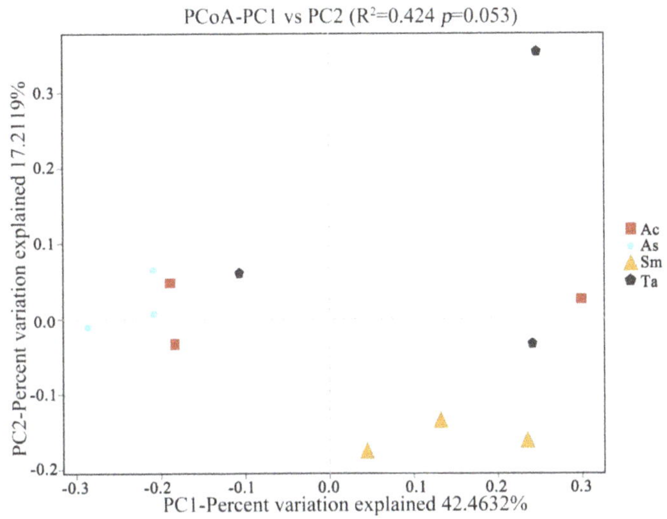

Figure 2. Principal coordinate analysis (PCoA) between four species of grasshoppers. Ac: red rectangle; As: blue circle; Sm: yellow triangle; Ta: gray polygon. The horizontal and vertical coordinates are the two eigenvalues that cause the largest difference between samples, and the main influence degree is reflected in the form of percentage. Ac, *Acridia chinensis*; As, *Atractomorpha sinensis*; Sm, *Sphingonotus mongolicus*; Ta, *Trilophidia annulate*.

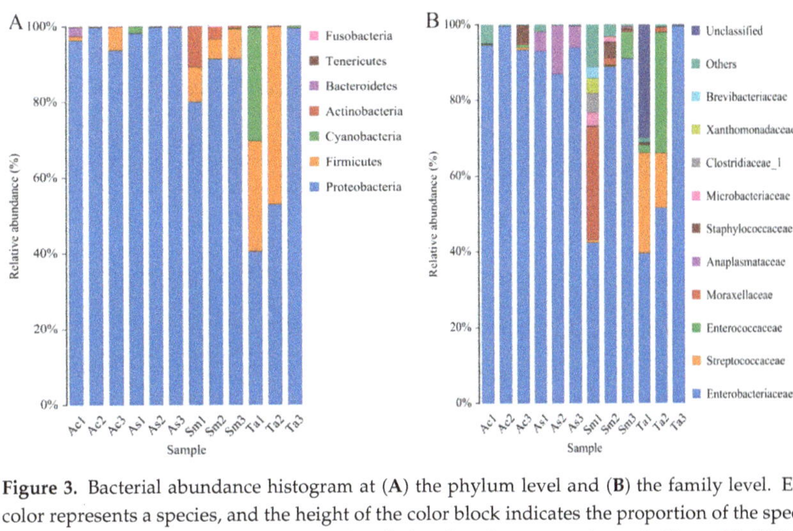

Figure 3. Bacterial abundance histogram at (**A**) the phylum level and (**B**) the family level. Each color represents a species, and the height of the color block indicates the proportion of the species in relative abundance. The top 10 genera in relative abundance were shown in Figure 3B. "Others" represented the remaining. "Unclassified" represented OTUs that were not commented. Ac, *Acridia chinensis*; As, *Atractomorpha sinensis*; Sm, *Sphingonotus mongolicus*; Ta, *Trilophidia annulate*.

Figure 4 combined the UPGMA cluster tree with the abundance histogram of each sample at the genus level. The similarity of species composition and abundance o among different samples could be intuitively judged. A total of 55 genera were annotated from 12 samples. The abundance histogram on the right showed the top ten genera with relative abundance greater than 1% in the intestines of four species of grasshoppers. However, according to the abundance histograms, the four species of grasshoppers differed at the genus level. There were *Klebsiella* and *Staphylococcus* in *Acridia chinensis* (Ac), *Klebsiella* and *Wolbachia* in *Atractomorpha sinensis* (As), *Klebsiella, Acinetobacter, Pantoea, Enterococcus, Staphylococcus, Stenotrophomonas, Microbacterium, Brevibacterium* and *Corynebacterium* in *Sphingonotus mongolicus* (Sm), *Klebsiella, Lactococcus* and *Enterococcus* in *Trilophidia annulate* (Ta). *Klebsiella* was the dominant genus shared by four species of grasshoppers. Compared with other genus, *Klebsiella* had the largest proportion, that is, the largest relative abundance. *Morganella* was unique to *Sphingonotus mongolicus* (Sm). Among the three samples of *Sphingonotus mongolicus* (Sm), it was only detected in Sm1, and the abundance was less than 0.01%. The cluster tree on the left shows the species composition in Figure 4 are most similar in samples of As1 and As3 and samples of Ac2 and Ta3. Ta1 and Ta2 had the obvious difference with other samples. Excluding Ta1 and Ta2, the difference between Sm1 and other samples was the most obvious. The difference in biodiversity among the three samples of the same species may be related to the difference in collection time and individual grasshopper.

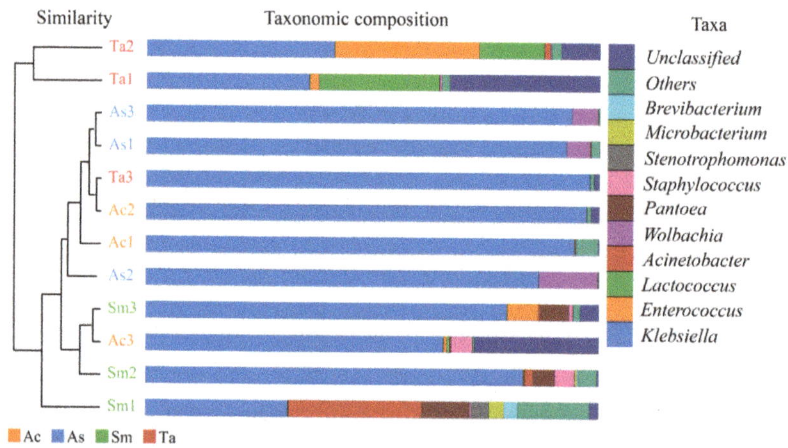

Figure 4. UPGMA cluster tree combined with histogram. The clustering tree and the abundance histogram are shown on the left and on the right, respectively. The top 10 genera in relative abundance were shown here. "Others" represented the remaining. "Unclassified" represented OTUs that were not commented. Ac, *Acridia chinensis;* As, *Atractomorpha sinensis;* Sm, *Sphingonotus mongolicus;* Ta, *Trilophidia annulate.*

The distance matrix was calculated by the weighted UniFrac method. A sample heat map was drawn by the R language tool. The heat map is a picture that uses color to represent the differences between samples. The color gradient from blue to red indicated that the distance between the samples was from close to far. Differences between two samples can be visually seen based on changes in the color gradient. The results were shown in Figure 5. The difference between Ta1 and other samples was red. The difference between Ta2 with Sm1 and other samples was between red and blue. The differences of the rest samples were blue, indicating that the microbial diversity and abundance were slight but insignificantly different among most samples. It was consistent with PCoA results ($p > 0.05$).

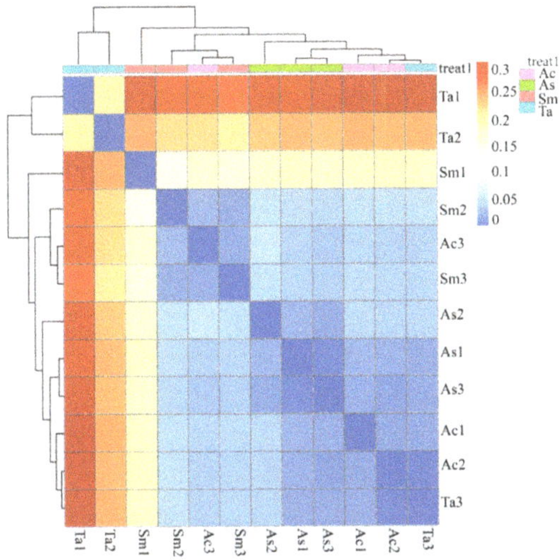

Figure 5. Heatmap of each sample at the OTU classification level. Dendrograms of hierarchical cluster analysis samples are shown on the left and at the top, respectively. The color gradient from blue to red indicated that the distance between the samples was from close to far. Ac, *Acridia chinensis*; As, *Atractomorpha sinensis*; Sm, *Sphingonotus mongolicus*; Ta, *Trilophidia annulate*.

In order to find biomarkers with statistical differences between different groups, we used linear discriminant analysis (LDA) effect size (LEfSe) to screen out different taxa at various levels (kingdom, phylum, class, order, family, genus, species) between different groups based on a standard LDA value greater than four (Figure 6). Biomarkers are molecules found in the body that indicate a specific biological condition. The biomarkers with LDA Scores greater than the set value of 4.0 were displayed and only screened in the guts of *Atractomorpha sinensis* (As) and *Sphingonotus mongolicus* (Sm). The LDA Scores of family Anaplasmataceae and genus *Wolbachia* selected from the guts of *Atractomorpha sinensis* (As) were similar. The biomarkers screened in the guts of *Sphingonotus mongolicus* (Sm) were genus *Acinetobacter*, order Actinobacteria, phylum Actinobacteria, order Micrococcales and genus *Pantoea*, all of which had LDA values greater than 4. Figure 7 shows the relative abundance of each Biomarker. As can be clearly seen in panel A, the relative abundance of biomarkers in *Sphingonotus mongolicus* (Sm) was obviously high. In the three samples of the same grasshopper species, the relative abundance of Biomarker was different, which was the result of the differences between the samples.

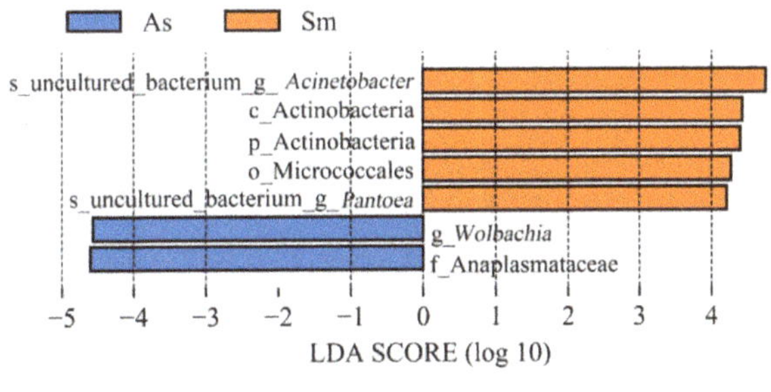

Figure 6. Bacterial taxa with linear discriminant analysis (LDA) score greater than four in the gut microbiota of different grasshopper. As, *Atractomorpha sinensis*; Sm, *Sphingonotus mongolicus*.

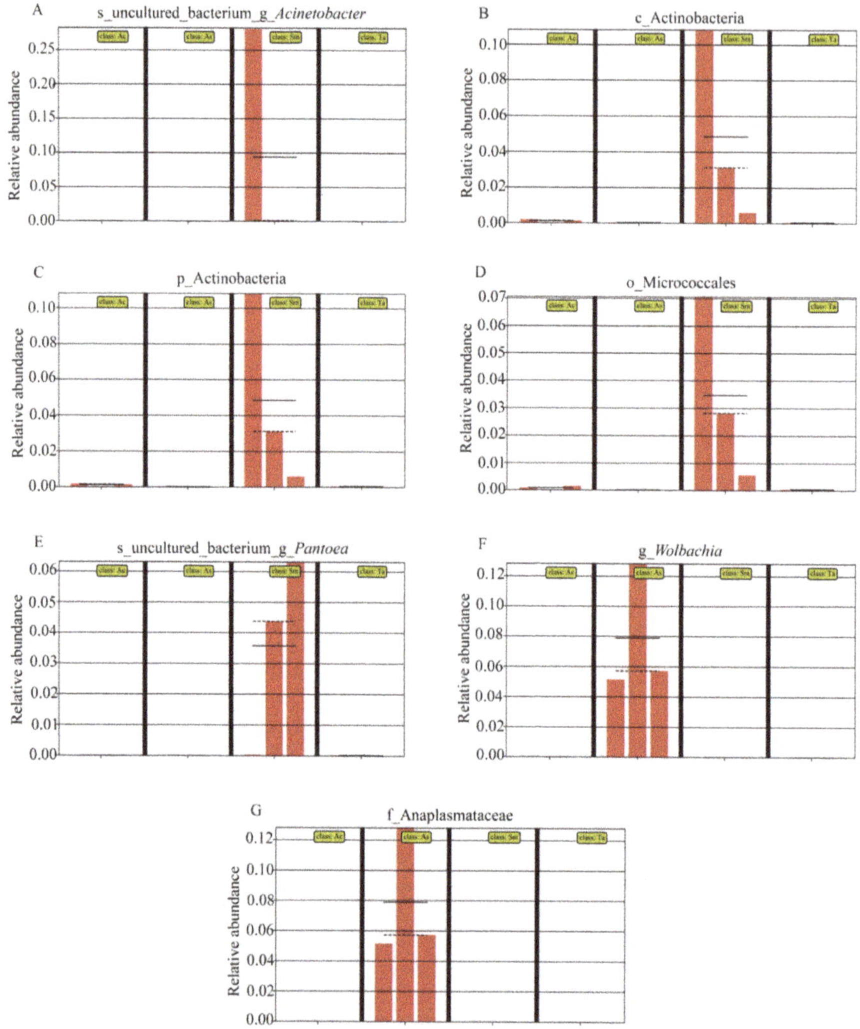

Figure 7. Abundance histogram of bacterial taxa with linear discriminant analysis (LDA) score greater than four in the gut microbiota. Different groups are separated by black bold solid lines. The solid line in the histogram of each group represents the average value of the expression amount of the reorganized sample, and the dotted line represents the median value of the expression amount of the group of samples. (**A**) s_uncultured_bacterium_g_*Acinetobacter*. (**B**) c_Actinobacteria. (**C**) p_ Actinobacteria. (**D**) o_Micrococcales. (**E**) s_uncultured_bacterium_g_*Pantoca*. (**F**) g_*Wolbachia*. (**G**) f_Anaplasmataceae.

3.2. *Correlation Analysis of Bacteria*

According to the abundance and change of each genus in each sample, Spearman rank correlation analysis was performed, and data of correlation > 0.1 and $p < 0.05$ were selected to construct a correlation network. Based on the analysis of the network diagram, the coexistence relationship of species in grasshopper intestinal samples could be obtained, and the interaction of species in the same environment and important model information could be obtained. Figure 8 shows the correlation analysis of the top 30 genera in abundances. *Klebsiella*, which has the highest relative abundance, had a significant positive correlation with *Enterobacter*, and had a significant negative correlation with *Wolbachia*, *Pantoea*, *Clostridium_sensu_stricto_1* and *Corynebacterium_1*.

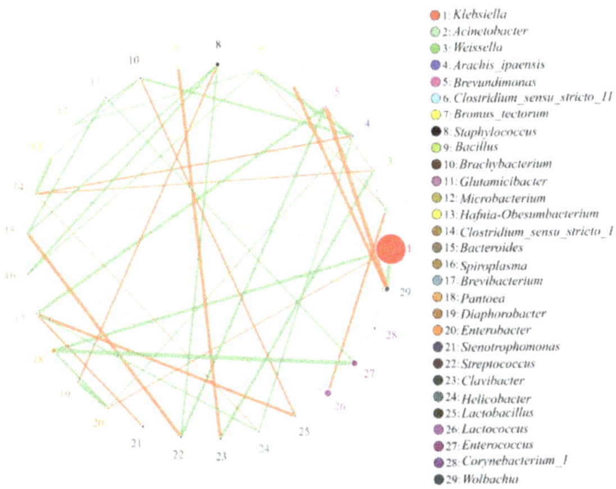

Figure 8. Correlation analysis of microorganisms at the genus level. The circles represented the genera, the size of the circles represented the average abundance, the lines represented the correlation between two species, the thickness and thinness of the lines represented the strength of the correlation, orange represented a positive correlation and green represented negative correlation. The correlation analysis of the top 30 genera in abundances are shown on the right.

3.3. Digestibility of Cellulose and Hemicellulose

After measurement and calculation, the cellulose content of wheat seedling was about 50.14%, hemicellulose content was about 8.39%. It was consistent with the cellulose content of Gramineae measured by Ye [24]. It was similar to the cellulose content of wheat straw, but significantly different to the hemicellulose content [25]. Table 4 shows the contents of cellulose and hemicellulose in the feces of four species of grasshoppers and the digestibility to cellulose and hemicellulose in wheat seedlings. The cellulose content of the feces of the *Sphingonotus mongolicus* (Sm) was 44.36% and *Trilophidia annulate* (Ta) was 41.54%. This indicated that *Trilophidia annulata* (Ta) had a slightly higher absorption of cellulose than that of the *Sphingonotus mongolicus* (Sm). Similarly, the absorption of cellulose by *Atractomorpha sinensis* (As) was higher than that by *Trilophidia annulata* (Ta), and by *Acrida cinerea* (Ac) was higher than that by *Atractomorpha sinensis* (As). In the same way, the content of hemicellulose in the feces of *Sphingonotus mongolicus* (Sm) reached 11.24%, ranking first among the four species of grasshoppers, followed by *Atractomorpha sinensis* (As) and *Trilophidia annulata* (Ta). Their fecal hemicellulose content was close. The hemicellulose content of the feces of *Acrida cinerea* (Ac) was 7.86%, ranking the last.

Table 4. The content and digestibility of cellulose and hemicellulose.

Species of Grasshopper	Sample ID	Cellulose Content in Feces (%)	Cellulose Digestibility (%)	Hemicellulose Content in Feces (%)	Hemicellulose Digestibility (%)
Acrida cinerea	Ac	33.28 ± 0.02	56.97 ± 0.09	7.86 ± 0.01	39.28 ± 0.12
Atractomorpha sinensis	As	37.29 ± 0.02	54.86 ± 0.06	11.37 ± 0.01	17.77 ± 0.10
Sphingonotus mongolicus	Sm	44.36 ± 0.03	67.91 ± 0.08	12.14 ± 0.01	47.51 ± 0.12
Trilophidia annulata	Ta	41.54 ± 0.04	49.87 ± 0.06	11.20 ± 0.02	19.25 ± 0.09

The cellulose digestibility of the four species of grasshoppers was higher than that of hemicellulose. Additionally, the digestibility of cellulose and hemicellulose of *Sphingonotus mongolicus* (Sm) were the highest, which were 67.91% and 47.51%, respectively. The results showed that the *Sphingonotus mongolicus* (Sm) had relatively high digestibility. The digestibility of cellulose from high to low were *Sphingonotus mongolicus* (67.91%), *Acrida cinerea* (56.97%), *Atractomorpha sinensis* (54.86%) and *Trilophidia annulata* (49.87%). The hemicellulose digestibility from high to low were *Sphingonotus mongolicus* (47.51%), *Acrida cinerea* (39.28%), *Trilophidia annulata* (19.25%) and *Atractomorpha sinensis* (17.77%). The digestibility of cellulose and hemicellulose had significant differences ($p < 0.01$), which may be due to the differences in the species and numbers of microorganisms.

3.4. Correlation Analysis of Intestinal Microorganism of Grasshopper with Digestibility of Cellulose and Hemicellulose

Cellulose and hemicellulose digestibility of four species of grasshoppers were determined, and Spearman correlations between them and gut microbes were analyzed. The results were shown in Figure 9, where CD represented cellulose digestibility and HD showed hemicellulose digestibility. Spearman correlation analysis showed that *Brevibacterium* ($p < 0.01$) and *Stenotrophomonas* ($p < 0.05$) were significantly correlated with cellulose digestibility. *Brevibacterium*, *Clavibacter*, *Microbacterium* and *Stenotrophomonas* were significantly correlated with the hemicellulose digestibility ($p < 0.05$). *Brevibacterium* can produce amylase [26]. Moreover, starch and cellulose were macromolecular polysaccharides composed of glucose. *Stenotrophomonas* could decompose xylan [27]. *Clavibacter* was a plant pathogen that destroyed plant cell walls by producing cellulases and pectinases [28]. This strain with cellulase activity isolated from insect guts included *Microbacterium* [29]. These also indirectly proved the reliability of the correlation analysis. The presence of these microorganisms helped grasshoppers digest plant cellulose and hemicellulose better.

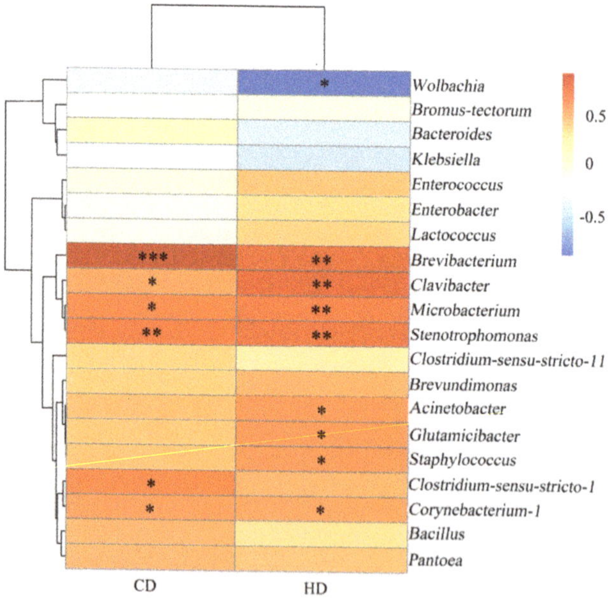

Figure 9. Heatmap of the correlation between digestibility and bacterial abundance. Dendrograms of hierarchical cluster analysis grouping genera is shown on the left. * There is a significant correlation of 5% between digestibility and bacteria. ** There is a significant correlation of 1% between digestibility and bacteria. *** There is a significant correlation of 0.1% between digestibility and bacteria. CD, cellulose digestibility; HD, hemicellulose digestibility.

4. Discussion

In this research, we used 16S rRNA gene high-throughput sequencing technology to analyze the bacterial diversity in the guts of four grasshopper species and determined the digestibility of cellulose and hemicellulose in those grasshoppers. We combined the analysis of the intestinal microbial diversity of *Acrida cinerea* (Ac), *Trilophidia annulata* (Ta), *Atractomorpha sinensis* (As) and *Sphingonotus mongolicus* (Sm) with their cellulose digestibility for the first time. This research showed that the composition of intestinal microorganisms of grasshoppers was diverse, which varied with different species, but there were still a small number of floras in common. There was a conserved core flora in different grasshopper species, which also indicated that the core flora had a symbiotic relationship with the grasshopper intestine and may play an important metabolic role in food digestion (cellulose degradation) and absorption. It laid a foundation for further research on the structure of the intestinal microorganism of grasshoppers, the relationship between microorganisms, the screening of microbial functional genes and the role of microorganisms in the life of grasshoppers.

Different living environments will lead to differences in the abundance and diversity of insect gut microbes [30,31]. Similarly, the gut microbial population of grasshoppers is also affected by relevant environmental conditions [5]. However, it is not clear how the living environment of grasshoppers affect their gut microbes. Yuan et al. confirmed that the gut bacterial structure of *G. molestacan* be influenced by the host plant [32]. Moro et al. showed that the diversity of gut microbes of the same species in regions was different [33]. Jesús M. et al. confirmed that different time scales strongly influence the diversity, composition and metabolic capabilities of *Brithys crini* gut microbial communities [34]. Huang et al. confirmed that both phylogeny and diet can impact the structure and composition of gut microbiomes [35]. The grasshoppers collected in this experiment were all adult, and the location and time were close to each other. To a large extent, the influence of time, environmental, climate and geographical conditions on the experimental results was avoided.

At the level of phylum, Proteobacteria accounted for the highest proportion, followed by Firmicutes. Muratore M. et al. found that there are bacterial phyla common to six grasshopper species from a coastal tallgrass prairie: Actinobacteria, Proteobacteria, Firmicutes, and to a lesser degree, Tenericutes [36]. Further, Wang et al. studied the gut microbial diversity of three species of grasshoppers, including *Oedaleus decorus asiaticus*, *Aiolopus tamulus* and *Shirakiacris shirakii*. Among them, Proteobacteria and Firmicutes were the most common. The intestinal microbial communities of the three species of grasshoppers are similar at the phylum level [22]. Mead et al. found that there were mainly four types of intestinal microbes, which were *Enterococcus* of Firmicutes, *Monserella*, *Pseudomonas* and *Enterobacteria* of Proteobacteria in the guts of *Melanoplus sanguinipes* [37]. In addition, using 16S rRNA gene sequencing, Schloss et al. found that the dominant intestinal phyla of *Saperda Vestita* was Proteobacteria [38]. Moreover, the largest relative proportion of the guts of the Mediterranean fruit fly was Enterobacteriaceae of Proteobacteria [39]. *Cnaphalocrocis medinalis* was the main pest of rice and the main dominant microflora in its larvae guts were Proteobacteria and Firmicutes [40]. Similarly, Kikuchi et al. found that the dominant microflora in the gut of *Riptortus cllavatus* were *Burkholderia* of Proteobacteria [41]. The above research results were consistent with this study. The abundance and structure of the intestinal microbes of these insects were different, but the dominant phyla were similar. At the genera level, *Klebsiella* accounted for the highest proportion in the intestinal microbes of the four species of grasshoppers, but the dominant genera were not the same. Barbosa et al. identified cellulase-producing bacteria by analyzing the 16S rDNA gene [42]. These strains were identified as *Klebsiella pneumoniae*, *Klebsiella* sp., and *Bacillus* sp. *Klebsiella pneumoniae* was the main cellulase-producing microorganism. In addition, Wang et al. found that *Klebsiella* accounted for the highest proportion of the microbial community in the three grasshopper species [22]. The specific role of *Klebsiella* in the guts of grasshoppers need to be further studied, but it was the common dominant bacteria in the guts of insects, and its important position cannot be ignored.

In this research, we found biomarkers with statistical differences between *Atractomorpha sinensis* (As) and *Sphingonotus mongolicus* (Sm) (Figure 7). The high abundance of *Acinetobacter*, *Pantoea*, and *Wolbachia* can be used as differential microorganisms to distinguish *Atractomorpha sinensis* (As) and *Sphingonotus mongolicus* (Sm). Hancock et al. found that *Wolbachia* could affect the reproduction of mosquitoes and reduce the spread of disease [43]. Further, a study on the brown planthopper showed that *Acinetobacter*, *Wolbachia* and *Staphylococcus* were significantly positively correlated with detoxification genes, that is, these symbiotic bacteria were involved in the metabolism of insecticides in the guts of the brown planthopper, which had positive significance for pest control [44]. *Acinetobacter* had esterase activity [45], which may also be related to nutrient metabolism of host insects [40]. *Pantoea* could provide vitamins and amino acids for host insects [46]. Therefore, gut microbes are closely related to the life activities of the host, differential microorganisms can be used in subsequent studies to explore their functions.

The main place where most bacteria in insect guts exist is the mid-hindgut [47]. In this research, we selected the mid-hindgut of grasshoppers as the experimental material, and the results proved that *Klebsiella* sp. were the common dominant bacteria in four species of grasshoppers. *Klebsiella* belongs to the Enterobacteriaceae. Smith et al. found that Enterobacteriaceae mainly reside in the hindgut and are involved in carbohydrate metabolism [48]. It has been reported that in the gut of *Bactrocera Oleae*, *Klebsiella* and *Enterobacter* were harmful for the host [49]. *Klebsiell oxytoca* in the gut of fruit fly delayed the emergence of parasitic wasps [50]. This suggested that *Klebsiella* and *Enterobacter* had some positive correlation and worked together in the hosts. However, Gao et al. found that *Klebsiella* can promote the growth and development of *Drosophila suzukii* to a certain extent [51]. Moreover, the *Klebsiella* isolated from the oral secretion of fall armyworm could down-regulate the activity of peroxidase and up-regulate the activity of trypsin inhibitor in tomato, thereby reducing the ability of tomato to resist pests [52]. The *Klebsiella* isolated from the larvae of *Dendrolimus kikuchii* could produce lipase [53].

In the determination of cellulose digestibility, adding sulfuric acid produced a large amount of heat, reducing the accuracy of the experimental results. The ice bath could effectively solve this problem and ensure the accuracy of the results. The results showed that the digestibility of cellulose was higher than that of hemicellulose. The digestibility of cellulose and hemicellulose varied greatly, which was related to the structure of cellulose and hemicellulose. The chemical structure of cellulose had high degree of polymerization, and the hydrogen bonding force between molecules determined that it was difficult to degrade [24]. Cellulose and hemicellulose had different decomposition products and different proportions in plants [54]. Consequently, the digestibility of grasshoppers was significantly different. The decomposition of cellulose and hemicellulose required the cooperation of a variety of microorganisms. The microorganisms that secreting cellulase may not participate in the breakdown of hemicellulose. Therefore, the number of microorganisms secreting cellulase and hemicellulose would affect the digestibility. The differences in the cellulose digestibility may be due to the differences in the amount of cellulase in the guts of different grasshoppers. The grasshoppers with high cellulose digestibility had a large number of microorganisms in their guts that can decompose cellulose and secrete a large amount of cellulase with high activity. The same was true for hemicellulose. Tian and Ba found that the cellulose digestibility of the rumen fluid of Tibetan sheep to the highland barley straw is 25.8% [55]. Li et al. studied the digestibility of sheep to corn stalks treated in different ways, and the results showed that the digestibility of crude fiber was 34.21–61.21% [56]. Further, Zhang et al. found that the digestibility of neutral detergent fiber and acid detergent fiber to wheat straw were 28.5–30.9% and 29.1–37.0%, respectively [57]. The cellulose digestibility in this research was close to that of mammals, and far higher than that of *Locusta migratoria manilensis*. This result may be due to different feeding materials. However, the digestion and utilization of cellulose and hemicellulose in the four species of grasshoppers were at high level, which might be related to the microorganisms in the gut. The breakdown of cellulose and hemicellulose requires the participation of enzymes. Additionally, the secretion of these enzymes requires the cooperation of a large number of microorganisms. However, in the guts of grasshoppers, which microorganism had the ability to decompose cellulose and hemicellulose and what was their specific roles in the decomposition process still need further research. *Bacillus licheniformis*, *O. intermedium*, and *M. paludicola* were isolated from the gut contents of termites (*Microcerotermes diversus*) as described previously [58]. They have high cellulose degradability. Kundu found 15 hemicellulolytic microbes in the guts of termites [59]. Similarly, Huang et al. isolated *Cellulomonas* sp. h9 from the intestinal tract of larvae of *Protaetia brevitarsis* [60]. It provides a research basis for the isolation of cellulose-degrading bacteria in the intestines of grasshoppers, which can be further studied. Some microorganisms related to cellulose and hemicellulose was obtained from Spearman correlation analysis. However, what role they play in the catabolism of cellulose and hemicellulose remains to be further verified. The digestibility of cellulose and hemicellulose of the four species of grasshoppers are high, and they have potential value as bioreactors for lignocellulose decomposition.

5. Conclusions

In conclusion, 16S rRNA gene sequences was used to determine the bacterial diversity of *Acrida cinerea*, *Trilophidia annulata*, *Atractomorpha sinensis* and *Sphingonotus mongolicus*, and correlation analysis was performed between the intestinal microbes of grasshoppers and the digestibility of cellulose and hemicellulose.

The diversity and abundance of intestinal microorganisms were different among all species, but there was no significant difference. *Acrida cinerea* had the highest bacterial species richness, and *Sphingonotus mongolicus* had the highest bacterial diversity. Proteobacteria and Firmicutes are the dominant bacteria in the intestinal microbial communities of the four grasshopper species. The dominant genera of different species of grasshoppers are different, and the common dominant species is *Klebsiella*. The intestinal microflora structure varied among the different species of grasshoppers, with the intestinal microflora structure of *Acrida cinerea* and *Atractomorpha sinensis* being the most similar. In addition, *Sphingonotus mongolicus* had the highest digestibility. The digestibility of cellulose was significantly different among species, as was the digestibility of hemicellulose. The digestibility of cellulose was higher than that of hemicellulose. Further, Spearman correlation analysis showed that *Brevibacterium* and *Stenotrophomonas* were significantly correlated with the cellulose digestibility. *Brevibacterium*, *Clavibacter*, *Microbacterium* and *Stenotrophomonas* were significantly correlated with the hemicellulose digestibility. The microorganisms mentioned above can be used as back-up to break down cellulose and hemicellulose.

Increasing the understanding of the structure and function of the grasshopper intestinal microflora will facilitate further research and the utilization of intestinal microorganisms in the future and contribute to the development of grasshoppers as a cellulose degradation bioreactors. Meanwhile, it provides a new idea for the decomposition and utilization of straw in agriculture and animal husbandry, which has important theoretical and practical significance.

Author Contributions: Conceptualization, X.-J.L., J.B. and Y.L.; methodology, J.B., Y.L. and W.-J.L.; software, Y.L. and L.W.; validation, J.B., Y.L. and X.-B.X.; formal analysis, Y.-Y.G.; investigation, F.-F.L.; resources, X.-J.L.; data curation, J.B. and Y.L.; writing—original draft preparation, J.B. and Y.L.; writing—review and editing, J.B., Y.L., W.-J.L., L.W. and X.-J.L.; visualization, J.B. and Y.L.; supervision, X.-J.L.; project administration, X.-J.L.; funding acquisition, X.-J.L. All authors have read and agreed to the published version of the manuscript.

Funding: This research was funded by the National Natural Science Foundation of China (No. 32070473 & No.31872274) and Natural Science Foundation of Hebei Province (No. C2018201139).

Institutional Review Board Statement: Not applicable.

Data Availability Statement: The data presented in this study are available in article here.

Acknowledgments: The authors would like to thank Xiang Li and Jing Wang for collecting specimens.

Conflicts of Interest: The authors declare no conflict of interest.

References

1. Basset, Y.; Cizek, L.; Cuénoud, P.; Didham, R.K.; Guilhaumon, F.; Missa, O.; Novotny, V.; Ødegaard, F.; Roslin, T.; Schmidl, J.; et al. Arthropod Diversity in a Tropical Forest. *Science* **2012**, *338*, 1481–1484. [CrossRef] [PubMed]
2. Hongoh, Y.; Deevong, P.; Inoue, T.; Moriya, S.; Trakulnaleamsai, S.; Ohkuma, M.; Vongkaluang, C.; Noparatnaraporn, N.; Kudo, T. Intra- and Interspecific Comparisons of Bacterial Diversity and Community Structure Support Coevolution of Gut Microbiota and Termite Host. *Appl. Environ. Microbiol.* **2005**, *71*, 6590–6599. [CrossRef] [PubMed]
3. Cao, L.; Ning, K. Metagenomics of insect gut: New borders of microbial big data. *Acta Microbiol. Sin.* **2018**, *58*, 964–984. [CrossRef]
4. Aziz, Z.; Nabil, R.; Said, E.; Houria, N.; Khadija, T.; Abderrahim, L.; Lahsen, E.G. Preliminary Study of the Intestinal Microbial Diversity of Three Acridoidae: *Oedipoda fuscocincta*, *Dociostaurus moroccanus*, and *Calliptamus barbarus* (Acrididae: Orthoptera), in the Moroccan Middle Atlas. *Indian J. Microbiol.* **2022**, *62*, 123–129. [CrossRef] [PubMed]
5. Lavy, O.; Gophna, U.; Gefen, E.; Ayali, A. Locust Bacterial Symbionts: An Update. *Insects* **2020**, *11*, 655. [CrossRef] [PubMed]
6. Dillon, R.J.; Dillon, V.M. The gut bacteria of insects: Nonpathogenic interactions. *Annu. Rev. Entomol.* **2004**, *49*, 71–92. [CrossRef]
7. Kwong, W.K.; Engel, P.; Koch, H.; Moran, N.A. Genomics and host specialization of honey bee and bumble bee gut symbionts. *Proc. Natl. Acad. Sci. USA* **2014**, *111*, 11509–11514. [CrossRef]
8. Ong, S.Y.; Kho, H.P.; Riedel, S.L.; Kim, S.W.; Gan, C.Y.; Taylor, T.D.; Sudesh, K. An integrative study on biologically re-covered polyhydroxyalkanoates (PHAs) and simultaneous assessment of gut microbiome in yellow mealworm. *J. Biotechnol.* **2018**, *265*, 31–39. [CrossRef]
9. Lee, F.J.; Rusch, D.B.; Stewart, F.J.; Mattila, H.R.; Newton, I.L.G. Saccharide breakdown and fermentation by the honey bee gut microbiome. *Environ. Microbiol.* **2014**, *17*, 796–815. [CrossRef]
10. Kaltenpoth, M.; Göttler, W.; Herzner, G.; Strohm, E. Symbiotic Bacteria Protect Wasp Larvae from Fungal Infestation. *Curr. Biol. CB* **2005**, *15*, 475–479. [CrossRef]
11. Chen, D.F.; Hou, L.; Wei, J.N.; Guo, S.; Cui, W.C.; Yang, P.C.; Kang, L.; Wang, X.H. Aggregation pheromone 4-vinylanisole promotes the synchrony of sexual maturation in female locusts. *eLife* **2022**, *11*, e74581. [CrossRef] [PubMed]
12. Arredondo-Santoyo, M.; Herrera-Camacho, J.; Vázquez-Garcidueñas, M.S.; Vázquez-Marrufo, G. Corn stover induces extracellular laccase activity in *Didymosphaeria* sp. (syn. = *Paraconiothyrium* sp.) and exhibits increased in vitro ruminal digestibility when treated with this fungal species. *Folia Microbiol.* **2020**, *65*, 849–861. [CrossRef] [PubMed]
13. Zhang, Q.; Zhang, J.; Zhao, S.; Song, P.; Chen, Y.; Liu, P.; Mao, C.; Li, X. Enhanced Biogas Production by Ligninolytic Strain *Enterobacter hormaechei* KA3 for Anaerobic Digestion of Corn Straw. *Energies* **2021**, *14*, 2990. [CrossRef]
14. Su, L.-J.; Liu, H.; Li, Y.; Zhang, H.-F.; Chen, M.; Gao, X.-H.; Wang, F.-Q.; Song, A.-D. Cellulolytic activity and structure of symbiotic bacteria in locust guts. *Genet. Mol. Res.* **2014**, *13*, 7926–7936. [CrossRef]
15. Luo, C.; Li, Y.; Chen, Y.; Fu, C.; Long, W.; Xiao, X.; Liao, H.; Yang, Y. Bamboo lignocellulose degradation by gut symbiotic microbiota of the bamboo snout beetle *Cyrtotrachelus buqueti*. *Biotechnol. Biofuels* **2019**, *12*, 70. [CrossRef]
16. Mori, H.; Maruyama, F.; Kato, H.; Toyoda, A.; Dozono, A.; Ohtsubo, Y.; Nagata, Y.; Fujiyama, A.; Tsuda, M.; Kurokawa, K. Design and Experimental Application of a Novel Non-Degenerate Universal Primer Set that Amplifies Prokaryotic 16S rRNA Genes with a Low Possibility to Amplify Eukaryotic rRNA Genes. *DNA Res.* **2014**, *21*, 217–227. [CrossRef]
17. Magoč, T.; Salzberg, S.L. FLASH: Fast length adjustment of short reads to improve genome assemblies. *Bioinformatics* **2011**, *27*, 2957–2963. [CrossRef]
18. Bolger, A.M.; Lohse, M.; Usadel, B. Trimmomatic: A flexible trimmer for Illumina sequence data. *Bioinformatics* **2014**, *30*, 2114–2120. [CrossRef]
19. Edgar, R.C.; Haas, B.J.; Clemente, J.C.; Quince, C.; Knight, R. UCHIME improves sensitivity and speed of chimera detection. *Bioinformatics* **2011**, *27*, 2194–2200. [CrossRef]
20. Bokulich, N.A.; Subramanian, S.; Faith, J.J.; Gevers, D.; Gordon, J.I.; Knight, R.; Mills, D.A.; Caporaso, J.G. Quality-filtering vastly improves diversity estimates from Illumina amplicon sequencing. *Nat. Methods* **2013**, *10*, 57–59. [CrossRef]
21. Edgar, R.C. UPARSE: Highly accurate OTU sequences from microbial amplicon reads. *Nat. Methods* **2013**, *10*, 996–998. [CrossRef] [PubMed]
22. Wang, J.-M.; Bai, J.; Zheng, F.-Y.; Ling, Y.; Li, X.; Wang, J.; Zhi, Y.-C.; Li, X.-J. Diversity of the gut microbiome in three grasshopper species using 16S rRNA and determination of cellulose digestibility. *PeerJ* **2020**, *8*, e10194. [CrossRef] [PubMed]
23. Kostic, A.; Gevers, D.; Siljander, H.; Vatanen, T.; Hyötyläinen, T.; Hämäläinen, A.-M.; Peet, A.; Tillmann, V.; Pöhö, P.; Mattila, I.; et al. The Dynamics of the Human Infant Gut Microbiome in Development and in Progression toward Type 1 Diabetes. *Cell Host Microbe* **2015**, *17*, 260–273. [CrossRef] [PubMed]
24. Ye, D.Y.; Huang, H.; Fu, H.Q.; Chen, H.Q. Advances in cellulose chemistry. *CIESC J.* **2006**, *57*, 1782–1791.
25. Zhang, H.; Fu, Q.; Liao, Q.; Xia, A.; Huang, Y.; Zhu, X.Q.; Zhu, X. Study on degradation kinetics of hemicellulose in wheat straw hydrothermal pretreatment. *CIESC J.* **2020**, *71*, 3098–3105. [CrossRef]

26. Andriani, A.; Bayuningsih, M.; Kusnadi, J.; Rahmani, N.; Juanssilfero, A.; Sari, L.; Ermayanti, T.; Yopi. Hydrolysis of Local Genotype Taro (*Colocasia esculenta*) Starch by Crude Amylase from *Brevibacterium* Sp. for Maltooligosaccharides Production. *IOP Conf. Ser. Earth Environ. Sci.* **2020**, *439*, 012065. [CrossRef]
27. Sipriyadi; Darwis, W.; Wibowo, R.H.; Farestiani, E. Characterization and Identification of Xylanolytic Bacteria Stenotrophomonas sp. EL-8 Isolated from Seagrass Substrates in Enggano Island. *IOP Conf. Ser. Earth Environ. Sci.* **2020**, *457*, 012068. [CrossRef]
28. Jahr, H.; Bahro, R.; Burger, A.; Ahlemeyer, J.; Eichenlaub, R. Interactions between *Clavibacter michiganensis* and its host plants. *Environ. Microbiol.* **1999**, *1*, 113–118. [CrossRef]
29. Shil, R.K.; Mojumder, S.; Sadida, F.F.; Uddin, M.; Sikdar, D. Isolation and Identification of Cellulolytic Bacteria from the Gut of Three Phytophagus Insect Species. *Braz. Arch. Biol. Technol.* **2014**, *57*, 927–932. [CrossRef]
30. Schmid, R.B.; Lehman, R.M.; Brözel, V.S.; Lundgren, J.G. Gut Bacterial Symbiont Diversity within Beneficial Insects Linked to Reductions in Local Biodiversity. *Ann. Èntomol. Soc. Am.* **2015**, *108*, 993–999. [CrossRef]
31. Donkersley, P.; Rhodes, G.; Pickup, R.W.; Jones, K.C.; Wilson, K. Bacterial communities associated with honeybee food stores are correlated with land use. *Ecol. Evol.* **2018**, *8*, 4743–4756. [CrossRef] [PubMed]
32. Yuan, X.; Zhang, X.; Liu, X.; Dong, Y.; Yan, Z.; Lv, D.; Wang, P.; Li, Y. Comparison of Gut Bacterial Communities of *Grapholita molesta* (Lepidoptera: Tortricidae) Reared on Different Host Plants. *Int. J. Mol. Sci.* **2021**, *22*, 6843. [CrossRef] [PubMed]
33. Moro, M.S.; Wu, X.; Wei, W.; Mendes, L.W.; Allen, K.C.; Pinheiro, J.B.; Clough, S.J.; Zucchi, M.I. Characterization and Comparison of Intestinal Bacterial Microbiomes of *Euschistus heros* and *Piezodorus guildinii* Collected in Brazil and the United States. *Front. Microbiol.* **2021**, *12*, 769965. [CrossRef] [PubMed]
34. Marín-Miret, J.; González-Serrano, F.; Rosas, T.; Baixeras, J.; Latorre, A.; Pérez-Cobas, A.E.; Moya, A. Temporal variations shape the gut microbiome ecology of the moth *Brithys crini*. *Environ. Microbiol.* **2022**. [CrossRef]
35. Huang, K.; Wang, J.; Huang, J.; Zhang, S.; Vogler, A.P.; Liu, Q.; Li, Y.; Yang, M.; Li, Y.; Zhou, X. Host Phylogeny and Diet Shape Gut Microbial Communities within Bamboo-Feeding Insects. *Front. Microbiol.* **2021**, *12*, 633075. [CrossRef]
36. Muratore, M.; Prather, C.; Sun, Y. The gut bacterial communities across six grasshopper species from a coastal tallgrass prairie. *PLoS ONE* **2020**, *15*, e0228406. [CrossRef]
37. Mead, L.J.; Khachatourians, G.G.; Jones, G.A. Microbial Ecology of the Gut in Laboratory Stocks of the Migratory Grasshopper, *Melanoplus sanguinipes* (Fab.) (Orthoptera: Acrididae). *Appl. Environ. Microbiol.* **1988**, *54*, 1174–1181. [CrossRef]
38. Schloss, P.D.; Delalibera, J.I.; Handelsman, J.; Raffa, K.F. Bacteria Associated with the Guts of Two Wood-Boring Beetles: *Anoplophora glabripennis* and *Saperda vestita* (Cerambycidae). *Environ. Èntomol.* **2006**, *35*, 625–629. [CrossRef]
39. Behar, A.; Yuval, B.; Jurkevitch, E. Gut bacterial communities in the Mediterranean fruit fly (*Ceratitis capitata*) and their impact on host longevity. *J. Insect Physiol.* **2008**, *54*, 1377–1383. [CrossRef]
40. Liu, X.G.; Yang, Y.J.; Liao, Q.J.; Xu, H.X.; Liu, Y.H.; Lv, Z.X. Analysis of the bacterial community structure and diversity in the intestine of *Cnaphalocrocis medinalis* (Lepidoptera: Pyralidae). *Acta Entomol. Sin.* **2016**, *59*, 965–976. [CrossRef]
41. Kikuchi, Y.; Hosokawa, T.; Fukatsu, T. Insect-Microbe Mutualism without Vertical Transmission: A Stinkbug Acquires a Beneficial Gut Symbiont from the Environment Every Generation. *Appl. Environ. Microbiol.* **2007**, *73*, 4308–4316. [CrossRef] [PubMed]
42. Barbosa, K.L.; Malta, V.R.D.S.; Machado, S.S.; Junior, G.A.L.; da Silva, A.P.V.; Almeida, R.M.R.G.; da Luz, J.M.R. Bacterial cellulase from the intestinal tract of the sugarcane borer. *Int. J. Biol. Macromol.* **2020**, *161*, 441–448. [CrossRef] [PubMed]
43. Hancock, P.A.; Sinkins, S.P.; Godfray, H.C.J. Strategies for Introducing *Wolbachia* to Reduce Transmission of Mosquito-Borne Diseases. *PLoS Negl. Trop. Dis.* **2011**, *5*, e1024. [CrossRef] [PubMed]
44. Tang, T.; Zhang, Y.; Cai, T.; Deng, X.; Liu, C.; Li, J.; He, S.; Li, J.; Wan, H. Antibiotics increased host insecticide susceptibility via collapsed bacterial symbionts reducing detoxification metabolism in the brown planthopper, *Nilaparvata lugens*. *J. Pest Sci.* **2020**, *94*, 757–767. [CrossRef]
45. Briones-Roblero, C.I.; Rodríguez-Díaz, R.; Santiago-Cruz, J.A.; Zúñiga, G.; Rivera-Orduña, F.N. Degradation capacities of bacteria and yeasts isolated from the gut of *Dendroctonus rhizophagus* (Curculionidae: Scolytinae). *Folia Microbiol.* **2017**, *62*, 1–9. [CrossRef]
46. Engel, P.; Moran, N.A. The gut microbiota of insects—Diversity in structure and function. *FEMS Microbiol. Rev.* **2013**, *37*, 699–735. [CrossRef]
47. MsangoSoko, K.; Bhattacharya, R.; Ramakrishnan, B.; Sharma, K.; Subramanian, S. Cellulolytic activity of gut bacteria isolated from the eri silkworm larvae, *Samia ricini*, (Lepidoptera: Saturniidae). *Int. J. Trop. Insect Sci.* **2021**, *41*, 2785–2794. [CrossRef]
48. Smith, C.C.; Srygley, R.B.; Healy, F.; Swaminath, K.; Mueller, U.G. Spatial Structure of the Mormon Cricket Gut Microbiome and its Predicted Contribution to Nutrition and Immune Function. *Front. Microbiol.* **2017**, *8*, 801. [CrossRef]
49. Koskinioti, P.; Ras, E.; Augustinos, A.A.; Beukeboom, L.W.; Mathiopoulos, K.D.; Caceres, C.; Bourtzis, K. Manipulation of insect gut microbiota towards the improvement of *Bactrocera oleae* artificial rearing. *Èntomol. Exp. Appl.* **2020**, *168*, 523–540. [CrossRef]
50. Koskinioti, P.; Ras, E.; Augustinos, A.A.; Beukeboom, L.W.; Mathiopoulos, K.D.; Caceres, C.; Bourtzis, K. The impact of fruit fly gut bacteria on the rearing of the parasitic wasp *Diachasmimorpha longicaudata*. *Èntomol. Exp. Appl.* **2020**, *168*, 541–559. [CrossRef]
51. Gao, H.H.; Qin, D.Y.; Dai, X.Y.; Liu, J.; Yu, Y. Effects of intestinal bacteria *Citrobacter freundi* and *Klebsiella oxytoca* on the development and substance metabolism of *Drosophila suzukii* (Diptera: Drosophilidae). *Acta Entomol. Sin.* **2020**, *63*, 84–91. [CrossRef]
52. Acevedo, F.E.; Peiffer, M.; Tan, C.-W.; Stanley, B.A.; Stanley, A.; Wang, J.; Jones, A.G.; Hoover, K.; Rosa, C.; Luthe, D.; et al. Fall Armyworm-Associated Gut Bacteria Modulate Plant Defense Responses. *Mol. Plant-Microbe Interact.* **2017**, *30*, 127–137. [CrossRef] [PubMed]

53. Sun, Y.H.; Zhou, K.Y.; Xiong, Z. Screening and Identification of Lipase-producing Bacteria from Intestinal Canal of *Dendrolimus kikuchii* and Preliminary Studies on Its Enzyme Properties. *J. Microbiol.* **2012**, *32*, 64–67.
54. Karimi, K.; Taherzadeh, M.J. A critical review of analytical methods in pretreatment of lignocelluloses: Composition, imaging, and crystallinity. *Bioresour. Technol.* **2016**, *200*, 1008–1018. [CrossRef] [PubMed]
55. Tian, F.Y.; Ping, B.A. Research on Cellulose-decomposing and Methanogenesis in Vitro within Rumen Fluid of Tibetan Sheep by Added Hydrolyzed Casein. *Chin. J. Vet. Med.* **2017**, *53*, 47–50.
56. Li, W.J.; Meng, L.; Luo, Q.J.; An, S.Z.; Chen, A.L.; Wang, J.Y. Effects of Different Utilization Methods of Corn Stalks on Digestion and Metabolism of Sheep. *Grass-Feed. Livest.* **2000**, *S1*, 55–62. [CrossRef]
57. Zhang, J.Q.; Ren, C.W.; Zhang, C.W. The effect of dry matter intake, milk production and digestibility of different wheat stalk chopped lengths and processing for dairy cows. *China Feed* **2019**, *6*, 17–21. [CrossRef]
58. Azizi-Shotorkhoft, A.; Mohammadabadi, T.; Motamedi, H.; Chaji, M.; Fazaeli, H. Isolation and identification of termite gut symbiotic bacteria with lignocellulose-degrading potential, and their effects on the nutritive value for ruminants of some by-products. *Anim. Feed Sci. Technol.* **2016**, *221*, 234–242. [CrossRef]
59. Kundu, P.; Manna, B.; Majumder, S.; Ghosh, A. Species-wide Metabolic Interaction Network for Understanding Natural Lignocellulose Digestion in Termite Gut Microbiota. *Sci. Rep.* **2019**, *9*, 16329. [CrossRef]
60. Huang, W.Q.; Shi, D.D.; Cai, H.Y.; Yu, D.L.; Meng, K.; Yang, P.L. Identification and Genome Analysis of a Cellulose Degrading Strain from the Intestinal Tract of *Protaetia brevitarsis* Larva. *J. Agric. Sci. Technol.* **2021**, *23*, 51–58. [CrossRef]

insects

MDPI

Article

Changes in the Host Gut Microbiota during Parasitization by Parasitic Wasp *Cotesia vestalis*

Shuaiqi Zhang [1,2,3], Jieling Huang [1,2,3], Qiuping Wang [1,2,3], Minsheng You [1,2,3,*] and Xiaofeng Xia [1,2,3,*]

1 State Key Laboratory of Ecological Pest Control for Fujian and Taiwan Crops, Institute of Applied Ecology, Fujian Agriculture and Forestry University, Fuzhou 350002, China

2 Ministerial and Provincial Joint Innovation Centre for Safety Production of Cross-Strait Crops, Fujian Agriculture and Forestry University, Fuzhou 350002, China

3 Joint International Research Laboratory of Ecological Pest Control, Ministry of Education, Fuzhou 350002, China

* Correspondence: msyou@fafu.edu.cn (M.Y.); xiaofengxia@fafu.edu.cn (X.X.)

Simple Summary: *Cotesia vestalis* is a larval endo-parasitoid of the diamondback moth (*Plutella xylostella*), which is a severe pest of cruciferous crops. The function of the gut microbiota of insects has been widely studied. However, it was unclear whether, and how, the gut microbiota of *P. xylostella* responds to its natural enemy, *C. vestalis*. In this study, a time-course experiment was performed to examine changes in the host–microbial community from the start of parasitization to the mature stage of the parasitoid larvae. Our results will provide a framework for studies of host-gut microbiota and parasitic wasp interactions.

Abstract: Parasites attack the host insects and possibly impact the host-gut microbiota, which leads to provision of a suitable host environment for parasites' development. However, little is known about whether and how the parasitic wasp *Cotesia vestalis* alters the gut microbiota of the host *Plutella xylostella*. In this study, 16S rDNA microbial profiling, combined with a traditional isolation and culture method, were used to assess changes in the bacterial microbiome of parasitized and non-parasitized hosts at different developmental stages of *C. vestalis* larvae. Parasitization affected both the diversity and structure of the host-gut microbiota, with a significant reduction in richness on the sixth day post parasitization (6 DPP) and significant differences in bacterial structure between parasitized and non-parasitized hosts on the third day. The bacterial abundance of host-gut microbiota changed significantly as the parasitization progressed, resulting in alteration of potential functional contribution. Notably, the relative abundance of the predominant family Enterobacteriaceae was significantly decreased on the third day post-parasitization. In addition, the results of traditional isolation and culture of bacteria indicated differences in the bacterial composition between the three DPP and CK3 groups, as with 16S microbial profiling. These findings shed light on the interaction between a parasitic wasp and gut bacteria in the host insect during parasitization.

Keywords: host gut microbiota; parasitoids; host-parasite-microbe interactions; host regulation

Citation: Zhang, S.; Huang, J.; Wang, Q.; You, M.; Xia, X. Changes in the Host Gut Microbiota during Parasitization by Parasitic Wasp *Cotesia vestalis*. *Insects* **2022**, *13*, 760. https://doi.org/10.3390/insects13090760

Academic Editors: Hongyu Zhang, Yin Wang and Xiaoxue Li

Received: 1 August 2022
Accepted: 22 August 2022
Published: 24 August 2022

Publisher's Note: MDPI stays neutral with regard to jurisdictional claims in published maps and institutional affiliations.

Copyright: © 2022 by the authors. Licensee MDPI, Basel, Switzerland. This article is an open access article distributed under the terms and conditions of the Creative Commons Attribution (CC BY) license (https://creativecommons.org/licenses/by/4.0/).

1. Introduction

In insects, the gut microbiota plays a substantial role in the host's life activities, which include digestion, nitrogen fixation [1], detoxification [2], development [3], pesticide resistance [4], behavior [5], and increasing host defenses against abiotic stress [6] and parasites [7]. Intestinal homeostasis is achieved by maintaining microbial populations at a specific density range to avoid excessive losses or to provide the required contribution to the host insect [8,9]. Therefore, characterization of the diversity and composition of gut microbiota in insects is essential for understanding the biology of the host insects [10].

There is increasing evidence that the diversity of the gut microbiome in host insects is influenced by host–parasite interactions, which provides a new perspective for understanding the co-evolution of host–parasite interactions. For the pathogenic fungus *Beauveria bassiana*–*Dendroctonus valens* association, the evenness, structure, and abundance of the host's bacterial community are substantially altered by infection with *B. bassiana*. The gut bacterium *Erwinia sp.* accelerates the mortality of the host [11]. For the tapeworm *Hymenolepis diminuta*–*Tenebrio molitor* association, considerable alteration in the host-gut bacteriome and mycobiome are found [12]. Recent studies have investigated the influences of parasitic wasp parasitization on the microbiome of their host insects. Changes in the host-gut microbiota caused by parasitization are observed in the host insects parasitized by the wasp *Cotesia flavipes* [13], *Cotesia glomerat* [14], *Lysiphlebia japonica* [15], and *Habrobracon hebetor* [16]. In contrast, trypanosomatid (*Lotmaria passim*) does not impact the general landscape of the honey bee (*Apis mellifera*)-gut microbiota [17]. Therefore, whether, and how, the parasitoid and host-microbiome interact needs to be analyzed specifically for each species.

For parasitic wasp–host interaction, parasitic wasps lay eggs in the hosts, regulate host physiology, and their larvae coexist with the host-gut microbiome in the host [18,19]. It has been indicated that host endosymbionts might influence host resistance to parasitoid wasps, and this has been mainly studied in aphids. Endosymbionts *Hamiltonella defensa* protect the pea aphid *Acyrthosiphon pisum* against the parasitoid wasp *Aphidius ervi* [20]. Endosymbionts *Regiella insecticola* provide vital protection for peach aphids *Myzus persicae* against wasps *Aphidius colemani* [21]. Meanwhile, a study showed that the differences in bacterial communities of *Drosophila melanogaster* influenced its resistance to parasitoids [22]. Furthermore, parasitic wasp embryos [23] and larvae [24,25] rely on nutrients from their hosts for development, and they regulate the metabolism of proteins, carbohydrates, and lipids in their hosts to satisfy their nutritional demands [26–28]. Besides these points, parasitic wasps modulate the host's immune system during adaptation, and studies have shown that they may suppress the expression of host antimicrobial peptide genes and Toll and IMD immune pathways [29–31], all of which are known to be important for maintaining host-gut microbial homeostasis [32–34]. These results indicate that parasitic wasps may influence the host microbiota by regulating host immunity. Collectively, the host microbiome and parasitic wasps are likely to interact.

The diamondback moth, *Plutella xylostella*, an important pest of cruciferous vegetable crops, causes severe economic losses worldwide [35]. The parasitic wasp, *Cotesia vestalis*, is a solitary endophagous parasitoid of *P. xylostella* larvae [36,37]. Several studies have investigated that the gut microbial diversity and composition of *P. xylostella* varied according to food type [38], insecticidal protoxins [39], insecticide resistance [40], and antibiotics [41]. However, changes in the gut microbiota of *P. xylostella* due to parasitism remain poorly understood. To explore whether the host-gut microbiota is involved in the interaction between host and parasite, 16S rDNA sequencing and traditional isolation and culture methods were performed to study the changes in the diversity and potential functions of gut microbiota in *P. xylostella* larvae when parasitized by *C. vestalis*. Our findings serve as a foundation for further studies into the association between the host-gut microbiota and parasitic wasps.

2. Materials and Methods

2.1. Insect Rearing and Sample Collection

Both *P. xylostella* and *C. vestalis* were initially collected from a cabbage-planting field in Fuzhou, China (25.95° N, 119.27° E) in May 2014. Then *P. xylostella* was reared on radish while *C. vestalis* was reared on the larvae of *P. xylostella*. Both insects and radishes were kept under controlled conditions (25 ± 2 °C, 60% ± 10% relative humidity, and 14 light:10 dark photoperiod) in the laboratory.

All samples were divided into two groups: parasitized larvae and non-parasitized larvae. The late second instar larvae of *P. xylostella* were individually exposed to mated *C.*

vestalis for parasitization to collect parasitized *P. xylostella*. The control groups were left unparasitized. For 16S rDNA sequencing, samples from parasitized larvae were collected on the first (1 DPP), third (3 DPP), and sixth day post-parasitization (6 DPP) according to the different development stages of the parasitic wasp [42]. Non-parasitized larvae were selected at the instar consistent with the parasitized larvae due to parasitized *P. xylostella* growing slower than non-parasitized ones (Figure 1). In addition, the samples of 3 DPP (the third day post-parasitization) and CK3 (unparasitized control group 3) were collected for the traditional isolation and culture of the gut bacteria. All larvae were soaked in 75% ethanol for 90 s and rinsed in sterile water three times. Then the surface-sterilized *P. xylostella* larvae were dissected in sterile 1% phosphate-buffered saline (PBS) under a microscope. For parasitized larvae, the gut of *P. xylostella* was collected when the eggs and larvae of *C. vestalis* were observed under a microscope. Eventually, each gut sample was stored at −80 °C until used. Four biological replicates per treatment were collected. Each biological replicate contained guts from 30 *P. xylostella* larvae.

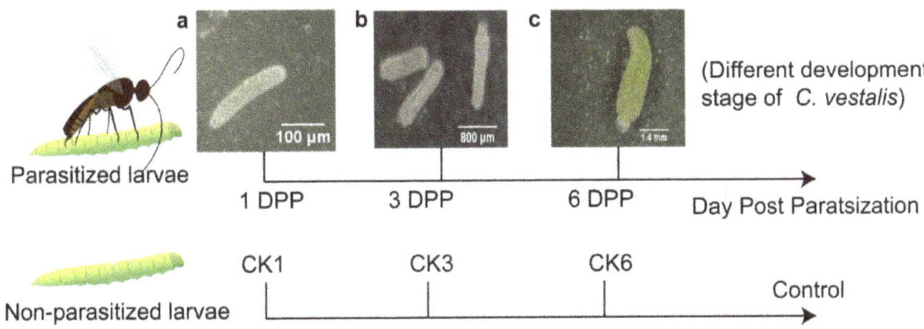

Figure 1. Experimental outline for exploring changes in the gut microbiota of parasitized *P. xylostella* and non-parasitized *P. xylostella*. (Different development stages of *C. vestalis* during sampling. (**a**) egg; (**b**) low instar larva; (**c**) mature larvae).

2.2. DNA Extraction and PCR Amplification of 16S rDNA Sequencing

Bacterial genomic DNA was extracted from the 24 gut samples using the E.Z.N.A.® soil DNA Kit (Omega Bio-Tek, Norcross, GA, USA). The 16S rRNA gene hypervariable region V3-V4 was amplified with primer set 338F/806R (Table S1) [43,44]. The PCR reaction was performed in a 20 μL volume including 4 μL 5 × Fast Pfu buffer, 2 μL 2.5 mM dNTPs, 0.8 μL each primer (5 μM), 0.4 μL Fast Pfu polymerase, 10 ng of template DNA, and appropriate ddH_2O. Cycling conditions were at 95 °C for 3 min, followed by 27 cycles at 95 °C for 30 s, 55 °C for 30 s, and 72 °C for 45 s, with a single extension at 72 °C for 10 min. All samples were amplified in triplicate. The PCR product obtained was purified using the AxyPrep DNA Gel Extraction Kit (Axygen Biosciences, Union City, CA, USA) and quantified using Quantus™ Fluorometer (Promega, Madison, WI, USA). Purified amplicons were paired-end sequenced on an Illumina MiSeq PE300 platform (Illumina, San Diego, CA, USA). The double-ended raw sequences were quality-filtered using fastp [45] and merged using FLASH [46], according to the following: (i) Reads of 300 bp were truncated at any site with an average quality score < 20 over a sliding window of 50 bp. Only reads ≥ 50 bp were retained. Reads containing N bases were removed. (ii) Overlapping sequences longer than 10 bp were assembled in which the maximum mismatch ratio was 0.2. Only assembled reads were used for the following analysis. (iii) Samples were distinguished according to the barcode (exact matching) and primers (2 nucleotide mismatch in matching).

Unique read sequences were identified from the optimized sequences (dereplication), singletons were discarded, and, then, these sequences were clustered into operational taxonomic units (OTUs) using UPARSE 7.1 at a 97% sequence similarity level [47]. Chimeras were removed during clustering. Chloroplast and mitochondrion sequences were removed for further analysis. The ribosomal database project (RDP) classifier (Version 2.11) was used to identify taxonomic groups based on the e SILVA 16S rRNA database [48] using a confidence threshold of 80% [49,50]. The raw data were submitted to the NCBI Sequence Read Archive (SRA) database (Accession Number: SAMN28027321- SAMN28027344).

2.3. Sequence Data Analysis

Based on the rarefied OTUs, rarefaction curves and alpha diversity indices were calculated with Mothur v1.30.1, including the observed richness (Sobs) and Shannon index [51]. The principal coordinate analysis (PCoA) based on Bray-Curtis dissimilarity was applied to determine the compositional difference of microbial communities, with ANOSIM (1000 permutations) testing the significance of the difference between samples. PICRUSt2 (Phylogenetic Investigations of Communities by Reconstruction of Unobserved States) was a bioinformatic tool for predicting and comparing functional attributes of microbial communities [52–57]. The potential function prediction of host-gut microbiota was analyzed by PICRUSt2 based on OTU representative sequences and abundances. All comparisons between two groups were analyzed by the Wilcoxon rank-sum test using Stats Package (R, version 3.3.1).

2.4. Isolation of Host Gut Bacteria

Thirty-five larvae from the 3 DPP and CK3 groups were randomly selected. The guts of surface-sterilized worms were separated and homogenized in sterile centrifuge tubes containing 1 mL 1% PBS solution. Ten-fold serial (10^{-1}, 10^{-2}, 10^{-3}, 10^{-4}, and 10^{-5}) dilutions of homogenized suspension were plated on four media, including Bile Aesculin Azide Agar (selective media for *Enterococcus*), Salmonella-Shigella Agar (selective media for *Salmonella*), Nutrient Agar (general media for bacteria), and Luria Bertani (general media for bacteria), and subsequently incubated at 37 °C. Plates were observed every 12 h to obtain the original bacterial strains. The isolates were categorized according to differences in colony size, color, and morphology. Then distinct morphological colonies were purified on LB plates for at least five generations to obtain monoclonal strains, followed by storing in 50% glycerol at −80 °C. The bacterial isolates obtained were grown in 500 μL liquid LB medium at 37 °C for 2–3 h. The 16S rRNA sequence was amplified by using universal primers 27F/1492R (Table S1) and the bacterial culture as a template. The PCR product was blasted in the NCBI database after sequencing. The 16S rRNA sequences of the bacteria isolated were deposited in the NCBI GenBank database with the accession number presented in Table S2. Furthermore, for evaluating the evolutionary relationships of all bacterial isolates and their closely related species, the phylogenetic tree was constructed by neighbor-joining analysis using MEGA 11.0 software [58].

3. Results

3.1. Effects of Parasitization on Host Gut Microbial Community Diversity and Structure by C. Vestalis

The 16S rDNA gene hypervariable region V3-V4 was sequenced in 24 samples of parasitized and non-parasitized *P. xylostella*, which yielded 2,173,198 sequences after standard quality filtering. The average length of the reads obtained from all samples was 428 bp. The sequences were clustered into 156 OTUs at 97% sequence identity using rarefied reads (64,327 reads per sample) for 1, 3, and 6 days post-parasitization (DPP), as well as for the control group. The rarefaction curves in all samples indicated adequate sampling and successful retrieval of OTUs. Rarefaction curves of all samples were flattened, showing that the actual bacterial diversity was effectively covered by sequencing (Figure 2a).

The bacterial community diversity and structure of parasitized and non-parasitized *P. xylostella* were analyzed using alpha diversity and beta diversity, respectively. The sobs index, reflecting microbial community richness, was significantly reduced on the 6 PPD compared with the other two parasitized groups (1 PPD and 3 PPD) (Wilcoxon rank-sum test, p = 0.03038). However, this difference at different developmental stages was not observed in the non-parasitized groups. Moreover, 6 PPD had a significantly lower value for the sobs index than the CK6 samples (p = 0.03038) (Figure 2b). In all time categories, however, there were no significant differences in community diversity evaluated by the Shannon index between parasitized and non-parasitized *P. xylostella* gut samples (Figure 2c). Taken together, the parasitization by *C. vestalis* decreased host bacterial community richness relative to that of non-parasitized *P. xylostella* on the sixth day after parasitization. Principle coordinate analysis (PCoA) of Bray-Curtis distances showed an apparent separation between the parasitized and control larvae on the third day after parasitization (ANOSIM, $p = 0.034$) (Figure 3b). By contrast, 1 and 6 DPP clustered closely with their respective controls (Figure 3a,c). In conclusion, the changes in gut bacterial structure between parasitized and control hosts were more apparent on the third day after parasitism than at the other two development times.

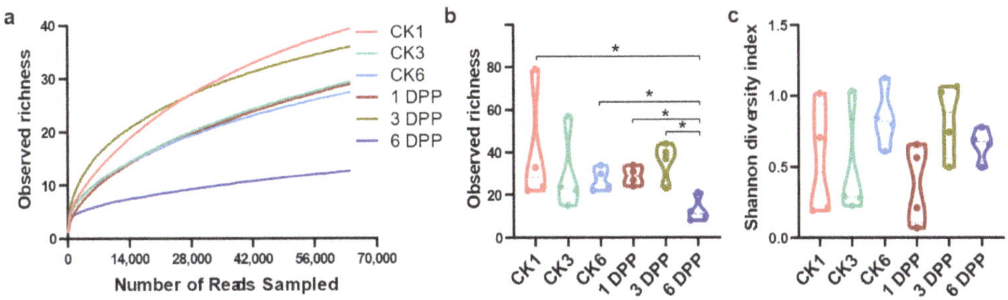

Figure 2. Alpha diversity of the host gut microbiome in the parasitized (CK1, CK3, CK6) and non-parasitized (1 DPP, 3 DPP, 6 DPP) groups at the OTU level. (**a**) Rarefaction curves based on Sobs values (the observed richness); (**b**,**c**) Violin plot showing sobs and Shannon values of bacterial communities in different samples. Wilcoxon rank-sum test between two independent samples was performed among treatments. The symbol "*" indicates statistically significant differences between the two groups being compared ($p < 0.05$).

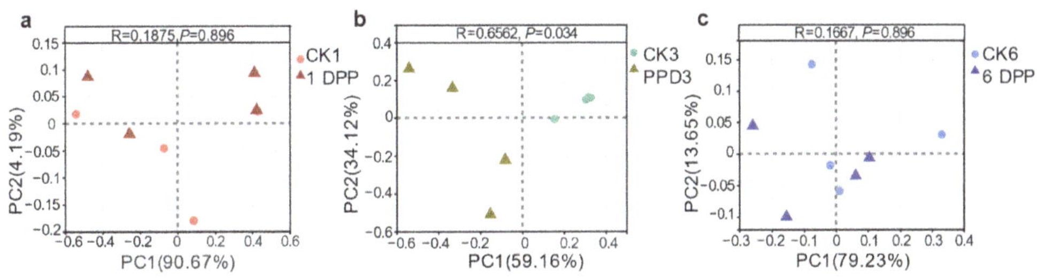

Figure 3. Principle coordinate analysis (PCoA) of the rarefied OTUs comparing the gut microbiota between parasitized and naïve control *P. xylostella* in different time categories with Bray-Curtis dissimilarity distance. (**a**) CK1 vs. 1 DPP, (**b**) CK3 vs. 3 DPP, (**c**) CK3 vs. 3 DPP. Analysis of similarities (ANOSIM) analyses revealed that the samples at 3 DPP were substantially different from those in the CK3 group ($p = 0.034$).

3.2. Impact of Parasitization on the Composition of Host-Gut Microbiota

Taxonomic analysis revealed that the major bacteria at the phylum level in all samples were Proteobacteria, Firmicutes, and Bacteroidetes, but these phyla did not significantly change between parasitized *P. xylostella* and their respective control groups (Tables S3 and S4). The host-gut bacterial community was dominated by four bacterial orders: Enterobacteriales, Lactobacillales, Pseudomonadales, and Flavobacteriales (Figure 4a). Among them, the proportion of Pseudomonadales was significantly reduced on 6 DPP compared to CK6 (Wilcoxon rank-sum test, $p = 0.02107$) (Table S4).

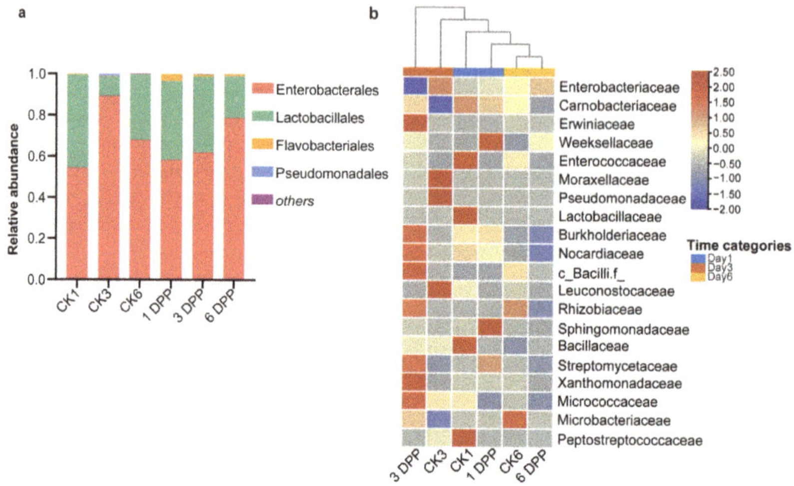

Figure 4. Impact of parasitization on the composition of host-gut bacterial community. (**a**) Relative abundance in the host-gut microbiome at the order level. "Others" included < 1% relative abundance taxa. (**b**) Heatmap of the family abundance in the *P. xylostella* gut microbiome in different time categories. Columns were clustered using the average method based on Euclidean distance, and rows were normalized.

A heatmap was plotted with the relative abundance of the top 20 shared families in six groups. The clustering of the gut samples at the family level indicated that the 3 DPP group showed dissimilarity from the other groups (Figure 4b). Among the top 20 families, in terms of abundance, the abundance of Enterobacteriaceae ($p = 0.03038$) and Leuconostocaceae ($p = 0.01771$) on the 3 DPP showed lower proportions compared with the CK3 group, whereas Xanthobacteraceae in 3 DPP was significantly more abundant than the control ($p = 0.03719$) (Figure 5a). The abundance of Nocardiaceae ($p = 0.04207$) and Rhizobiaceae ($p = 0.02558$) decreased on the 6 DPP compared with CK6 group (Figure 5b).

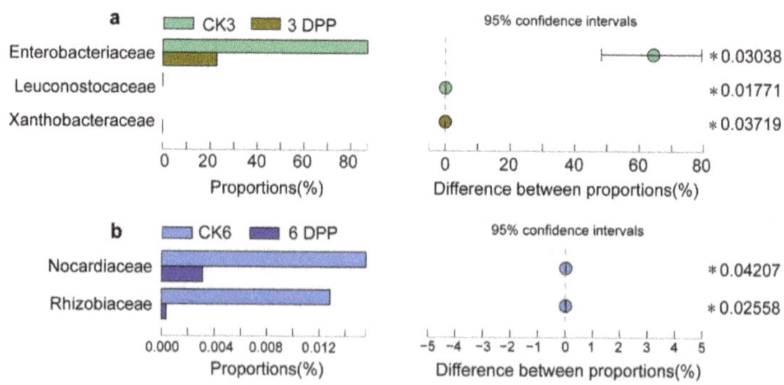

Figure 5. The difference in relative proportion (%) between parasitized and non-parasitized larvae at different sampling times at the family level. (**a**) CK3 vs. 3 DPP, (**b**) CK6 vs. 6 DPP. Statistical analysis was performed by the Wilcoxon rank-sum test. The symbol * indicates $p < 0.05$.

At the genus level, the gut bacterial community was dominated by *Enterobacter*, *Carnobacterium*, *Pantoea*, an unidentified genus of Enterobacteriaceae, and *Chryseobacterium*, with at least 1% relative abundance (Figure S1). Alterations in bacterial proportions were seen at the genus level, which was consistent with the family level. In particular, significant reductions in the genus *Enterobacter* were observed on the 3 DPP compared with non-parasitized larvae (Figure S2) (p = 0.03038). *Pantoea* was one of the dominant bacteria enriched mainly in 3 DPP with a mean relative abundance of 39.29% (Table S3). However, no significant change in the bacterial proportions was observed on the third day after parasitization compared to the control, as one replicate of the 3 DPP sample had lower values than the others (Table S4).

3.3. *Effects of Parasitization on Host-Gut Microbial Function by C. vestalis*

The different functional contribution of host-gut microbiota was predicted using the top thirty shared Kyoto Encyclopedia of Genes and Genomes (KEGG) level 3 inferred by PICRUSt2 in all samples. The roles of parasitized and non-parasitized host-gut microbes mostly comprised Metabolism, Genetic Information Processing, Environmental Information Processing, Cellular Processes, and Human diseases. In the most prevalent metabolism category, pathways related to the biosynthesis of secondary metabolites, biosynthesis of amino acids, pentose phosphate pathway, purine/starch and sucrose/cysteine and methionine metabolism predominated on 3 DPP. In contrast, metabolic pathways, microbial metabolism in diverse environments, oxidative phosphorylation, pyrimidine/fructose and mannose/propanoate/glyoxylate and dicarboxylate/butanoate/sulfur/porphyrin and chlorophyll metabolism were significantly reduced. In other functional categories, ABC transporters, ribosome, quorum sensing and flagellar assembly were increased significantly in the 3 DPP group, while the two-component system was more predominant in CK3. However, no significant difference was observed between parasitized and non-parasitized hosts on the first and sixth days (Table S5). Above all, the 3 DPP group showed the most obvious changes in the relative abundance of bacterial functions compared to CK3 among all-time categories, similar to the differences in the structure and composition of host-gut microbiota (Figure 6).

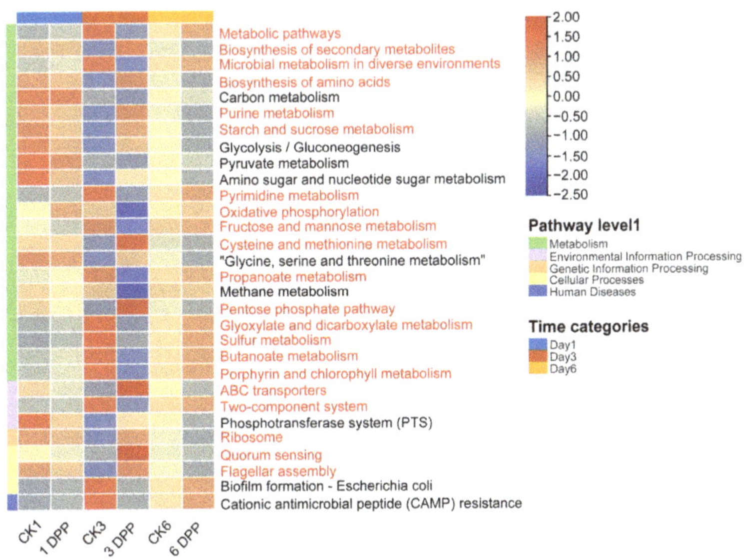

Figure 6. Relative abundance (%) of host-gut microbiota functions between parasitized and non-parasitized larvae at different time categories at the Kyoto Encyclopedia of Genes and Genomes (KEGG) level 3. The heatmap plot was normalized by row. The red letter indicates that the special function at KEGG pathway level 3 significantly differed from the control group during the parasitism (p < 0.05). Group color bars on the left indicate that the functions were grouped according to pathway level 1.

3.4. Isolation and Culture of Bacteria from Parasitized and Non-Parasitized Host Gut

As indicated in the high throughput sequencing results, the beta-diversity, composition, and specific function of host-gut bacteria were more variable than the other two development stages on the 3rd day post-parasitization compared to control. According to these changes, the gut samples from 3 DPP and CK3 were chosen to explore the difference in gut microbiota between parasitized and non-parasitized *P. xylostella* using the traditional isolation and culture methods.

The 16S rDNA gene sequencing analysis resulted in the identification of 7 species from 3 DPP and 8 species from CK3. The bacterial isolates identified as *Cedecea lapagei* (CK3-6, 3 DPP-4), *Carnobacterium maltaromaticum* (CK3-7, 3 DPP-8), and *Enterococcus termitis* (CK3-4, 3 DPP-2) were present in both groups. Four bacterial isolates from the genera *Stenotrophomonas* (CK3-1), *Acinetobacter* (CK3-2), *Enterobacter* (CK3-3), and *Bacillus* (CK3-5) were uniquely found in the unparasitized control group. Moreover, there were five strains specific to the 3 DPP group, containing the genus *Neisseria* (3 DPP-1), *Klebsiella* (3 DPP-3, 3 DPP-5), *Citrobacter* (3 DPP-6), and *Staphylococcus* (3 DPP-7) (Table S2). Phylogenetic analysis of all isolates with the closest relatives showed that the prevalent phyla were Proteobacteria and Firmicutes in both groups, consistent with the high throughput sequencing results (Figure 7).

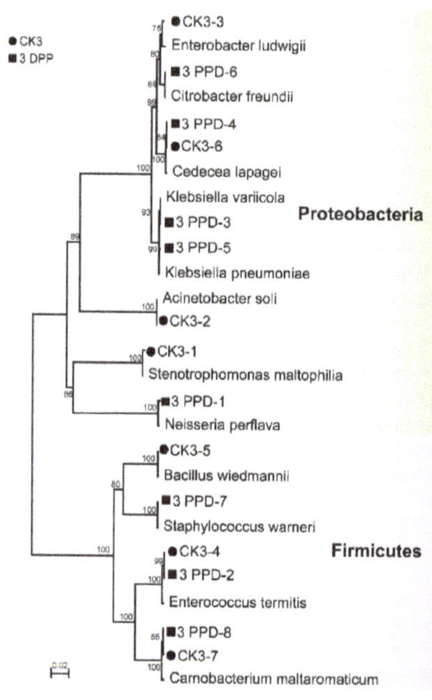

Figure 7. Neighbor-joining tree of bacterial isolates from parasitized and non-parasitized *P. xylostella* and their closely related species based on sequencing of the 16S rDNA gene. The nodes' bootstrap values were based on 1000 replicates. The scaled bar represents 0.02 estimated phylogenetic divergence.

4. Discussion

How the gut microbiota of *P. xylostella* change due to parasitization by *C. vestalis* at different development stages was investigated in this work. In terms of alpha diversity, we discovered that the bacterial community richness index (sobs) decreased in the late stage of parasitization (6 DPP), whether compared to the early phase of the parasitization process or the non-parasitized group. Interestingly, all the microbial diversity in aphids (*Aphis gosypii*) parasitized by *Lysiphlebia japonica* was lower than that in non-parasitized aphids at 8 h, 16 h, 1 day, 2 days, and 3 days [15]. Additionally, rare microbial taxa have been proven to contribute to community stability and persistence [59,60], so we retained the low-abundance OTUs. The existence of low-abundance OTUs in the other groups was most likely responsible for the Sobs index decreasing in the 6 DPP group and the Shannon value remaining similar to that of the other groups. However, the beta diversity showed that the gut bacterial structure of the host altered significantly compared with the control only on the third day. As previously demonstrated for *C. flavipes*, whereas alpha-diversity analysis revealed changes in the richness of gut microbiota at different stages (1, 5, and 9 "days after parasitization, DAP") of *D. saccharalis* parasitization by *C. flavipes*, the beta-diversity analysis revealed that the parasitoid influenced the host-gut microbiota only on 5 DAP [13]. The findings suggest that the response mode of host-gut microbiota to parasitoid varies at different phases of parasitization. It has been shown that the nutritional physiology [61] and immune response capacity [62] of the host are different at various stages of parasitoid larval development and may influence the dynamics of microbial diversity in the host.

According to the taxonomic analysis, the bacterial microbiome of non-parasitized *P. xylostella* was dominated by Enterobacteriaceae, followed by Carnobacteriaceae. A previous study has also shown that these two families are the most abundant in the gut of *P. xylostella* [63]. However, significant declines in Enterobacteriaceae of samples at 3 DPP were reported in our investigation, resulting in the Enterobacteriaceae no longer being the most abundant family in the bacterial microbiome on the third day post-parasitization. A previous study has also observed that parasitoid envenomation led to a predominant shift of gut bacterial composition in *Galleria mellonella* [16]. This suggests that *C. vestalis* may significantly disturb the composition of host-gut microbiota in the middle phase of parasitization. The declines in Enterobacteriaceae appear to have been caused by the genus *Enterobacter*, with a similar change in proportions at the genus level. The *Enterobacter* sp. isolated from the gut of *Bactrocera oleae* significantly reduced parasitism rate and fecundity of *Diachasmimorpha longicaudata*. This suggests that the reduction of *Enterobacter* from the *P. xylostella* gut may impact the suitability of the host environment for the *C. vestalis*. Furthermore, the abundance of Enterobacteriaceae recovered to the highest family on 6 DPP, while PCoA analyses showed a similar bacterial structure to CK6, reflecting that the greater impact of parasitic wasps on the microbial community in the host at the 3 DPP was temporary. According to a recent study, the total count of hemocytes in *Diatraea saccharalis* was lowest on the third day after parasitization by *Cotesia flavipes*, while hemocyte viability was significantly higher at 5 DAP for parasitized larvae compared with non-parasitized larvae over 0–10 DAP [64]. The dynamic of the host bacterial community in our study may be due to the immune regulation of the host insect by the parasitic wasp during its development. Previously, researchers considered that alternations in the structure of the gut microbiome could contribute to the variations in the susceptibility to pathogenic microorganisms [65]. Alterations in the host-gut microbiome generated by parasitoid envenomation were found to enhance fungal infection [16]. Whether the interaction between *P. xylostella* and *C. vestalis* leads to similar results remains to be further studied.

It is worth noting that, due to the reduced relative abundance of *Pantoea* in one sample of 3DPP compared to the other replicates, the difference in the abundance of *Pantoea* between 3DPP and CK3 was not significant. Nevertheless, *Pantoea* became the most dominant genus in the host-gut microbiome on the third day after parasitization. *Pantoea* strains are commonly found in the guts of insects [66]. *Pantoea agglomerans* was previously found to produce antifungal phenols, which may play a role in host defense and have an important impact on the composition of the gut flora [67]. Based on its high abundance in 3DPP, it is worth continuing to pay attention to the changes and functions of this kind of flora in future studies.

The unique structure and physicochemical environment of the insect gut result in a complex and functionally diverse gut microbial community [1]. In the current study, the main functional groups of gut microbiota in parasitized and non-parasitized larvae were similar, and it is assumed that fixed groups play a role in the host, which may be the result of their co-evolution with the host. Additionally, functional KEGG pathway analysis revealed significant differences between samples from the 3 DDP group and CK3, with specific pathways increasing or decreasing in relative abundance. A previous report also suggested that *C. flavipes* might alter the potential function of its host-gut microbiota [13]. Significant differences in gut microbiota functional profiles between parasitized and non-parasitized hosts were mainly enriched to several metabolism-related pathways. These differences suggest that the gut bacteria may affect nutrient replenishment and food digestion in the parasitized *P. xylostella*. Previous studies found that parasitic wasps could regulate the host's metabolic levels to provide a suitable environment for the development of wasps [68,69], and gut microbiota may play a role in this regulation. However, considering the limitations of PICRUSt2, the analysis to predict the function of gut microbiota only provided some preliminary results. Based on these results, the functional shifts of the host-gut microbiota during parasitization might be determined by combining metabonomics and metagenomics in the future. In addition, hosts in the mid-stage of parasitization could be chosen as study objects in future experiments.

In this study, Proteobacteria and Firmicutes were the most common phyla that could be cultured in *P. xylostella*. A previous investigation also found that cultured bacterial strains isolated from *P. xylostella* were dominated by these two phyla [70]. Furthermore, the results of traditional isolation and culture of bacteria also indicated differences in the host-gut bacteriome during parasitization. The original strains isolated from *P. xylostella* gut provide valuable resources for the future study of their functions in the interaction between *P. xylostella* and *C. vestalis*. Besides, the bacterial isolates from genus *Neisseria*, *Klebsiella*, and *Citrobacter* obtained using culture methods on the 3 DPP were not detected by high-throughput sequencing, which may be due to the methodological nature of OTU picking and the limitations of taxonomic databases inserting important biases in community analyses. There were still many limitations in this study. Our selection of media types is not yet comprehensive, and the culture was only conducted in an aerobic environment. Further exploration of the culturable bacteria in *P. xylostella*, with a broader range of media and culture methods, is still required.

5. Conclusions

To the best of our knowledge, this study provides the first comprehensive description of shifts in the gut bacteriome of *P. xylostella* during parasitization by *C. vestalis*. The degree of changes in bacterial community structure and composition caused by *C. vestalis* varied at the different larval developmental stages of wasps according to the time-series experiments. The most obvious alterations in the structure and composition of host-gut microbiota at 3 DPP affect the potential functional contribution of the gut bacterial community. These alterations suggest that *C. vestalis* larvae may adapt and regulate their host environment by changing the balance of host-gut microbiota. However, the specific biological significance of bacteria cultured from parasitized *P. xylostella*, as well as the mechanisms causing changes in the host microbial community, remain to be tested. In conclusion, our results provide a framework of interactions among *P. xylostella*, its symbionts, and its parasitic enemy, *C. vestalis*, wherein regulation of the host by the parasitic wasp is associated with host-gut bacteria, which could help in understanding the regulation of host by parasitic wasp associated with host-gut bacteria.

Supplementary Materials: The following supporting information can be downloaded at: https://www.mdpi.com/article/10.3390/insects13090760/s1, Table S1: Primer sequences for this study; Table S2: Online blast-based alignment of 16S rDNA gene for cultured gut bacteria; Table S3: Relative abundance in the host-gut microbiome at the phylum, order, family, and genus level; Table S4: Comparison of relative abundance of host-gut microbiota at the phylum, order, family, and genus level between parasitized and non-parasitized groups; Table S5: Comparison of relative abundance of host-gut microbiota function at KEGG pathway level 3 with significant differences between parasitized and control groups; Figure S1: Relative abundance in the host-gut microbiome at the genus level; Figure S2: The difference in relative proportion (%) between parasitized and non-parasitized larvae at different sampling times at genus level.

Author Contributions: Data curation, S.Z. and J.H.; Software, S.Z.; Visualization, S.Z. and Q.W. Writing-original draft, S.Z.; Writing—review and editing, all authors; Supervision, M.Y. and X.X.; Project Administration, M.Y. and X.X. All authors have read and agreed to the published version of the manuscript.

Funding: This research was funded by the project of the National Key Research and Development Program of China (2017YFE0122000), the project of the National Natural Science Foundation of China (Nos. 31871968), and the open project of Fujian Key Laboratory of crop pest monitoring and control (MIMCP-201902).

Institutional Review Board Statement: Not applicable.

Informed Consent Statement: Not applicable.

Data Availability Statement: Not applicable.

Acknowledgments: We are very grateful to Bowen Feng (Majorbio Bio-Pharm Technology) for his help with the 16S rDNA sequencing data analysis.

Conflicts of Interest: The authors declare no conflict of interest.

References

1. Engel, P.; Moran, N.A. The gut microbiota of insects—Diversity in structure and function. *FEMS Microbiol. Rev.* **2013**, *37*, 699–735. [CrossRef] [PubMed]
2. Watanabe, H.; Tokuda, G. Cellulolytic Systems in Insects. *Annu. Rev. Entomol.* **2009**, *55*, 609–632. [CrossRef] [PubMed]
3. Zheng, H.; Powell, J.E.; Steele, M.I.; Dietrich, C.; Moran, N.A. Honeybee gut microbiota promotes host weight gain via bacterial metabolism and hormonal signaling. *Proc. Natl. Acad. Sci. USA* **2017**, *114*, 4775. [CrossRef] [PubMed]
4. Eterovic, M.; Kirfel, P.; Grotmann, J.; Vilcinskas, A. Fitness costs of infection with *Serratia symbiotica* are associated with greater susceptibility to insecticides in the pea aphid *Acyrthosiphon pisum*: *Serratia symbiotica* correlates with susceptibility to insecticides in the pea aphid. *Pest Manag. Sci.* **2018**, *74*, 1829–1836.
5. Jia, Y.; Jin, S.; Hu, K.; Geng, L.; Han, C.; Kang, R.; Pang, Y.; Ling, E.; Tan, E.; Pan, Y.; et al. Gut microbiome modulates Drosophila aggression through octopamine signaling. *Nat. Commun.* **2021**, *12*, 2698. [CrossRef]
6. Raza, M.; Wang, Y.; Cai, Z.; Bai, S.; Awan, U.; Zhang, Z.-Y.; Zheng, W.; Zhang, H. Gut microbiota promotes host resistance to low-temperature stress by stimulating its arginine and proline metabolism pathway in adult *Bactrocera dorsalis*. *PLoS Pathog.* **2020**, *16*, e1008441. [CrossRef]
7. Gao, H.; Bai, L.; Jiang, Y.; Huang, W.; Wang, L.; Li, S.; Zhu, G.; Wang, D.; Huang, Z.; Li, X.; et al. A natural symbiotic bacterium drives mosquito refractoriness to Plasmodium infection via secretion of an antimalarial lipase. *Nat. Microbiol.* **2021**, *6*, 806–817. [CrossRef]
8. Charroux, B.; Royet, J. Gut-Microbiota interactions in non-mammals: What can we learn from Drosophila? *Semin. Immunol.* **2012**, *24*, 17–24. [CrossRef]
9. Buchon, N.; Broderick, N.; Lemaitre, B. Gut homeostasis in a microbial world: Insights from Drosophila melanogaster, Nature reviews. *Microbiology* **2013**, *11*, 615–626.
10. Pernice, M.; Simpson, S.J.; Ponton, F. Towards an integrated understanding of gut microbiota using insects as model systems. *J. Insect Physiol.* **2014**, *69*, 12–18. [CrossRef]
11. Xu, L.; Deng, J.; Zhou, F.; Cheng, C.; Lu, M. Gut microbiota in an invasive bark beetle infected by a pathogenic fungus accelerates beetle mortality. *J. Pest Sci.* **2019**, *92*, 343–351. [CrossRef]
12. Fredensborg, B.L.; Kálvalí, I.; Johannesen, T.B.; Stensvold, C.R.; Kapel, C. Parasites modulate the gut-microbiome in insects: A proof-of-concept study. *PLoS ONE* **2020**, *15*, e0227561. [CrossRef] [PubMed]
13. Cavichiolli de Oliveira, N.; Cônsoli, F. Beyond host regulation: Changes in gut microbiome of permissive and nonpermissive hosts following parasitization by the wasp *Cotesia flavipes*. *FEMS Microbiol. Ecol.* **2019**, *96*, fiz206. [CrossRef] [PubMed]
14. Gloder, G.; Bourne, M.E.; Verreth, C.; Wilberts, L.; Bossaert, S.; Crauwels, S.; Dicke, M.; Poelman, E.H.; Jacquemyn, H.; Lievens, B. Parasitism by endoparasitoid wasps alters the internal but not the external microbiome in host caterpillars. *Anim. Microbiome* **2021**, *3*, 73. [CrossRef]
15. Gao, X.; Niu, R.; Xiangzhen, Z.; Wang, L.; Ji, J.; Niu, L.; Wu, C.; Zhang, S.; Luo, J.; Cui, J. Characterization and comparison of the bacterial microbiota of *Lysiphlebia japonica* parasitioid wasps and their aphid host *Aphis gosypii*. *Pest Manag. Sci.* **2021**, *77*, 2710–2718. [CrossRef]
16. Polenogova, O.; Kabilov, M.; Maksim, T.; Rotskaya, U.N.; Krivopalov, A.; Morozova, V.; Mozhaitseva, K.; Kryukova, N.; Alikina, T.Y.; Kryukov, V.; et al. Parasitoid envenomation alters the *Galleria mellonella* midgut microbiota and immunity, thereby promoting fungal infection. *Sci. Rep.* **2019**, *9*, 4012. [CrossRef]
17. Liu, Q.; Lei, J.; Darby, A.C.; Kadowaki, T. Trypanosomatid parasite dynamically changes the transcriptome during infection and modifies honey bee physiology. *Commun. Biol.* **2020**, *3*, 51. [CrossRef]

18. Beckage, N.; Gelman, D. Wasp parasitoid disruption of host development: Implications for new biologically based strategies for insect control. *Annu. Rev. Entomol.* **2004**, *49*, 299–330. [CrossRef]
19. Schafellner, C.; Marktl, R.C.; Nussbaumer, C.; Schopf, A. Parasitism-Induced effects of *Glyptapanteles liparidis* (Hym., Braconidae) on the juvenile hormone titer of its host, *Lymantria dispar*: The role of the parasitoid larvae. *J. Insect Physiol.* **2004**, *50*, 1181–1189. [CrossRef]
20. Oliver, K.; Russell, J.; Moran, N.; Hunter, M. Facultative bacterial symbionts in aphids confer resistance to parasitic wasps. *Proc. Natl. Acad. Sci. USA* **2003**, *100*, 1803–1807. [CrossRef]
21. Christoph, V.; Lukas, G.; Paula, R. A strain of the bacterial symbiont Regiella insecticola protects aphids against parasitoids. *Biol. Lett.* **2010**, *6*, 109–111.
22. Chaplinska, M.; Gerritsma, S.; Dini-Andreote, F.; Salles, J.; Wertheim, B. Bacterial communities differ among *Drosophila melanogaster* populations and affect host resistance against parasitoids. *PLoS ONE* **2016**, *11*, e0167726. [CrossRef] [PubMed]
23. Mark Jervis, P.F. Towards a general perspective on life-history evolution and diversification in parasitoid wasps. *Biol. J. Linn. Soc.* **2011**, *104*, 443–461. [CrossRef]
24. Grimaldi, A.; Caccia, S.; Congiu, T.; Ferrarese, R.; Eguileor, M.D. Structure and function of the extraembryonic membrane persisting around the larvae of the parasitold *Toxoneuron nigriceps*. *J. Insect Physiol.* **2006**, *52*, 870–880. [CrossRef] [PubMed]
25. Thompson, N.S. Host nutrition determines blood nutrient composition and mediates parasite developmental success: *Manduca sexta* L. parasitized by *Cotesia congregata (Say)*. *J. Exp. Biol.* **2005**, *208*, 625–635. [CrossRef]
26. Jiang, J.; Ji, X.; Yin, Y.; Wan, N. The effect of nucleopolyhedrovirus infection and/or parasitism by *Microplitis pallidipes* on hemolymph proteins, sugars and lipids in *Spodoptera exigua* larvae. *BioControl* **2013**, *58*, 777–788. [CrossRef]
27. Wang, Y.; Wu, X.; Wang, Z.; Chen, T.; Zhou, S.; Chen, J.; Pang, L.; Ye, X.; Shi, M.; Huang, J.; et al. Symbiotic bracovirus of a parasite manipulates host lipid metabolism via tachykinin signaling. *PLoS Pathog.* **2021**, *17*, e1009365. [CrossRef]
28. Siebert, A.; Doucette, L.; Simpson-Haidaris, P.J.; Werren, J. Parasitoid wasp venom elevates sorbitol and alters expression of metabolic genes in human kidney cells. *Toxicon* **2019**, *161*, 57–64. [CrossRef]
29. Thoetkiattikul, H.; Beck, M.; Strand, M. Inhibitor B-like proteins from a polydnavirus inhibit NF- B activation and suppress the insect immune response. *Proc. Natl. Acad. Sci. USA* **2005**, *102*, 11426–11431. [CrossRef]
30. Strand, M.R. Polydnaviruses: Abrogation of invertebrate immune systems. In *Encyclopedia of Virology*, 3rd ed.; Mahy, B.W.J., Van Regenmortel, M.H.V., Eds.; Academic Press: Oxford, UK, 2008; pp. 250–256.
31. Lu, Z. A metalloprotease homolog venom protein from a parasitoid wasp suppresses the toll pathway in host hemocytes. *Front. Immunol.* **2018**, *9*, 2301.
32. Ha, E.-M.; Oh, C.-T.; Ryu, J.-H.; Bae, Y.-S.; Kang, S.-W.; Jang, I.-H.; Brey, P.; Lee, W.-J. An antioxidant system required for host protection against gut infection in drosophila. *Dev. Cell* **2005**, *8*, 125–132. [CrossRef] [PubMed]
33. Ponton, F.; Morimoto, J.; Robinson, K.; Kumar, S.; Cotter, S.; Wilson, K.; Simpson, S. Macronutrients modulate survival to infection and immunity in *Drosophila*. *J. Anim. Ecol.* **2019**, *89*, 460–470. [CrossRef] [PubMed]
34. Bai, S.; Yao, Z.; Raza, M.F.; Cai, Z.; Zhang, H. Regulatory mechanisms of microbial homeostasis in insect gut. *Insect Sci.* **2020**, *28*, 286–301. [CrossRef] [PubMed]
35. Furlong, M.; Wright, D.; Dosdall, L. Diamondback moth ecology and management: Problems, progress and prospects. *Annu. Rev. Entomol.* **2013**, *58*, 517–541. [CrossRef]
36. Talekar, N.S.; Shelton, A.M. Biology, ecology and management of the Diamondback Moth. *Annu. Rev. Entomol.* **2003**, *38*, 275–301. [CrossRef]
37. Li, Z.; Xia, F.; Liu, S.S.; You, M.; Furlong, M.J. Biology, ecology and management of the Diamondback Moth in China. *Annu. Rev. Entomol.* **2015**, *61*, 277–296. [CrossRef]
38. Yang, F.; Saqib, H.; Chen, J.; Ruan, Q.; Vasseur, L.; He, W.; You, M. Differential profiles of gut microbiota and metabolites associated with host shift of *Plutella xylostella*. *Int. J. Mol. Sci.* **2020**, *21*, 6283. [CrossRef]
39. Li, S.; Xu, X.; De Mandal, S.; Shakeel, M.; Hua, Y.; Shoukat, R.; Fu, D.; Jin, F. Gut microbiota mediate *Plutella xylostella* susceptibility to Bt Cry1Ac protoxin is associated with host immune response. *Environ. Pollut.* **2021**, *271*, 116271. [CrossRef]
40. Xia, X.; Zheng, D.; Zhong, H.; Qin, B.; Gurr, G.; Vasseur, L.; Lin, H.; Bai, J.; He, W.; You, M. DNA sequencing reveals the midgut microbiota of diamondback moth, *Plutella xylostella* (L.) and a possible relationship with insecticide resistance. *PLoS ONE* **2013**, *8*, e68852.
41. Lin, X.L.; Kang, Z.W.; Pan, Q.J.; Liu, T.X. Evaluation of five antibiotics on larval gut bacterial diversity of *Plutella xylostella* (Lepidoptera: Plutellidae). *Insect Sci.* **2015**, *22*, 619–628. [CrossRef]
42. Alizadeh, M.; Rassoulian, G.; Karimzadeh, J.; Hosseini-Naveh, V.; Farazmand, H. Biological study of *Plutella xylostella* (L.) (Lep: Plutellidae) and it's solitary endoparasitoid, *Cotesia vestalis (Haliday)* (Hym. Braconidae) under laboratory conditions. *Pak. J. Biol. Sci. PJBS* **2011**, *14*, 1090–1099. [CrossRef]
43. Liu, C.; Zhao, D.; Ma, W.; Guo, Y.; Wang, A.; Wang, Q.; Lee, D.J. Denitrifying sulfide removal process on high-salinity wastewaters in the presence of *Halomonas* sp. *Appl. Microbiol. Biotechnol.* **2016**, *100*, 1421–1426. [CrossRef]
44. Mori, H.; Maruyama, F.; Kato, H.; Toyoda, A.; Dozono, A.; Ohtsubo, Y.; Nagata, Y.; Fujiyama, A.; Tsuda, M.; Kurokawa, K. Design and experimental application of a novel non-degenerate universal primer set that amplifies prokaryotic 16S rRNA genes with a low possibility to amplify eukaryotic rRNA genes. *DNA Res.* **2014**, *21*, 217–227. [CrossRef]
45. Chen, S.; Zhou, Y.; Chen, Y.; Gu, J. Fastp: An ultra-fast all-in-one FASTQ preprocessor. *Bioinformatics* **2018**, *34*, i884–i890. [CrossRef]

46. Magoc, T.; Salzberg, S. FLASH: Fast Length Adjustment of Short Reads to Improve Genome Assemblies. *Bioinformatics* **2011**, *27*, 2957–2963. [CrossRef]
47. Edgar, R.C. UPARSE: Highly accurate OTU sequences from microbial amplicon reads. *Nat. Methods* **2013**, *10*, 996–998. [CrossRef]
48. Quast, C.; Pruesse, E.; Yilmaz, P.; Gerken, J.; Glckner, F.O. The SILVA ribosomal RNA gene database project: Improved data processing and web-based tools. *Nucleic Acids Res.* **2012**, *41*, D590–D596. [CrossRef]
49. Wang, Q.; Garrity, G.; Tiedje, J.; Cole, J.R. Naive Bayesian classifier for rapid assignment of rRNA sequences into the new bacterial taxonomy. *Appl. Environ. Microbiol.* **2007**, *73*, 5264–5267. [CrossRef]
50. Claesson, M.; O'Sullivan, O.; Wang, Q.; Nikkilä, J.; Marchesi, J.; Smidt, H.; de Vos, W.; Ross, R.; O'Toole, P. Comparative analysis of pyrosequencing and a phylogenetic microarray for exploring microbial community structures in the human distal intestine. *PLoS ONE* **2009**, *4*, e6669. [CrossRef]
51. Schloss, P.D.; Westcott, S.L.; Ryabin, T.; Hall, J.R.; Hartmann, M.; Hollister, E.B.; Lesniewski, R.A.; Oakley, B.B.; Parks, D.H.; Robinson, C.J. Introducing mothur: Open-Source, platform-independent, community-supported software for describing and comparing microbial communities. *Appl. Environ. Microbiol.* **2009**, *75*, 7537. [CrossRef]
52. Douglas, G.; Maffei, V.; Zaneveld, J.; Yurgel, S.; Brown, J.; Taylor, C.; Huttenhower, C.; Langille, M. PICRUSt2: An improved and extensible approach for metagenome inference. *bioRxiv* **2019**. [CrossRef]
53. Douglas, G.; Maffei, V.; Zaneveld, J.; Yurgel, S.; Brown, J.; Taylor, C.; Huttenhower, C.; Langille, M. PICRUSt2 for prediction of metagenome functions. *Nat. Biotechnol.* **2020**, *38*, 685–688. [CrossRef]
54. Hao, X.; Liu, X.; Chen, J.; Wang, B.; Li, Y.; Ye, Y.; Ma, W.; Ma, L. Effects on community composition and function *Pinus massoniana* infected by *Bursaphelenchus xylophilus*. *BMC Microbiol.* **2022**, *22*, 157. [CrossRef]
55. Bletz, M.; Goedbloed, D.; Sanchez, E.; Reinhardt, T.; Tebbe, C.; Bhuju, S.; Geffers, R.; Jarek, M.; Vences, M.; Steinfartz, S. Amphibian gut microbiota shifts differentially in community structure but converges on habitat-specific predicted functions. *Nat. Commun.* **2016**, *7*, 13699. [CrossRef]
56. Yuan, X.; Zhang, X.; Liu, X.; Dong, Y.; Yan, Z.; Lv, D.; Wang, P.; Li, Y. Comparison of Gut Bacterial Communities of Grapholita molesta (Lepidoptera: Tortricidae) Reared on Different Host Plants. *Int. J. Mol. Sci.* **2021**, *22*, 6843. [CrossRef]
57. Moo, P.S.; Rosales, M.; Ibarra-Laclette, E.; Desgarennes, D.; Huerta, C.; Lamelas, A. Diversity and Composition of the Gut Microbiota in the Developmental Stages of the Dung Beetle *Copris incertus* Say (Coleoptera, Scarabaeidae). *Front. Microbiol.* **2020**, *11*, 1698.
58. Koichiro, T.; Glen, S.; Sudhir, K. MEGA11: Molecular evolutionary genetics analysis version 11. *Mol. Biol. Evol.* **2021**, *7*, 3022–3027.
59. Li, P.; Xue, Y.; Shi, J.; Pan, A.; Tang, X.; Ming, F. The response of dominant and rare taxa for fungal diversity within different root environments to the cultivation of Bt and conventional cotton varieties. *Microbiome* **2018**, *6*, 184. [CrossRef]
60. Xiong, C.; He, J.-Z.; Singh, B.; Wang, J.; Li, P.-P.; Zhang, Q.-B.; Han, L.-L.; Shen, J.-P.; Ge, A.-H.; Wu, C.-F.; et al. Rare taxa maintain the stability of crop mycobiomes and ecosystem functions. *Environ. Microbiol.* **2020**, *23*, 1907–1924. [CrossRef]
61. Kaeslin, M.; Pfister-Wilhelm, R.; Lanzrein, B. Influence of the parasitoid *Chelonus inanitus* and its polydnavirus on host nutritional physiology and implications for parasitoid development. *J. Insect Physiol.* **2005**, *51*, 1330–1339. [CrossRef]
62. Mahmoud, A.M.A.; De Luna-Santillana, E.J.; Rodríguez-Perez, M.A. Parasitism by the endoparasitoid, *Cotesia flavipes* induces cellular immunosuppression and enhances susceptibility of the sugar cane borer, *Diatraea saccharalis* to *Bacillus thuringiensis*. *J. Insect Sci.* **2011**, *11*, 119. [CrossRef] [PubMed]
63. Xia, X.; Gurr, G.M.; Vasseur, L.; Zheng, D.; Zhong, H.; Qin, B.; Lin, J.; Wang, Y.; Song, F.; Li, Y.; et al. Metagenomic sequencing of diamondback moth gut microbiome unveils key holobiont adaptations for herbivory. *Front. Microbiol.* **2017**, *8*, 663. [CrossRef]
64. Guidotti Pinto, C.; Walker, A.; Robinson, S.; King, G.; Rossi, G. Proteotranscriptomics reveals the secretory dynamics of teratocytes, regulators of parasitization by the endoparasitoid wasp *Cotesia flavipes*. *J. Insect Physiol.* **2022**, *139*, 104395. [CrossRef]
65. Wei, G.; Lai, Y.; Wang, G.; Chen, H.; Li, F.; Wang, S. Insect pathogenic fungus interacts with the gut microbiota to accelerate mosquito mortality. *Proc. Natl. Acad. Sci. USA* **2017**, *114*, 5994–5999. [CrossRef]
66. White, J.A.; Richards, N.K.; Laugraud, A.; Saeed, A.; Curry, M.M.; McNeill, M.R. Endosymbiotic Candidates for Parasitoid Defense in Exotic and Native New Zealand Weevils. *Microb. Ecol.* **2015**, *70*, 274–286. [CrossRef]
67. Dillon, R. Chemical barriers to gut infection in the desert locust: In vivo production of antimicrobial phenols associated with the bacterium *Pantoea agglomerans*. *J. Invertebr. Pathol.* **1995**, *66*, 72–75. [CrossRef]
68. Nakamatsu, Y.; Kuriya, K.; Harvey, J.A.; Tanaka, T. Influence of nutrient deficiency caused by host developmental arrest on the growth and development of a koinobiont parasitoid. *J. Insect Physiol.* **2006**, *52*, 1105–1112. [CrossRef]
69. Becchimanzi, A.; Avolio, M.; Di Lelio, I.; Marinelli, A.; Varricchio, P.; Grimaldi, A.; Eguileor, M.; Pennacchio, F.; Caccia, S. Host regulation by the ectophagous parasitoid wasp *Bracon nigricans*. *J. Insect Physiol.* **2017**, *101*, 73–81. [CrossRef]
70. Lin, X.; Pan, Q.; Tian, H.; Douglas, A.; Liu, T. Bacteria abundance and diversity of different life stages of *Plutella xylostella* (Lepidoptera: Plutellidae), revealed by bacteria culture-dependent and PCR-DGGE methods. *Insect Sci.* **2015**, *22*, 375–385. [CrossRef]

insects

MDPI

Article

Topical Fungal Infection Induces Shifts in the Gut Microbiota Structure of Brown Planthopper, *Nilaparvata lugens* (Homoptera: Delphacidae)

Zhengliang Wang †, Yiqing Cheng †, Yandan Wang and Xiaoping Yu *

Zhejiang Provincial Key Laboratory of Biometrology and Inspection and Quarantine, College of Life Sciences, China Jiliang University, Hangzhou 310018, China; zhengliang.w0234@163.com (Z.W.); chengyiqing11@163.com (Y.C.); wydxcjlu@163.com (Y.W.)

* Correspondence: yuxiaoping19630306@163.com

† These authors contributed equally to this work.

Simple Summary: Fungal entomopathogens are important natural enemies of insect pests and widely applied for biocontrol. Gut microbiota play important roles in mediating insect physiology and behavior. There is growing evidence that alteration of gut microbial communities due to pathological and environmental exposure can have detrimental impacts on host health and pathogen resistance. Here, we investigated the effects of topical infection with *Metarhizium anisopliae* fungus on the gut microbial community structure of the brown planthopper (*Nilaparvata lugens*, BPH), a destructive insect pest of rice. Our results demonstrated dramatic changes of gut bacterial community structure after topical fungal infection in BPH, as indicated by a significant increase in bacterial load, a significant decrease in bacterial community evenness and significant shifts in dominant bacterial abundance at the taxonomic level below the class. The dysbiosis of the gut bacteria might partly be due to the suppression of gut immunity caused by topical fungal infection. Our results highlighted the importance of the gut microbial community in fungal pathogenesis in insects.

Abstract: The brown planthopper (*Nilaparvata lugens*, BPH) is a destructive insect pest posing a serious threat to rice production. The fungal entomopathogen *Metarhizium anisopliae* is a promising alternative that can be used for BPH biocontrol. Recent studies have highlighted the significant involvement of gut microbiota in the insect–fungus interactions. In the presented study, we investigated the effects of topical fungal infection on the gut microbial community structure in BPH. Our results revealed that topical infection with *M. anisopliae* increased the bacterial load and altered the bacterial community structure in the gut of BPH. The relative abundances of the dominant gut bacteria at the order, family and genus level were significantly different between fungus-infected and uninfected groups. At the genus level, the uninfected BPH harbored high proportions of *Pantoea* and *Enterobacter* in the gut, whereas the fungus-infected BPH gut was absolutely dominated by *Acinetobacter*. Moreover, topical fungal infection significantly inhibited the expressions of immune-related genes encoding anti-microbial protein and dual oxidase that were involved in the maintenance of gut microbiota homeostasis, indicating that gut bacteria imbalance might be attributed in part to the suppression of gut immunity caused by fungal pathogen. Our results highlighted the importance of the gut microbial community during interactions between fungal pathogens and insect hosts.

Keywords: brown planthopper; gut microbiota; fungal entomopathogen; gut immunity

Citation: Wang, Z.; Cheng, Y.; Wang, Y.; Yu, X. Topical Fungal Infection Induces Shifts in the Gut Microbiota Structure of Brown Planthopper, *Nilaparvata lugens* (Homoptera: Delphacidae). *Insects* **2022**, *13*, 528. https://doi.org/10.3390/insects13060528

Academic Editors: Hongyu Zhang, Yin Wang and Xiaoxue Li

Received: 28 April 2022
Accepted: 6 June 2022
Published: 8 June 2022

Publisher's Note: MDPI stays neutral with regard to jurisdictional claims in published maps and institutional affiliations.

Copyright: © 2022 by the authors. Licensee MDPI, Basel, Switzerland. This article is an open access article distributed under the terms and conditions of the Creative Commons Attribution (CC BY) license (https://creativecommons.org/licenses/by/4.0/).

1. Introduction

The brown planthopper (BPH), *Nilaparvata lugens* Stål (Hemiptera: Delphacidae), is one of the most destructive piercing-sucking pests on rice (*Oryza sativa* L.). It can cause damages directly through sucking sap from the phloem of rice plants and indirectly via transmitting plant-pathogenic viruses, resulting in substantial yield and economic losses every year in rice producing areas [1]. Chemical control is the primary method for controlling this pest. However, BPH has evolved heavy resistance to a variety of conventional chemical insecticides [2]. Additionally, the long-term and unreasonable using of synthetic chemicals has also caused severe environmental pollution and ecological damage [2]. Thus, environmentally friendly alternative methods for BPH control are urgently required.

Numerous practical research evidenced that an effective alternative approach for BPH control is to make use of fungal entomopathogens [3–7]. Entomopathogenic fungi infect their host insects via attachment of conidia to the host cuticle, then invade into host body by conidia germination and hyphal penetration, and finally proliferate in insect haemocoel and kill the hosts [8]. At present, many fungal strains with high virulence to BPH have been screened based on laboratory and/or filed bioassays, and most of them belonging to *Beauveria*, *Metarhizium* and *Isaria* [5–7]. For example, *Metarhizium anisopliae* CQMa421 has shown high control efficiency against BPH during both the nymphal and adult stages, but without adverse effects on natural enemies [7]. Nevertheless, fungal pesticidal agents suffer the disadvantage of having a relatively slower killing speed when compared with chemical insecticides, which hampered their widespread application [9]. To develop approaches to enhance fungal pesticidal efficacy, a better understanding of the insect–fungus interaction is required.

Fungal infection often triggers the insect innate immune system, including cellular and humoral responses [8]. In the insects–fungi interaction model, immune aspects in insects have been extensively studied in individuals or tissues (hemolymph and fat body) that are traditionally attributed to immune responses [10–12]. A large number of immune-related genes have been found to be up-regulated in the whole BPH body after topical infection with *M. anisopliae* based on comparative transcriptomic analysis [13,14]. Nowadays, accumulating studies have revealed that the gut also plays an important role in shaping insect immunity.

The insect gut is a complex ecosystem consisting of diverse communities of microbes that play important roles in host physiology including nutrition metabolism, immunity modulation and pathogen defense [15]. Recent studies have shown that insect gut microbiota could affect to the pathogenic process and the pesticidal efficiency of insect pathogens [16–21]. For example, gut microbiota of the beet armyworm *Spodoptera exigua* could enhance baculovirus virulence by modulating gut immunity [16]. Similar effects were found for the cotton bollworm *Helicoverpa armigera* [17], the gypsy moth *Lymantria dispar* [18] and the malaria mosquito *Anopheles stephensi* [19] in response to viral, bacterial and fungal challenged, respectively. In contrast, gut bacteria in the honey bee *Apis mellifera* [20] and cockroach *Blattella germanica* [21] could protect their hosts against invading pathogens by up-regulated the host immune response or by producing antimicrobial compounds. To date, the interaction between insect gut microbiota and pathogen infection was mainly based on models of "insect-virus" or "insect-bacteria" [16–18,20]. However, unlike viral and bacterial pathogens, which invade insects through their oral cavity and/or gut, insect fungal pathogens infect insects primarily through the cuticle [8]. Hence, the interactions between insect gut microbiota and fungi might be more complex and fascinating. However, few reports have explored the interplay between fungal infection and insect gut microbial associates.

In this study, we aim to investigate how gut microbitoa in BPH respond to topical infection with a fungal entomopathogen, *M. anisopliae*, which has great potential for BPH biocontrol [3,14]. Building on the literature and preliminary research, we hypothesized that the gut microbial community homeostasis would be disturbed after fungal infection by topical route in BPH. To test this hypothesis, we determined the changes in the composition of gut bacterial community after topical fungal infection by in vitro culture, quantitative polymerase chain reaction (qPCR) and high-throughput 16S rRNA amplicon sequencing. Moreover, the expression patterns of gut-homeostasis-related genes in the gut of BPH during the course of fungal infection were assessed by quantitative real time polymerase chain reaction (qRT-PCR) analysis.

2. Materials and Methods

2.1. Insect and Fungal Entomopathogen

The BPH population was originally collected from rice fields in a paddy field in Yuyao, Zhejiang province of China (121°33 E, 29°99 N) and maintained on the rice variety TN1 in the insectary greenhouse for more than 20 generations under controlled conditions (27 ± 1 °C, 70 ± 10% relative humidity and a 14:10 h light/dark photoperiod). The entomopathogenic fungal strain *M. anisopliae* ARSEF456 (designated as Ma456 herein) was grown on the plates of Potato dextrose agar (PDA) at the regime of 28 °C and 12:12 h (light/dark cycle).

2.2. Topical Fungal Infection and Gut Dissection

Aerial conidia of Ma456 produced on PDA plates were washed with 0.02% Tween 80 solution and adjusted to a final concentration of 1×10^8 conidia/mL using a hemocytometer. BPH nymphs (24 h after molting at the fifth instar) were prepared for topical fungal infection following previous protocol [14]. Briefly, batches of 30–40 nymphs on 3-cm high rice seedlings in uncaged cups were sprayed with 1 mL conidial suspension using a handheld micro sprayer. The amount of conidia deposited onto the nymphs was measured as number of conidia mm^{-2} using microscopic counts of conidia collected onto four glass slips (20 × 20 mm) under each spray. Control nymphs (CK group) were treated with an equal-volume of 0.02% Tween 80. In order to test whether 0.02% Tween 80 has detrimental impacts on the gut microbial community structure, BPH nymphs were also sprayed with 1 mL ddH_2O. All sprayed nymphs were reared in situ in a growth chamber at 25 °C and a 14:10 h light/dark photoperiod and BPH mortality was recorded daily for 10 days. Fresh rice seedlings were supplied every 3 days for their feeding during the period of rearing.

For sampling of BPH gut, samples of 400 surviving nymphs after post-infection of 4 days were collected from the fungal treatment group and the control group, respectively. Prior to gut dissection, all nymphs were surface sterilized by washing them three rinses with 75% ethanol for 1 min each time, followed by three rinses with ddH_2O for 1 min each time. Gut samples were gently dissected by using sterile forceps under a stereomicroscope and then homogenized in 1 mL phosphate-buffered saline (1× PBS) and frozen at −80 °C. All gut samples were divided into four sets for in vitro microbial culture, DNA and RNA extraction, respectively.

2.3. Quantification of Gut Bacteria by CFU Counting Assay

To check the quantity changes of BPH gut microbiota in response to fungal infection, gut samples suspended in 1× PBS were diluted to a suitable concentration (10^{-2}–10^{-5}) and 100 μL aliquots of suspension of each gut sample were spread onto the surface of Luria-Bertani (LB) agar plates, followed by 24–48 h incubation at 37 °C. Then, the colony forming units (CFUs) of gut bacteria were counted and calculated for each gut. At least five replicates for each sample were used for analysis.

2.4. Quantification of Gut Bacteria by 16S rRNA Gene qPCR Assay

Total DNA of BPH gut was extracted using the DNeasy Tissue Kit (Qiagen, Hilden, Germany) according to the manufacturer's protocol. The quality and quantity of the extracted DNA were determined by a Nanodrop 2000 spectrophotometer (Thermo Fisher Scientific, Waltham, MA, USA) and 1% agarose gel electrophoresis. The total gut bacterial load was quantified by qPCR with a pair of universal primers 1114F (5′-CGGCAACGAGCGCAACCC-3′) and 1275R (5′-CCATTGTAGCACGTGTGTAGCC-3′) targeting 16S rRNA gene. All qPCR reactions were carried out in triplicate with SYBR® Premix Ex TaqTM (Takara, Kusatsu, Japan) according to the manufacturer's instructions. The BPH housekeeping 18S ribosomal protein gene (18S) was used as the internal standard.

2.5. DNA Extraction, PCR Amplification and High-Throughput Sequencing

Total gut DNA from Ma456-infected and 0.02% Tween 80-treated nymphs was extracted and qualified in preparation for 16S rRNA gene amplicon sequencing as the method described as above. Bacterial V3-V4 region of the 16S rRNA gene was amplified using specific barcoded primers 338F (5′-barcode-ACTCCTACGGGAGGCAGCAG-3′) and 806R (5′-barcode-GGACTACHVGGGTWTCTAAT-3′). The barcode fragments were used to sort multiple samples in a single sequencing run. PCR reactions were performed in a total volume of 25 μL, containing 2.5 μL 5× FastPfu Buffer, 2 μL dNTPs (2.5 mM), 0.5 μL each primer (10 μM), 0.5 μL FastPfu Polymerase (5 U/μL), 1 μL template DNA (about 50 ng) and 13 μL ddH_2O. The PCR procedures were as follows: an initial denaturation for 3 min at 94 °C, followed by 30 cycles of denaturation for 30 s at 94 °C, annealing for 30 s at 55 °C, elongation for 30 s at 72 °C, and a final extension step for 5 min at 72 °C. PCR reactions were conducted in triplicate for each sample and the PCR products were pooled to minimize the PCR bias. After evaluation by 2% agarose gel electrophoresis, the high-quality amplicons from each sample were adjusted to an equal concentration and subsequently sent for sequencing on an Illumina NovaSeq platform according to the standard protocols at LC-Bio Technology (Hangzhou, China). Each experiment was repeated with three independently isolated DNA samples (biological replicates).

2.6. 16S rRNA Gene Amplicon Sequence Analysis

Paired-end reads were assigned to appropriate samples based on unique barcodes and truncated by cutting off the primer and barcode sequence, and then assembled using FLASH software (Columbia, MD, USA) [22]. Raw sequences were quality-filtered using QIIME version 1.8.0 [23]. Low complexity sequences, sequences with ambiguous bases and sequences with length below 250 bp were discarded. Operational units (OTUs) were clustered with a 97% similarity cut-off using UPARSE version 7.1 [24]. The taxonomic classification of each bacterial OTU was assigned by RDP Classifier (http://rdp.cme.msu.edu/, accessed on 20 February 2022) against the Silva (SSU115) 16S rRNA database using a confidence threshold of 70%. The alpha diversity and beta diversity indices were calculated using QIIME V1.8.0. Alpha diversity analysis is used to analyze complexity of species diversity for a sample, including Chao1 index, Simpson index and Shannon estimator. Beta diversity was applied to evaluate structural variation of bacterial community among samples using the weighted UniFrac distance metric, and visualized using principal coordinate analysis (PCoA). Linear discriminant analysis coupled with effect size measurements (LEfSe) was applied to identify differentially abundant bacterial taxa among groups. Only those taxa that obtained a log linear discriminant analysis (LDA) score > 3.0 and p-value < 0.05 were ultimately considered. The raw data have been deposited into NCBI Sequence Read Archive database under the accession number PRJNA832103.

2.7. qRT-PCR Analysis of Gut-Homeostasis-Related Genes

Total gut RNA from Ma456-infected and 0.02% Tween 80-treated nymphs was extracted using TRIzol® Reagent and treated with DNase I (New England Biolabs, Ipswich, UK) according to the manufacturer's instructions. After evaluation by RNase-free agarose gel electrophoresis and a NanoDrop 2000 spectrophotometer, all RNA samples were reversely transcribed into cDNAs with PrimeScriptTM RT kit (Takara, Kusatsu, Japan) and then assessed for the transcript levels of ten gut-homeostasis-related genes via qRT-PCR with paired primers (Table S1). All qRT-PCR experiments were performed with SYBR® Premix Ex Taq™ (Takara, Kusatsu, Japan) under the following conditions: an initial denaturation for 30 s at 95 °C, followed by 40 cycles of 5 s at 95 °C and 30 s at 60 °C, and a final step for generation of melting curves. The 18S ribosomal protein gene (18S) of BPH was used as the internal standard. The relative expression level of each gene in each group was estimated using the $2^{-\Delta\Delta Ct}$ method [25]. qRT-PCR analysis was conducted in the triplicate assays, each of which contained three technical replicates.

2.8. Statistical Analysis

Statistical analyses were performed using DPS software v7.05 [26]. Differences between the fungal treatment group and control group in bacterial CFU counts, bacterial *16S rRNA* gene quantification, alpha diversity, bacterial taxa abundance at different taxonomic levels and immune-related gene expression level were analyzed by unpaired two-tailed Student's t-test or one way analysis of variance (ANOVA), followed by Tukey's honestly significant difference (HSD) test. Differences were considered significant if p-value < 0.05.

3. Results

3.1. Fungal Infection Caused High Mortality of BPH Nymphs

A laboratory bioassay was conducted to verity the virulence of the fungal conidia to the and to fifth-instar nymphs of BPH. The concentrations of Ma456 conidia deposited onto BPH nymphs were 952 $\pm$ 84 conidia mm^{-2} and showed no statistically significant difference among the bioassay replicates. Cumulative mortality of BPH nymphs during a 10-day observation period after exposure to fungal spray was illustrated in Figure 1. Corrected mortality of 27.7% and 54.2% for sprayed nymphs were observed at 4 and 6 days post infection (dpi), respectively, and reached 69.2% at 10 dpi.

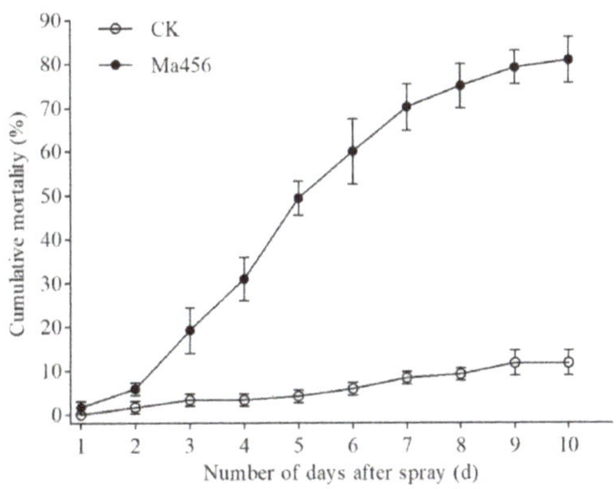

Figure 1. Cumulative mortality rate of BPH fifth-instar nymphs infected by Ma456. Error bars: SD of the mean from three replicates.

3.2. Fungal Infection Enhanced Bacterial Load in BPH Gut

The cultivable bacteria loads in the gut of BPH at 4 dpi were determined by the CFU counting assays. As a result, the load of the culturable bacteria was significantly increased in topical fungal infected groups compared with uninfected controls. The number of bacterial CFUs in the gut of BPH at 4 dpi was 2.15 $\pm$ 0.53 $\times$ 10^4 per gut, which were 2.1-fold higher than that in the 0.02% Tween 80-treated group, in which 1.04 $\pm$ 0.09 $\times$ 10^4 CFUs per gut was detected (Figure 2A). The qPCR result also showed that the total bacterial load in the fungus-infected group was significantly higher (about 3.2-fold) than that in the control groups (Figure 2B). No significant differences in both bacterial CFUs count and relative bacterial 16S rRNA level were observed between 0.02% Tween 80-treated and ddH_2O-treated groups.

3.3. Fungal Infection Decreased Bacterial Community Evenness in BPH Gut

16S rRNA gene amplicon sequencing generated a total number of 494,588 raw reads from six gut samples. After quality filtering and chimera removal, a total of 445,848 clean reads were remained, including 187,059 reads from the fungus-infected group and 258,789 reads from the control group, which then clustered into a total of 192 and 169 bacterial OTUs at a 97% similarity level, respectively (Table 1). The taxonomy of all gut bacterial OTUs was presented in Table S2 (online only). The alpha diversity indices were estimated using three measurements, including Chao1, Simpson and Shannon indices. As a result, there were no significant differences between the fungus-infected BPH and the control BPH in terms of Chao1 index that reflected microbial community richness. However, the fungus-infected group shown a markedly lower Simpson and Shannon diversity indices when compared with the control group ($p < 0.05$), indicating that topical fungal infection decreased the bacterial community evenness (Table 1).

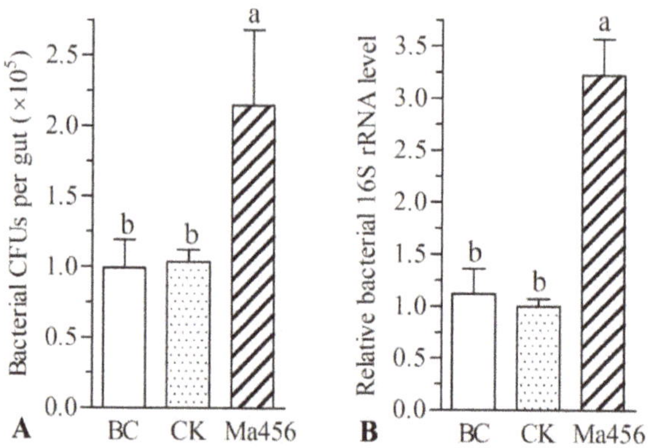

Figure 2. Bacterial CFUs count (**A**) and relative quantification of bacterial 16S rRNA (**B**) in the gut of BPH treated by topical spraying fungal conidia (Ma456), 0.02% Tween 80 (CK) and ddH_2O (BC), respectively. Different letters on the bars of each group denote significant differences among BPH gut samples ($p < 0.05$). Error bars: SD from three repeated assays.

Table 1. Bacterial community alpha-diversity characteristics in the gut of BPH infected and uninfected with fungal pathogen.

Sample	Raw Tags	Valid Tags	OTUs	Shannon †	Simpson †	Chao1 †	Coverage †
CK	219,128	187,059	192	3.03 (±0.32) [a]	0.76 (±0.01) [a]	76.67 (±20.54) [a]	0.999 (±0.001)
Ma456	275,460	258,789	169	1.91 (±0.41) [b]	0.59 (±0.06) [b]	77.11 (±7.52) [a]	1.000 (±0.000)

† Numbers represent mean (±standard error) and different lowercase letters on the same row indicate differences for $p < 0.05$. CK and Ma456 refer to the gut sample from 0.02% Tween 80-treated group and fungus-infected group, respectively.

3.4. *Fungal Infection Altered Bacterial Community Composition in BPH Gut*

Based on the OTU classification, a total of 13 bacterial phyla consisting of 18 classes, 44 orders, 69 families and 111 genera were assigned in the fungus-infected gut samples, while the total bacterial OTUs in the control gut samples were annotated into 11 phyla, 17 classes, 43 orders, 61 families and 101 genera. The bacteria with the relative abundance over 1.00% in at least one group at the levels of phylum, class and order in BPH gut were shown in Figure 3. No significant differences were observed in the relative abundance of bacteria at the level of phylum and class. The most dominant gut bacteria in both BPH groups belonged to the phylum Proteobacteria (Figure 3A) and the class Gammaproteobacteria (Figure 3B), accounting for more than 90% in each group. At the order level, the bacteria from Pseudomonadales were the most predominant in the fungus-infected gut samples. However, the most dominant bacteria in the control gut samples were represented by Enterobacterales (Figure 3C). The relative abundance of Pseudomonadales in the fungus-infected group was 69.16 ± 8.47%, which was significant higher than that in the control group (2.37 ± 1.08%). By contrast, the relative abundance of Enterobacterales in the fungus-infected group (27.30 ± 9.36%) was significant lower than that in the control group (90.98 ± 7.00%). Among the bacterial families, Moraxellaceae was the most dominant family in the fungus-infected gut samples (69.15 ± 8.47%), followed by

Erwiniaceae (15.88 ± 8.73%) and Enterobacteriaceae (11.24 ± 3.61%). However, significant lower abundance of Moraxellaceae (2.19 ± 0.69%), and significant higher abundance of Erwiniaceae (65.43 ± 4.37%) and Enterobacteriaceae (23.85 ± 7.38%) were observed in the control gut samples (Figure 3D). The variations in bacterial community compositions at the genus level were visualized on the heat map of top 15 abundant bacteria (Figure 4). Remarkably, bacterial communities from the fungus-infected gut samples were dominated by members of the genus *Acinetobacter*, with the relative abundance of 68.48 ± 7.32%, which was extremely higher than that in the control gut samples (about 2%). In contrast, the dominant genus of bacteria in the control gut samples was represented by *Pantoea* and *Enterobacter*, with the relative abundance of 64.76 ± 3.50% and 25.21 ± 8.42%, respectively.

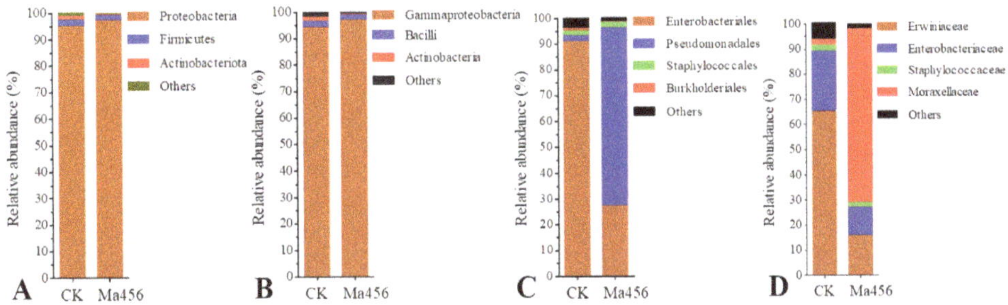

Figure 3. The bacteria with the relative abundance over 1.00% in at least one BPH population at the levels of phylum (**A**), class (**B**), order (**C**) and family (**D**) in the gut of BPH that treated by fungal infection (Ma456) and 0.02% Tween 80 (CK), respectively.

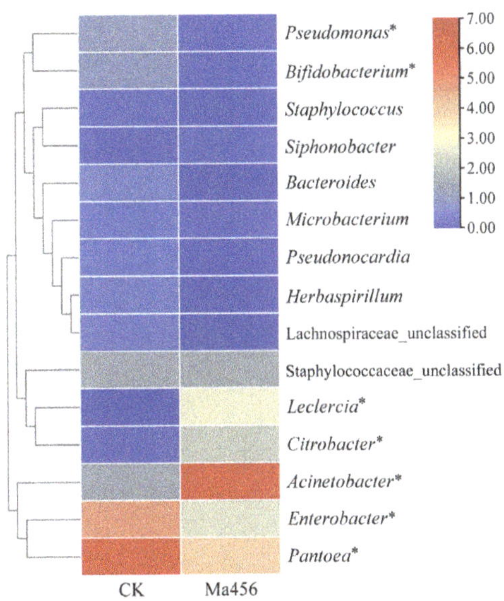

Figure 4. Heatmap of the relative abundance of the top 15 predominant bacterial genera in the gut of BPH that were treated by fungal infection (Ma456) and 0.02% Tween 80 (CK). The color scale represents values of relative abundances (%) normalized by $\log_2$. Zero values were added as 1 and $\log_2$ transformed. Asterisked species differ significantly in the relative abundance between two BPH gut samples ($p < 0.05$).

To more rigorously compare the bacterial community structure between the fungus-infected and control group, a PCoA analysis plot of samples using the weight UniFrac distance metric was performed. As shown in Figure 5, the bacterial communities from the fungus-infected gut samples clustered independently and distinctly from the control gut samples based on the weighted UniFrac PCoA plot, as significant differences were observed between them along the PC1 axis ($p < 0.05$). Based on this analysis the BPH gut bacterial community following exposure to Ma456 was distinct from untreated guts. The observed differences in beta-diversity were directly reflected by the strong shifts in the taxonomic composition of the gut bacterial community in BPH after fungal infection that was described above. For instance, the relative abundance of Moraxellaceae was noticeably higher in the fungus-infected group, whereas Erwiniaceae were significantly enriched in the control group. At the genus level, more than one-third of the bacterial genera had significant differences in the relative abundance between the fungus-infected and control group, such as *Pantoea, Acinetobacter, Enterobacter, Leclercia, Citrobacter* and *Pseudomonas* (Figure 3).

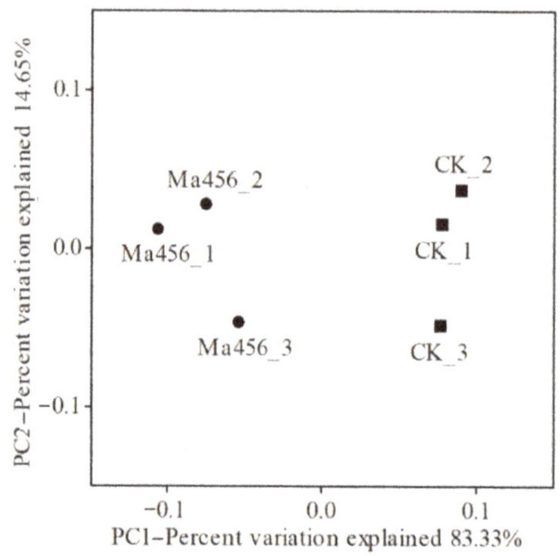

Figure 5. Principal coordinate analysis (PCoA) of beta diversity based on the weighted UniFrac distance metric for gut bacterial communities in the fungus-infected (Ma456) and 0.02% Tween 80-treated (CK) BPH.

LEfSe analysis was also applied to identify gut bacterial taxa that differed significantly in abundance between the fungus-infected and control group. A total of 21 differentially abundant taxa were detected between the two groups, all of which had a log LDA score > 3.0 (Figure 6). For instance, at the genus level, bacteria from the fungus-infected group were enriched with *Acinetobacter* from family Moraxellaceae, *Leclercia* and *Citrobacter* from family Enterobacteriaceae, while the relative abundances of the bacterial taxa from *Pantoea* belonging to family Erwiniaceae and Enterobacter belonging to family Enterobacteriaceae were higher in the control group. The LEfSe analysis was consistent with the results from the comparative analysis of bacterial composition community as presented above.

3.5. Fungal Infection Modulated Expressions of Gut-Homeostasis-Related Genes in BPH

Ten gut-homeostasis-related genes were assessed for the transcript levels in the gut of BPH after topical fungal infection through qRT-PCR. As illustrated in Figure 7, nine gut-homeostasis-related genes were differentially expressed in the fungus-infected gut samples when compared to the control. Among them, two antimicrobial peptide (AMP) encoding genes (*defA* and *defB*) and three immune responsive effector genes encoding i-type lyzysome (*iLys1, iLys2* and *iLys3*) were significantly down-regulated their transcript levels after fungal infection. For instance, the transcript levels of *defA* and *defB* were significantly repressed by 61.4% and 85.3% in the gut of fungal-challenged BPH, respectively. A pattern recognition receptor gene encoding peptidoglycan recognition protein LC (*PGRP-LC*) involved in the immune deficiency (Imd) signal pathway and a dual oxidase (DUOX) encoding gene (*Duox1*) linked to the production of reactive oxygen species (ROS) were also significantly down-regulated their transcript levels in the gut of BPH when suffering fungal infection. However, two immune responsive effector genes encoding i-type lyzysomes (*iLys6* and *iLys7*) were significantly up-regulated their transcript levels in the fungus-infected gut samples relative to the control.

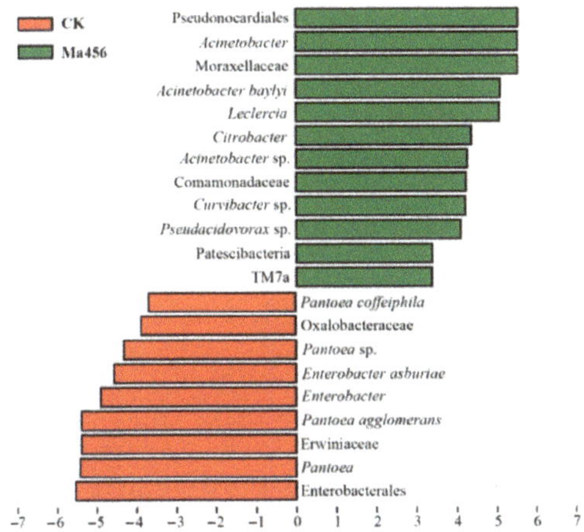

Figure 6. Different structures of gut bacteria between the fungus-infected (Ma456) and 0.02% Tween 80-treated (CK) groups identified by LEfSe analysis with linear discriminant analysis (LDA) scores of 3.0. Differentially abundant taxa are represented by histograms with LDA scores (red, CK; green, Ma456).

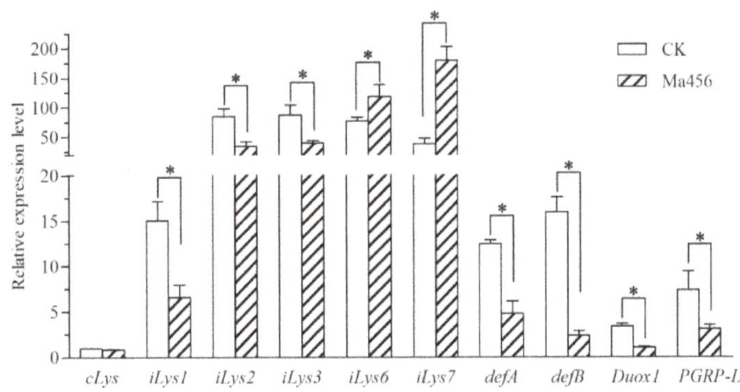

Figure 7. Relative expression levels of ten gut-homeostasis-related genes in the gut of BPH that treated by fungal infection (Ma456) and 0.02% Tween 80 (CK). Asterisk indicates significant difference between two BPH gut samples ($p < 0.05$). Error bars: SD from three repeated assays.

4. Discussion

Insect gut microbiota play vital roles in host ecology and physiology, particular in provision of nutrition necessary that essential for host growth and modulation of host immune defense against their pathogens [10,27,28]. Many factors, including the developmental stage, sex and phylogeny of the host, stressful condition and environmental habitat, have been shown to affect gut microbial community structure in insects [29]. Recently, the effects of insecticide and host rice varieties on the gut bacterial composition of BPH were assessed by high-throughput amplicon sequencing [30–32]. However, the influence of pathogen infection on the gut microbial composition and structure of BPH is still unclear. To fill this gap, we characterized and compared the gut bacterial community in BPH sprayed with and without the conidia suspension of an insect fungal pathogen.

Numerous studies have confirmed that *per os* infection by microbial pathogens (e.g., bacteria and virus) could influence the gut microbial community structure in insects, and the changes of gut microbial composition could in turn to alter host susceptibility to pathogen infection [16–18]. Insect fungal pathogen attack and kill insects primarily by penetrating the host integument and proliferating in the hemocoel cavity by exhausting host nutrients and producing toxins, *per os* infection occurs occasionally or even rarely [8]. In view of this, previous studies focusing on the interactions between fungal pathogens and host insects paid little attention to the status of gut microbiota, especially when the host was a homopteran insect with piercing and sucking mouthparts. Our results revealed that topically fungal infection could also cause a dramatic alteration in gut bacterial community structure in the sap-sucking homopteran rice pests BPH, as indicated by a significant increase in gut bacterial load and a significant decrease in bacterial community evenness. However, the changes in bacterial load and evenness did affect the status of the dominant gut bacterial phylum. According to the bacterial OTUs classification analysis, bacteria affiliated with the phylum Proteobacteria was found to be the most predominant in both fungus-infected and control gut samples, in consistent with the data reported in previous studies on the gut microbiota of diverse insect groups, including data for BPH populations with different virulence levels and insecticide-resistant levels [29–32]. This seems to imply that Proteobacteria are widely present in the gut in insects and play a vital role in host fitness and environmental adaptability.

Although the most dominant gut bacteria at higher taxonomic levels (phylum and class) showed no significant differences in BPH after fungal challenge, the relative abundances of the dominant order, family and genus of gut bacterial community were significantly different. Remarkably, at the genus level, the bacteria from *Pantoea* were the most prevalent in the control gut samples, whereas the fungus-infected gut samples was absolutely dominated by *Acinetobacter* (about 70%). *Acinetobacter* is a bacterial genus commonly found in the insect gut samples, including species of a symbiotic nature which could provide their hosts with essential nutrients [33], and species with insecticidal potential that could serve as a pathogen against their hosts [34]. The shift in dominant bacteria from *Pantoea* to *Acinetobacter* in the gut of BPH highly suggested that *Acinetobacter* might play an important role in the course of fungal infection. Recently, a significant increase in the abundance of *Serratia* and *Erwinia* was also observed in the guts of *Anopheles* mosquitos and *Dendroctonus* beetles after topical fungal infection, respectively, and furthermore, these gut bacteria overgrown in the gut reciprocally could promote the killing speed of fungal pathogen against their host insects [19,35]. Whether *Acinetobacter* have the ability to promote the fungal killing of BPH is still unclear and warrants further investigation.

Accumulating studies have proved that homeostasis in the gut bacterial community is partly determined by gut immunity [36–38]. The Imd signal pathway regulating the production of AMPs and the DUOX-ROS system leading to the production of ROS are considered to be the two major pathways for insects to maintain the homeostasis of gut microbiota [39–41]. In this study, a pattern recognition receptor encoding gene (*PGRP-LC*) and two AMP encoding genes (*defA* and *defB*) involved in the Imd pathway and a DUOX encoding gene (*Duox1*) involved in DUOX-ROS system were observed significantly repressed in the gut of BPH when suffering Ma456 challenge, suggesting that topical fungal infection could cause a level of immune suppression in the gut of BPH, which might subsequently lead to dysbiosis of its gut bacterial community. Recent studies have also shown that the suppression of immune responses by fungal invasion was attributed to the shifts of bacterial community structure in the insect gut [19,42,43]. For instance, the expressions of one DUOX and five AMPs encoding genes in the midgut of *A. stephensi* were significantly down-regulated after topical infection with an insect fungal pathogen

Beauveria bassiana, which subsequently cause a significant increase in gut bacterial load and a significant decrease in bacterial diversity [19]. A similar phenomenon was also observed in the Colorado potato beetle when topically infected by *M. robertsii*, in which a gene regulating the activity of the DUOX was significantly inhibited along with the proliferation of *Serratia* in the gut during fungal infection [42]. Gut microbiota imbalance might also be caused by host physiological reactions under stressful conditions, including pathogen infection and insecticide exposure [44,45]. The micro-ecology of the insect gut might be altered by toxic compounds released by fungal pathogen during infection, such as oosporein secreted by *Beauveria*, which was proved to be effective in changing the gut bacterial community in mosquito via weakening the host DUOX-ROS system [19]. *M. anisopliae* also produce bioactive metabolite destruxins that are toxic to insect hosts [46]. The interaction among fungal destruxin, insect gut immunity and gut microbiota could be the subject of future studies.

In summary, we demonstrated dramatic changes in BPH gut bacterial community structure after topical fungal infection, as expressed by a significant increase in bacterial load, a significant decrease in bacterial community evenness and significant changes in dominant bacterial abundance at the taxonomic level below the class. The suppression of gut immunity might partly account for the gut microbiota imbalance. Our results highlighted the importance of considering the gut microbial community when determined the interactions between fungal pathogen and insect host.

Supplementary Materials: The following supporting information can be downloaded at: https://www.mdpi.com/article/10.3390/insects13060528/s1; Table S1: Specific primer pairs of the gut-homeostasis-related genes for qRT-PCR.; Table S2: Bacterial OTUs in the gut of BPH infected and uninfected by fungal pathogen.

Author Contributions: Z.W. and X.Y., conceived and designed the experiments; Z.W. and Y.C., performed the experiments; Z.W., Y.C. and Y.W., analyzed the data; Z.W., Y.C. and Y.W., wrote the paper; X.Y., reviewed and edited the paper. All authors have read and agreed to the published version of the manuscript.

Funding: This work was financially supported by the Natural Science Foundation of Zhejiang Province (Grant No. LR19C140001), the National Natural Science Foundation of China (Grant no. 31972347; U21A20223) and the Fundamental Research Funds for the Provincial Universities of Zhejiang (Grant No. 2022YW82; 2020YW27).

Institutional Review Board Statement: Not applicable.

Informed Consent Statement: Not applicable.

Data Availability Statement: The data presented in this study are available in the article and Supplementary Materials.

Conflicts of Interest: The authors declare no conflict of interest.

References

1. Cheng, X.; Zhu, L.; He, G. Towards understanding of molecular interactions between rice and the brown planthopper. *Mol. Plant* **2013**, *6*, 621–634. [CrossRef] [PubMed]
2. Jena, K.K.; Kim, S.M. Current status of brown planthopper (BPH) resistance and genetics. *Rice* **2010**, *3*, 161–171. [CrossRef]
3. Jin, S.F.; Feng, M.G.; Ying, S.H.; Mu, W.J.; Chen, J.Q. Evaluation of alternative rice planthopper control by the combined action of oil-formulated *Metarhizium anisopliae* and low-rate buprofezin. *Pest Manag. Sci.* **2011**, *67*, 36–43. [CrossRef] [PubMed]
4. Tang, J.F.; Liu, X.Y.; Ding, Y.C.; Jiang, W.J.; Xie, J.Q. Evaluation of *Metarhizium anisopliae* for rice planthopper control and its synergy with selected insecticides. *Crop Prot.* **2019**, *121*, 132–138. [CrossRef]
5. Zhao, Q.; Ye, L.; Wang, Z.; Li, Y.; Zhang, Y.; Keyhani, N.O.; Huang, Z. Sustainable control of the rice pest, *Nilaparvata lugens*, using the entomopathogenic fungus *Isaria javanica*. *Pest Manag. Sci.* **2021**, *77*, 1452–1464. [CrossRef]
6. Li, M.; Li, S.; Xu, A.; Lin, H.; Chen, D.; Wang, H. Selection of *Beauveria* isolates pathogenic to adults of *Nilaparvata lugens*. *J. Insect Sci.* **2014**, *14*, 1–12. [CrossRef]
7. Peng, G.; Xie, J.; Guo, R. Long-term field evaluation and large-scale application of a *Metarhizium anisopliae* strain for controlling major rice pests. *J. Pest Sci.* **2021**, *94*, 969–980. [CrossRef]

8. Islam, W.; Adnan, M.; Shabbir, A.; Naveed, H.; Abubakar, Y.S.; Qasim, M.; Tayyab, M.; Noman, A.; Nisar, M.S.; Khan, K.A.; et al. Insect-fungal-interactions: A detailed review on entomopathogenic fungi pathogenicity to combat insect pests. *Microb. Pathog.* **2021**, *159*, 105122. [CrossRef]
9. Lu, H.L.; St Leger, R.J. Insect immunity to entomopathogenic fungi. *Adv. Genet.* **2016**, *94*, 251–285.
10. Vertyporokh, L.; Wojda, I. Expression of the insect metalloproteinase inhibitor IMPI in the fat body of *Galleria mellonella* exposed to infection with *Beauveria bassiana*. *Acta Biochim. Pol.* **2017**, *64*, 273–278. [CrossRef]
11. Chakraborty, M.; Mandal, S.M.; Basak, A.; Ghosh, A.K. Identification of a novel humoral antifungal defense molecule in the hemolymph of tasar silkworm *Antheraea mylitta*. *Biochem. Biophys. Res. Commun.* **2019**, *519*, 121–126. [CrossRef] [PubMed]
12. Al Souhail, Q.; Hiromasa, Y.; Rahnamaeian, M.; Giraldo, M.C.; Takahashi, D.; Valent, B.; Vilcinskas, A.; Kanost, M.R. Characterization and regulation of expression of an antifungal peptide from hemolymph of an insect, *Manduca sexta*. *Dev. Comp. Immunol.* **2016**, *61*, 258–268. [CrossRef] [PubMed]
13. Peng, Y.; Tang, J.; Xie, J. Transcriptomic analysis of the brown planthopper, *Nilaparvata lugens*, at different stages after *Metarhizium anisopliae* challenge. *Insects* **2020**, *11*, 139. [CrossRef] [PubMed]
14. Wang, Z.L.; Pan, H.B.; Li, M.Y.; Wu, W.; Yu, X.P. Comprehensive insights into host-pathogen interaction between brown planthopper and a fungal entomopathogen by dual RNA sequencing. *Pest Manag. Sci.* **2021**, *77*, 4903–4914. [CrossRef]
15. Schmidt, K.; Engel, P. Mechanisms underlying gut microbiota-host interactions in insects. *J. Exp. Biol.* **2021**, *224*, jeb207696. [CrossRef]
16. Jakubowska, A.K.; Vogel, H.; Herrero, S. Increase in gut microbiota after immune suppression in baculovirus-infected larvae. *PLoS Pathog.* **2013**, *9*, e1003379. [CrossRef]
17. Yuan, C.; Xing, L.; Wang, M.; Hu, Z.; Zou, Z. Microbiota modulates gut immunity and promotes baculovirus infection in *Helicoverpa armigera*. *Insect Sci.* **2021**, *28*, 1766–1779. [CrossRef]
18. Broderick, N.A.; Robinson, C.J.; McMahon, M.D.; Holt, J.; Handelsman, J.; Raffa, K.F. Contributions of gut bacteria to *Bacillus thuringiensis*-induced mortality vary across a range of Lepidoptera. *BMC Biol.* **2009**, *7*, 11. [CrossRef]
19. Wei, G.; Lai, Y.; Wang, G.; Chen, H.; Li, F.; Wang, S. Insect pathogenic fungus interacts with the gut microbiota to accelerate mosquito mortality. *Proc. Natl. Acad. Sci. USA* **2017**, *114*, 5994–5999. [CrossRef]
20. Steele, M.I.; Motta, E.V.S.; Gattu, T.; Martinez, D.; Moran, N.A. The gut microbiota protects bees from invasion by a bacterial pathogen. *Microbiol. Spectr.* **2021**, *9*, e0039421. [CrossRef]
21. Zhang, F.; Sun, X.X.; Zhang, X.C.; Zhang, S.; Lu, J.; Xia, Y.M.; Huang, Y.H.; Wang, X.J. The interactions between gut microbiota and entomopathogenic fungi: A potential approach for biological control of *Blattella germanica* (L.). *Pest Manag. Sci.* **2018**, *74*, 438–447. [CrossRef] [PubMed]
22. Magoč, T.; Salzberg, S.L. FLASH: Fast length adjustment of short reads to improve genome assemblies. *Bioinformatics* **2011**, *27*, 2957–2963. [CrossRef] [PubMed]
23. Caporaso, J.G.; Kuczynski, J.; Stombaugh, J.; Bittinger, K.; Knight, R. QIIME allows analysis of high-throughput community sequencing data. *Nat. Methods* **2010**, *7*, 335–336. [CrossRef]
24. Edgar, R.C. UPARSE: Highly accurate OTU sequences from microbial amplicon reads. *Nat. Methods* **2013**, *10*, 996–998. [CrossRef] [PubMed]
25. Livak, K.J.; Schmittgen, T.D. Analysis of relative gene expression data using real-time quantitative PCR and the 2(-Delta Delta C(T)) method. *Methods* **2011**, *25*, 402–408. [CrossRef]
26. Tang, Q.Y.; Zhang, C.X. Data processing system (DPS) software with experimental design, statistical analysis and data mining developed for use in entomological research. *Insect Sci.* **2013**, *20*, 254–260. [CrossRef] [PubMed]
27. Liberti, J.; Engel, P. The gut microbiota-brain axis of insects. *Curr. Opin. Insect Sci.* **2020**, *39*, 6–13. [CrossRef]
28. Ramirez, J.L.; Souza-Neto, J.; Torres, C.R.; Rovira, J.; Ortiz, A.; Pascale, J.M.; Dimopoulos, G. Reciprocal tripartite interactions between the *Aedes aegypti* midgut microbiota, innate immune system and dengue virus influences vector competence. *PLoS Negl. Trop. Dis.* **2012**, *6*, e1561. [CrossRef]
29. Yun, J.H.; Roh, S.W.; Whon, T.W.; Jung, M.J.; Kim, M.S.; Park, D.S.; Yoon, C.; Nam, Y.D.; Kim, Y.J.; Choi, J.H.; et al. Insect gut bacterial diversity determined by environmental habitat, diet, developmental stage, and phylogeny of host. *Appl. Environ. Microbiol.* **2014**, *80*, 5254–5264. [CrossRef]
30. Vijayakumar, M.M.; More, R.P.; Rangasamy, A.; Gandhi, G.R.; Muthugounder, M.; Thiruvengadam, V.; Samaddar, S.; Jalali, S.K.; Sa, T. Gut Bacterial diversity of insecticide-susceptible and -resistant nymphs of the brown planthopper *Nilaparvata lugens* Stål (Hemiptera: Delphacidae) and elucidation of their putative functional roles. *J. Microbiol. Biotechnol.* **2018**, *28*, 976–986. [CrossRef]
31. Zhang, Y.; Tang, T.; Li, W.; Cai, T.; Li, J.; Wan, H. Functional profiling of the gut microbiomes in two different populations of the brown planthopper, *Nilaparvata lugens*. *J. Asia-Pac. Entomol.* **2018**, *21*, 1309–1314. [CrossRef]
32. Wang, Z.L.; Pan, H.B.; Wu, W.; Li, M.Y.; Yu, X.P. The gut bacterial flora associated with brown planthopper is affected by host rice varieties. *Arch. Microbiol.* **2021**, *203*, 325–333. [CrossRef] [PubMed]
33. Briones-Roblero, C.I.; Rodríguez-Díaz, R.; Santiago-Cruz, J.A.; Zúñiga, G.; Rivera-Orduña, F.N. Degradation capacities of bacteria and yeasts isolated from the gut of *Dendroctonus rhizophagus* (Curculionidae: Scolytinae). *Folia Microbiol.* **2017**, *62*, 1–9. [CrossRef] [PubMed]
34. Eski, A.; Demir, İ.; Güllü, M.; Demirbağ, Z. Biodiversity and pathogenicity of bacteria associated with the gut microbiota of beet armyworm, *Spodoptera exigua* Hübner (Lepidoptera: Noctuidae). *Microb. Pathog.* **2018**, *121*, 350–358. [CrossRef]

35. Xu, L.; Deng, J.; Zhou, F.; Cheng, C.; Lu, M. Gut microbiota in an invasive bark beetle infected by a pathogenic fungus accelerates beetle mortality. *J. Pest Sci.* **2019**, *92*, 343–351. [CrossRef]
36. You, H.; Lee, W.J.; Lee, W.J. Homeostasis between gut-associated microorganisms and the immune system in *Drosophila*. *Curr. Opin. Immunol.* **2014**, *30*, 48–53. [CrossRef]
37. Pang, X.; Xiao, X.; Liu, Y.; Zhang, R.; Liu, J.; Liu, Q.; Wang, P.; Cheng, G. Mosquito C-type lectins maintain gut microbiome homeostasis. *Nat. Microbiol.* **2016**, *1*, 16023. [CrossRef]
38. Bai, S.; Yao, Z.; Raza, M.F.; Cai, Z.; Zhang, H. Regulatory mechanisms of microbial homeostasis in insect gut. *Insect Sci.* **2021**, *28*, 286–301. [CrossRef]
39. Xiao, R.; Wang, X.; Xie, E.; Ji, T.; Li, X.; Muhammad, A.; Yin, X.; Hou, Y.; Shi, Z. An IMD-like pathway mediates the intestinal immunity to modulate the homeostasis of gut microbiota in *Rhynchophorus ferrugineus* Olivier (Coleoptera: Dryophthoridae). *Dev. Comp. Immunol.* **2019**, *97*, 20–27. [CrossRef]
40. Jang, S.; Mergaert, P.; Ohbayashi, T.; Ishigami, K.; Shigenobu, S.; Itoh, H.; Kikuchi, Y. Dual oxidase enables insect gut symbiosis by mediating respiratory network formation. *Proc. Natl. Acad. Sci. USA* **2021**, *118*, e2020922118. [CrossRef]
41. Ha, E.M.; Lee, K.A.; Seo, Y.Y.; Kim, S.H.; Lim, J.H.; Oh, B.H.; Kim, J.; Lee, W.J. Coordination of multiple dual oxidase-regulatory pathways in responses to commensal and infectious microbes in *Drosophila* gut. *Nat. Immunol.* **2009**, *10*, 949–957. [CrossRef] [PubMed]
42. Kryukov, V.Y.; Rotskaya, U.; Yaroslavtseva, O.; Polenogova, O.; Kryukova, N.; Akhanaev, Y.; Krivopalov, A.; Alikina, T.; Vorontsova, Y.L.; Slepneva, I.; et al. Fungus *Metarhizium robertsii* and neurotoxic insecticide affect gut immunity and microbiota in Colorado potato beetles. *Sci. Rep.* **2021**, *11*, 1299. [CrossRef] [PubMed]
43. Ramirez, J.L.; Dunlap, C.A.; Muturi, E.J.; Barletta, A.B.F.; Rooney, A.P. Entomopathogenic fungal infection leads to temporospatial modulation of the mosquito immune system. *PLoS Negl. Trop. Dis.* **2018**, *12*, e0006433. [CrossRef] [PubMed]
44. Li, F.; Li, M.; Mao, T.; Wang, H.; Chen, J.; Lu, Z.; Qu, J.; Fang, Y.; Gu, Z.; Li, B. Effects of phoxim exposure on gut microbial composition in the silkworm, *Bombyx mori*. *Ecotoxicol. Environ. Saf.* **2020**, *189*, 110011. [CrossRef] [PubMed]
45. Wu, C.Y.; Meng, J.; Merchant, A.; Zhang, Y.X.; Li, M.W.; Zhou, X.G.; Wang, Q. Microbial response to fungal infection in a fungus-growing termite, *Odontotermes formosanus* (Shiraki). *Front. Microbiol.* **2021**, *12*, 723508. [CrossRef]
46. Ríos-Moreno, A.; Garrido-Jurado, I.; Resquín-Romero, G.; Arroyo-Manzanares, N.; Arce, L.; Quesada-Moraga, E. Destruxin A production by *Metarhizium brunneum* strains during transient endophytic colonisation of *Solanum tuberosum*. *Biocontrol Sci. Technol.* **2016**, *26*, 1574–1585. [CrossRef]

insects

MDPI

Article

Observation of the Antimicrobial Activities of Two Actinomycetes in the Harvester Ant *Messor orientalis*

Yiyang Wu [1,2], Yaxuan Liu [3], Jinyong Yu [4], Yijuan Xu [1,*] and Siqi Chen [1,*]

1 Guangdong Laboratory for Lingnan Modern Agriculture, Red Imported Fire Ant Research Center, South China Agricultural University, Guangzhou 510642, China; yiyang_hongling@163.com
2 Sendelta International Academy, Shenzhen 518000, China
3 Department of Material Science and Engineering, College of Engineering, Carnegie Mellon University, Pittsburgh, PA 15213-2683, USA; yaxuanl2@andrew.cmu.edu
4 College of Agronomy and Biotechnology, Hebei Normal University of Science & Technology, Qinhuangdao 066600, China; yujing_75211@163.com
* Correspondence: xuyijuan@scau.edu.cn (Y.X.); chensq@stu.scau.edu.cn (S.C.)

Simple Summary: Observations in the animal room have shown that the seeds stored by harvester ants, although in a damp environment, are less likely to mold. It was hypothesized that harvester ants may use actinomycetes to protect their seed stores, given that leafcutter ants use actinomycetes as producers of defensive substances. Two actinomycetes were isolated from the harvester ant *Messor orientalis*. The fermentation broth of the actinomycetes showed significant inhibitory effects on the three indicator fungi. Coculture experiments supported the observed inhibitory effects. The antifungal activities of actinomycetes in harvester ants were revealed. This research provides a significant theoretical reference for the abovementioned hypothesis and for the potential agricultural applications of these actinomycetes for multiple crops.

Abstract: Observations have shown that seeds collected by harvester ants are less likely to mold. Based on evolutionary analysis and other research, it was hypothesized that harvester ants could apply actinomycetes to protect seeds, similar to the protection of mutualistic fungi by leafcutter ants. Two actinomycetes were successfully isolated from the harvester ant *Messor orientalis*. The taxonomic status of the actinomycetes was determined by 16S rRNA sequence analysis and biochemical experimental observations. Their inhibitory effects on plant pathogens were measured. One of the bacteria was identified as *Brachybacterium phenoliresistens* and denoted as *B. phenoliresistens MO*. The other belonged to the genus *Microbacterium*. It was named *Microbacterium* sp. Growth rate determination and coculture experiments were performed to explore the inhibitory effect of actinomycetes on indicator plant pathogens. The inhibition rates of the actinomycetes toward *Peronophythora litchii* and *Rhizoctonia solani* were 100% in media containing 30% or more fermentation broth, and they also showed an inhibitory effect on *Colletotrichum siamense*. The coculture experiment supported this result by showing that the growth of *P. litchii* and *R. solani* was inhibited in the presence of actinomycetes. Therefore, the results of this study show the agricultural application potential of these bacteria and may provide a reference for research on the symbiosis of harvester ants with actinomycetes.

Keywords: *Messor orientalis*; harvester ants; actinomycetes; plant pathogens; fungicide

Citation: Wu, Y.; Liu, Y.; Yu, J.; Xu, Y.; Chen, S. Observation of the Antimicrobial Activities of Two Actinomycetes in the Harvester Ant *Messor orientalis*. *Insects* **2022**, *13*, 691. https://doi.org/10.3390/insects13080691

Academic Editor: Brian T. Forschler

Received: 2 July 2022
Accepted: 28 July 2022
Published: 31 July 2022

Publisher's Note: MDPI stays neutral with regard to jurisdictional claims in published maps and institutional affiliations.

Copyright: © 2022 by the authors. Licensee MDPI, Basel, Switzerland. This article is an open access article distributed under the terms and conditions of the Creative Commons Attribution (CC BY) license (https://creativecommons.org/licenses/by/4.0/).

1. Introduction

Symbiotic microbes exist in a vast majority of insects [1], with actinomycetes accounting for a large proportion of the microbes found in insects [2]. Actinomycetes can help insects adapt to their habitats and resist natural enemies, and they even play a dominant role in the food digestion process [3–5]. Various metabolites of symbiotic actinomycetes derived from insects have shown bacteriostatic activity. Lu et al. isolated a wide range of metabolites from the actinomycete *Streptomyces violaceoruber BYC-01*, which showed inhibitory effects on fungi obtained from termite nests. A single compound, fogacin, was extracted from the fermentation broth of this strain via distillation and ethyl acetate extraction [6–8]. Insect-derived *Streptomyces* species exhibit high inhibitory activity [9]. *Streptomyces* species symbiotically associated with *Dendroctonus frontalis* produce the secondary metabolites frontalamides and mycangimycin [9–11]. Mycangimycin inhibits the beetle's antagonistic fungus *Ophiostoma minus*, while frontalamides have general antifungal activity [9–11]. Sceliphrolactam, an antifungal compound isolated from *Streptomyces*, was found to be associated with a mud dauber wasp [9,12]. These results are promising, and there is a wide variety of insects in the world, so research on insect-associated microorganisms has high application potential, and several abundant sources of active metabolites have yet to be explored. Ants play an indispensable regulatory role in the ecosystem [13]. Studies have shown that leafcutter ants place parasitic fungal spores near symbiotic actinomycetes, *e.g.*, *Nocardiopsis*, until these spores lose their infectivity [14]. The coexistence of *Streptomyces* sp. with strong bacteriostatic properties has been reported in invasive fire ants (*Solenopsis invicta Buren*) [15]. These studies showed that actinomycetes in insects may provide abundant resources for the development of antimicrobial agents. These symbioses are best exemplified in fungus-growing ants [5,8,16], carpenter ant [17–19], solitary digger wasps [12], and southern pine beetles [11].

A study by Tarsa et al. revealed a negative correlation between the occurrence of seed-collecting ants and that of plant pathogens [20]. Harvester ants, i.e., *Messor Forel*, collect and store plant seeds in their nests, which may affect microbial composition [21]. Observations in the animal room have shown that the seeds stored by harvester ants are less likely to mold, but the underlying mechanism remains unclear. Based on analyses from previous studies, actinomycetes in harvester ants may play an important role in this phenomenon.

According to Kang, 10–15% of agricultural production in the world is lost due to improper storage and diseases, among other reasons [22], and 70–80% of the total loss is attributed to plant pathogenic fungi. Thus, to mitigate the loss of grain to plant pathogenic fungi, effective fungicides must be developed. The recent development of fungicides involves the use of plant extracts and the isolation of new compounds from microbial metabolites [23]. Actinomycetes are highly applicable in biological control because their metabolites possess strong bacteriostatic properties [24,25].

In this study, actinomycetes were isolated from *Messor orientalis*, and two actinomycete species belonging to the genera *Microbacterium* and *Brachybacterium* were analyzed. The antimicrobial activity of symbiotic actinomycetes against fungi such as *Colletotrichum siamense* was assayed. The results obtained in this work may provide a scientific reference for the development of new fungicides and aid future research on seed protection mechanisms.

2. Materials and Methods

2.1. Collection of Ant Samples and Indicator Fungi

The ant samples were collected by local collectors in Wujiaqu, Xinjiang, China, in 2020. The ants were kept in test tubes in a storage box, which was placed in an animal culture room at a constant temperature of 27 °C. The ants were morphologically identified as *M. orientalis*.

Three plant pathogens, i.e., *C. siamense* (collection code: CSGD18001), *Peronophythora litchii* (collection code: PLGD18001), and *Rhizoctonia solani* (collection code: RSGD18001), were kindly provided by the Plant Fungi Laboratory, South China Agricultural University.

2.2. Isolation of Actinomycetes

The collected ants were provided with water and *Phalarit canariensis* seeds as food every day. To avoid the influence of disturbance during collection on the community of symbiotic actinomycetes, we let the ant colony stabilize indoors for two weeks before the isolation of actinomycetes. Twenty-five harvester ants from 4 different colonies (6–7 ants per colony) were washed with sterilized water and 70% alcohol. Then, the ants in each group were rinsed twice with sterilized water. The ants were ground with 1 mL water to obtain a liquid ground-ant sample.

Dilutions of the liquid ground-ant sample (10^{-2}, 10^{-3}, and 10^{-4}) were plated on sterilized Gauze's No. 1 agar medium (every 150 mL was amended with 25 mg of cyclohex-imide and 8 mg of nalidixic acid). Each dilution was repeated 3 times. Then, the samples were sealed and kept at 28 °C for 7 days. The actinomycete colonies obtained were then inoculated on sterilized Gauze's No. 1 medium and cultured at 28 °C for 7 days.

2.3. 16S rRNA Gene Sequencing and Biochemical Identification

A single colony of each actinomycete was used to extract the pure genomic DNA using the Bacterial DNA Extraction Kit (Tiangen Biotech (Beijing) Co., Ltd.) according to the manufacturer's instructions. Primers 27F (5′-AGAGTTTGATCCTGGCTCAG-3′) and 1492R (5′-GGTTACCTTGTTACGACT-3′) were used as universal primers for bacterial 16S rRNA amplification. The total PCR volume was 25 μL, including 12.5 μL of 2×Taq PCR Mastermix (Tiangen Biotech (Beijing) Co., Ltd., Beijing, China), 1 μL of each primer (10 μM), 1.5 μL of DNA template, and 9 μL of ddH_2O. The PCR protocol is shown in Table 1.

Table 1. Actinomycete 16S rRNA amplification PCR protocol.

Step	Reaction Temperature (°C)	Reaction Time (min)
Initialization	94	15
Denaturing	94	0.5 [a]
Annealing	55	0.5 [a]
Elongation	72	1 [a]
Stop	72	10

[a] Step was repeated 30 times.

The PCR products were sent to RuiBiotech for Sanger sequencing. Reads greater than 1400 bp in length were used for the database analysis. The sequencing results were compared using NCBI nucleotide BLAST (https://blast.ncbi.nlm.nih.gov/Blast.cgi, accessed on 20 July 2022). Then, 16S rRNA sequences with the highest similarity to those of the isolated strains and typical strains in the same genus were obtained for phylogenetic analysis. The phylogenetic trees evaluated with 1000 bootstrap replications were inferred using the maximum likelihood method based on the Tamura 3-parameter model in MEGA7 (GenBank accession number: OM665406; OM665407).

Inositol, maltose, dextrose, rhamnose, sucrose, Neisser-fructose, hydrogen sulfide production, xylose, mannitol, and raffinose identification tubes purchased from Huancai Microbial Technology were used to test the substrate utilization of the isolated strains. A total of 50 μL of activated broth was added to the identification tubes and then cultured at 37 °C for 24 h. Substrate utilization data for other species in similar genera were obtained for comparison with the data for the isolated strains. Gram staining of the colonies was conducted and then observed microscopically.

2.4. Bioassay of the Fungal Inhibition Effect

A single colony of actinomycetes was activated in liquid BHI medium (37 °C ± 1 °C, 160 r/min) for 24 h. Then, 1 mL of the activated broth was added to a 250 mL flask that contained 100 mL of soybean powder fermentation broth. The mixture was cultured in a shaker (28 °C ± 1 °C, 160 r/min) for 7 days. The fermentation broth was centrifuged (4 °C, 8000 r/min, 20 min) and filtered using a 0.22 μm filter membrane to obtain the aseptic filtrate. Various amounts of filtrate (1 mL, 1.5 mL, and 2 mL) were added to sterilized PDA medium (4 mL, 3.5 mL, and 3 mL) on a plate with a diameter of 5.5 cm. Then, the medium was allowed to cool to 55 °C. The control PDA plate was supplemented with various amounts of sterile water (1 mL, 1.5 mL, and 2 mL). Plant pathogenic fungal plugs with a diameter of 0.5 cm were placed on PDA medium. For each actinomycete sample, the experiment was repeated three times on media containing each fermentation broth filtrate dilution for each fungus. The average diameters of the fungal colonies in the experimental and control groups were recorded. The bacteriostatic rate of the actinomycete against the three-indicator plant pathogenic fungi was calculated according to the following equation:

$$\text{Inhibition rate\%} = \frac{(\text{ADc} - \text{D}) - (\text{ADt} - \text{D})}{(\text{ADc} - \text{D})} \times 100\%$$

where ADc represents the average colony diameter in the control group, ADt represents the average colony diameter in the treatment group, and D represents the diameter of the fungal plugs.

The inhibitory effect on phytopathogens was shown directly by the confrontation culture method. Circular plugs of fungi (diameter = 0.5 cm) were placed in one-quarter of the plates after the actinobacterial plugs were placed in the other quarter of the plates for 7 days at 28 °C.

The Kruskal–Wallis (KW) nonparametric analysis of variance was used to compare the different treatments. The Mann–Whitney U test for multiple comparisons among the different groups was used if the results of the Kruskal–Wallis test showed significant differences at a significance level of 0.05.

3. Results

3.1. Identification of Actinomycetes

Two strains of bacteria, A and B, were identified in *M. orientalis* based on morphological observation and 16S rRNA sequencing. Figure 1 shows the results obtained from the 1.5% agarose gel electrophoresis of the PCR products. Clear bands corresponding to a length of 1500 bp were observed.

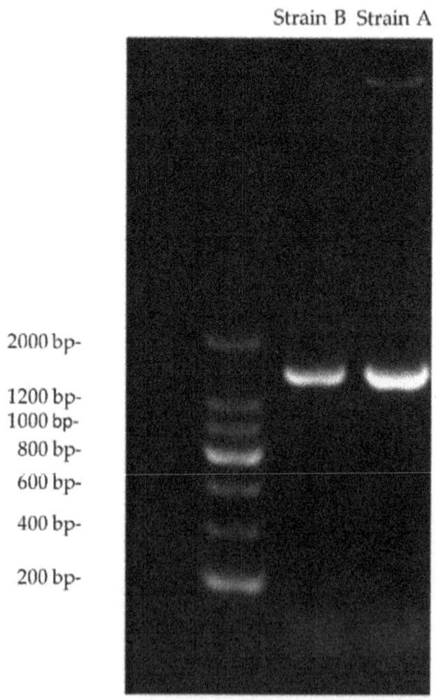

Figure 1. PCR amplification and gel electrophoresis of strains A and B.

The 16S rRNA sequences (GenBank accession number: OM665406; OM665407) of the two bacteria were compared using NCBI nucleotide BLAST. Based on the results, these two bacteria were identified as actinomycetes belonging to the genera *Microbacterium* and *Brachybacterium*. The similarity between strain A and *Brachybacterium phenoliresistens* was 99.86%. The similarity between strain B and *Microbacterium barkeri* was 99.34%. The physiological and biochemical characteristics of strains A and B are shown in Tables 2 and 3, respectively. The results indicated that these two actinomycetes were successfully identified. However, it is notable that the rhamnose utilization capacity of strain B differed from that of *M. barkeri*.

Table 2. Physiological and biochemical characteristics of strain A. Strains: 1, *B. phenoliresistens* LMG 23707^T; 2, *B. saceli* DSM 14566^T; 3, *B. alimentarium* CCM 4520^T; 4, *B. freconis* DSM 14564^T; 5, *B. paraconglomeratum* DSM 46361^T; 6, *B. faecium* CCM 4372^T. Abbreviations: +, positive; (+), weakly positive; −, negative; ND, not determined.

Characteristic	Strain A	1	2	3	4	5	6
H_2S production	−	−	−	+	+	+	−
Acid production from:							
D-fructose	+	ND	+	−	+	+	−
Maltose	+	+	+	−	+	+	(+)
D-mannose	+	ND	(+)	(+)	+	+	−
L-rhamnose	+	+	(+)	(+)	+	+	−
Sucrose	+	+	−	+	(+)	−	−
D-xylose	+	+	−	−	−	+	−
Galactose	+	ND	+	+	+	+	+

Table 3. Physiological and biochemical characteristics of strain B. Strains: 1, *M. barkeri* DSM 20145^T; 2, *M. chocolatum* IFO 3758^T; 3, *M. hominis* IFO 15708^T; 4, *M. thalassium* IFO 16060^T, IFO 16061; 5, *M. halophilum* IFO 16062^T; 6, *M. laevaniformans* IFO 15709^T. Abbreviations: +, positive; −, negative; ND, not determined.

Characteristic	Strain B	1	2	3	4	5	6
H_2S production	−	ND	+	+	−	−	+
Utilization of:							
Maltose	+	ND	+	+	+	+	+
D-mannose	−	ND	+	+	+	+	+
Acid production from:							
L-rhamnose	−	+	−	−	−	+	−
Sucrose	+	ND	+	+	+	−	−
D-xylose	−	ND	−	−	−	+	−
Galactose	−	+	−	+	−	−	+

Figure 2 shows the phylogenetic tree that was constructed based on 20 known strains of *Brachybacterium* and strain A (shown as *Brachybacterium* sp.). As shown in the phylogenetic tree, *B. phenoliresistens* and strain A were closely related, with high repeatability. Figure 3 shows the phylogenetic tree that was constructed based on 11 strains with the highest similarity to strain B (shown as *Microbacterium* sp.), according to BLAST. The findings showed that strain B and *M. barkeri* were closely related.

Based on the analysis, strain A was preliminarily considered to be a *Brachybacterium* strain, probably a strain of *B. phenoliresistens*. Hence, strain A is denoted *Brachybacterium* sp. *MO*. Strain B belongs to the *Microbacterium* genus. However, strain B has yet to be identified by multiphase classification and identification, so it is denoted as *Microbacterium* sp.

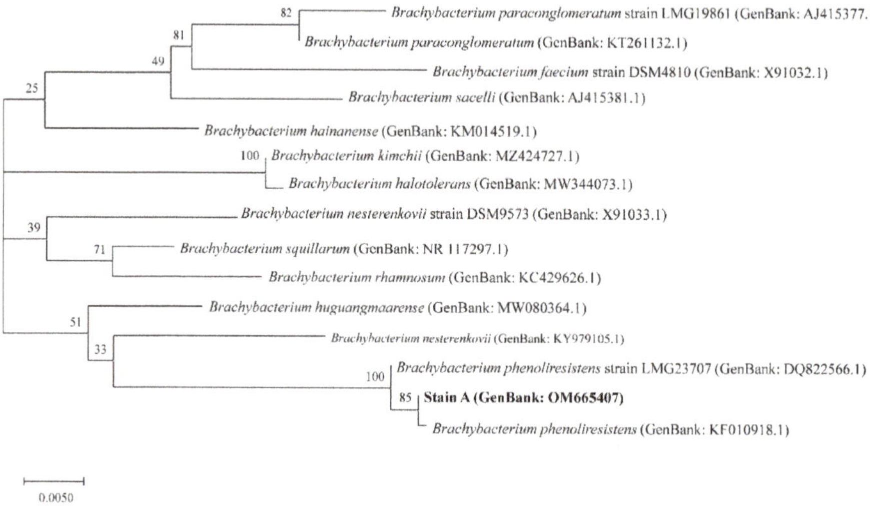

Figure 2. Molecular phylogenetic analysis of strain A (*Brachybacterium* sp.) using the maximum likelihood method.

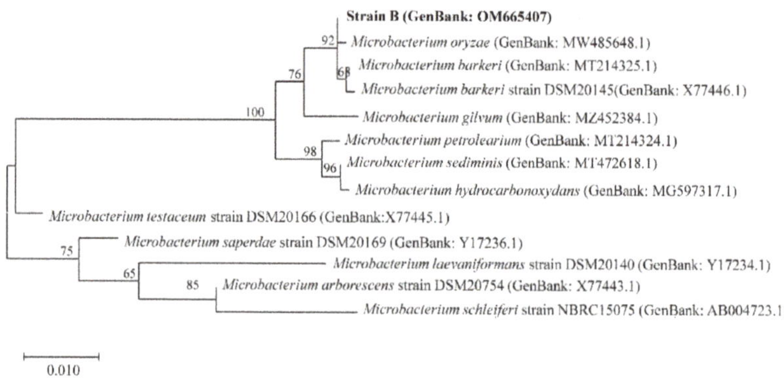

Figure 3. Molecular phylogenetic analysis of strain B (*Microbacterium* sp.) using the maximum likelihood method.

3.2. *Study of Fungal Inhibitory Activity*

3.2.1. Fungal Inhibitory Activity of the Fermentation Broth

The inhibitory effects of *B. phenoliresistens MO* and *Microbacterium* sp. on three types of plant pathogenic fungi with different fermentation broth concentrations are shown in Figures 4 and 5, respectively. Significant inhibitory effects were exhibited by these two actinomycete fermentation broths on the three plant pathogens. As the fermentation broth filtrate concentration increased, the inhibitory effect of the bacteria on plant pathogenic fungi increased. As shown in Figures 4 and 5, the inhibitory rates exhibited by these two actinomycetes on *R. solani* and *P. litchii* were both 100%. However, the inhibitory effects of both actinomycetes on *C. siamense* were weaker. To illustrate these results visually, digital photographs of the pathogenic fungi on the fermentation broth filtrate plate are presented in Figures S1 (*C. siamense*), S2 (*R. solani*), and S3 (*P. litchii*) in the Supporting material. The growth of these plant pathogenic fungi on the fermentation filtrate plate was inhibited.

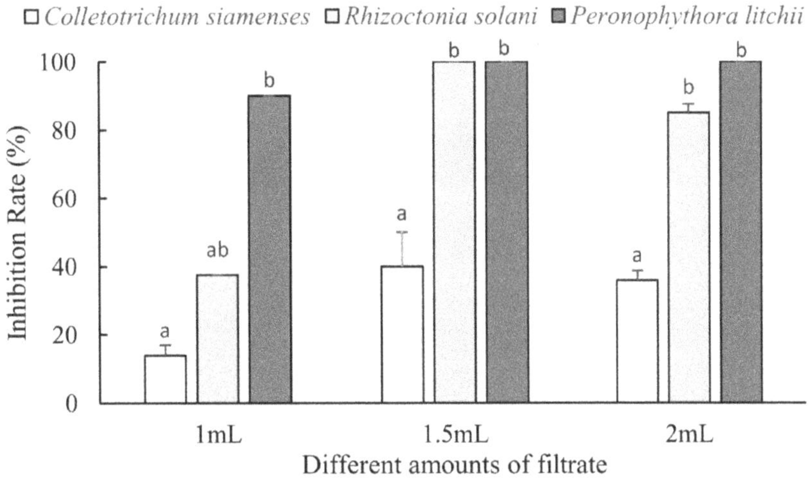

Figure 4. Inhibitory effect of *Brachybacterium phenoliresistens* against three plant pathogens with different amounts of filtrate. For each amount of filtrate, bars with the same letter are not significantly different ($p > 0.05$, Mann–Whitney U test).

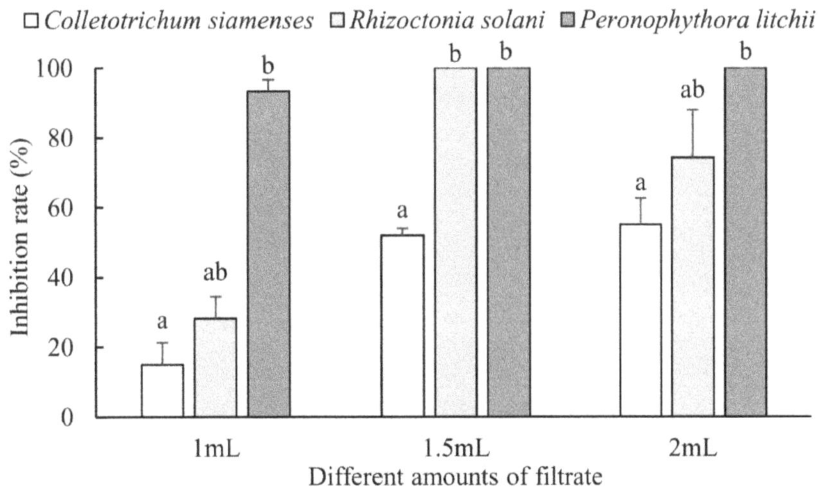

Figure 5. Inhibitory effect of *Microbacterium* sp. against three plant pathogens with different amounts of filtrate. For each amount of filtrate, bars with the same letter are not significantly different ($p > 0.05$, Mann–Whitney U test).

3.2.2. Inhibitory Activity of Actinomycete Colonies

On the bacterial plate, the growth of the pathogen was almost unaffected by live *B. phenoliresistens MO* and *Microbacterium* sp., which was consistent with the results obtained from the fermentation broth inhibition experiment. *P. litchii* growth was significantly inhibited by strains A and B. Mycelial growth around live strain B was relatively unaffected, while germination could not occur around live strain A, as shown in Figure 6.

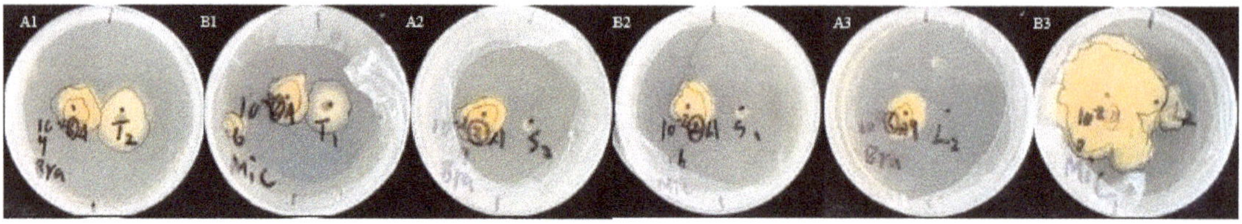

Figure 6. Cogrowth of actinomycetes and fungi. (**A**) *Brachybacterium phenoliresistens* MO. (**B**) *Microbacterium* sp. (**1**) *Colletotrichum siamense,* (**2**) *Rhizoctonia solani*, and (**3**) *Peronophthora litchii.*

4. Conclusions

Two actinomycete strains were isolated from the harvester ants. Based on the 16S rRNA and substrate utilization analysis, strain A belongs to the *Brachybacterium* genus, and it is very likely that *B. phenoliresistens*. Hence, strain A is denoted *B. phenoliresistens MO*. Strain B belongs to the *Microbacterium* genus. Because there were differences in the substrate utilization results, strain B must be further classified. It is denoted here as *Microbacterium* sp.

The inhibitory effects of the 2 actinomycetes on plant pathogenic fungi were assayed. When the actinomycete fermentation broth concentration was higher than 30%, the inhibition rates on the indicator fungi, i.e., *P. litchii* and *R. solani,* were significantly high, wherein a 100% inhibition rate was recorded. The inhibitory effects of these two actinomycetes on *C. siamense* were also notable, indicating prospects for agricultural applications. Through coculturing live bacteria, we obtained results consistent with a previous conclusion, showing that the two strains exhibited strong inhibitory effects on *P. litchii* and *R. solani*.

5. Discussion

Rice sheath blight caused by *R. solani* is one of the three major rice diseases and is the major disease in rice-producing areas in Asia. As a result, the damage caused by *R. solani* has led to significant agricultural losses in China each year [26]. In addition, infestation by *P. litchii* in litchi during storage and transport can result in great losses each year [27]. In this work, two strains of actinomycetes were used on these two types of plant pathogenic fungi at certain concentrations, and an inhibition rate of 100% was observed; therefore, the inhibitory effect of the live bacterial cultures on the pathogenic fungi was confirmed. However, the exact mechanism remains unclear. This result is possibly related to the bacterial fermentation products. In general, the inhibitory rates exhibited by the actinomycetes on *P. litchii* were higher than those of the other two pathogenic fungi, which suggests that *P. litchi* is more sensitive to actinomycetes [15]. The 100% pathogenic fungus inhibition rate may indicate the potential agricultural applications of these actinomycetes for multiple crops. Multiple methods—HPLC, metabolomic profiling, and gas chromatography—can be used to study the chemical characteristics, including toxicity, light degradability, and stability, of antifungal metabolites to determine whether actinomycetes can be used as fungicides against *R. solani* and *P. litchii*. As such, this work indicates a potential means of reducing the economic loss caused by damage to rice and litchi by *R. solani* and *P. litchii*.

Microbacterium is an abundant component of bacterial communities in the soil, insects, and leaf material of plants [8]. In previous studies, *M. testaceum* KU313 isolated from stored rice grains was antagonistic to *Aspergillus flavus* and *Penicillium* spp. [28]. *Microbacterium* sp. LGMB471 isolated from the medicinal plant *Vochysia divergens* inhibited the development of *Phyllosticta citricarpa* [29]. *Microbacterium* sp. isolated from tomato plants inhibited the growth of *Alternaria alternata*, *Corynespora cassiicola,* and *Stemphylium lycopersici* [30]. Among these, antifungal compounds, 5-methyl-2-phenyl-1H-indole produced by strain KU143, 7-O-β-D-glucosyl-genistein and 7-O-β-D-glucosyl-daidzein produced by strain LGMB471, were also discovered [28,29]. *Brachybacterium* exists in various environments, such as soil (poultry deep litter, contaminated sand), roots, fermented food, and animals [31]. It was reported that *B. paraconglomeratum* YEBPT2 isolated from banana contributed to antifungal activity against *Fusarium oxysporum* f.sp. *cubense* (Foc), and nine bioactive metabolites were identified as diethyl hydrazine, carbonic acid, nitrosopyrrolidine, 4H-pyran, valeric acid, butanoic acid, trioxsalen, deoxy-d-mannoic acid, and amino caprylic acid [32].

Ants are one of the most successful terrestrial species [33]. Many researchers devoted to screening the high biotechnological potential of ant-associated microorganisms, as well as the significant ecological impact of microbial secondary metabolites [5,8,9,13–18,34–36]. Nonetheless, most studies have mainly focused on ant–fungus–actinomycete tripartite mutualism evolved by leaf-cutting ants, which use antifungal microbial secondary metabolites produced by actinobacteria (*Streptomyces* spp., *Nocardia* spp., *Pseudonocardia* spp., *Amycolatopsis* spp., etc.) to control pathogens in their fungal gardens [5,9,16,35–39]. The rich diversity of antimicrobial secondary metabolites plays a driving role in shaping the ecosystems of leaf-cutting ants. Only by revealing the chemical nature of antibiotics can we begin to fully understand the complex interactions between multi-organismic partners. In this work, *B. phenoliresistens MO.* and *Microbacterium* sp. exhibited pronounced antifungal properties; however, the possible symbiotic relationship between harvester ants and actinomycetes remains unclear. The genera *Microbacterium* isolated from gardens and starter cultures of *Atta* could play disease-suppressing or other unknown roles [8], while no specific function was shown for *Brachybacterium* isolated from the abdomen of *Leucocoprinus gongylophorus* [40]. To better understand the ecological role of microorganisms associated with *Messor orientalis*, it is crucial to analyze the chemical composition and evaluate the biological activity of their metabolites. In further work in this area, we wish to identify more species of functional actinomycetes from *Messor orientalis* by using diverse isolation methods and media. In addition, we will further examine the relationship between harvester ants and actinomycetes to understand whether the ants and microbes have a mutually beneficial relationship.

Supplementary Materials: The following supporting information can be downloaded at: https://www.mdpi.com/xxx/s1, Figure S1: Growth of *Colletotrichum siamense* on various inhibitory media. (A) *Brachybacterium phenoliresistens* MO, (B) *Microbacterium* sp., (D) control; Figure S2: Growth of *Rhizoctonia solani* on various inhibitory media. (A) *Brachybacterium phenoliresistens* MO, (B) *Microbacterium* sp., (D) control; Figure S3: Growth of *Peronophthora litchii* on various inhibitory media. (A) *Brachybacterium phenoliresistens* MO, (B) *Microbacterium* sp., (D) control.

Author Contributions: Conceptualization, Y.X., Y.W., and S.C.; methodology, Y.W., J.Y., Y.L., and S.C.; formal analysis, Y.X., S.C., J.Y., Y.L., and Y.W.; writing—original draft preparation, Y.X., S.C., and Y.W.; writing—review and editing, Y.X., S.C., J.Y., Y.L., and Y.W.; funding acquisition, Y.X. All authors have read and agreed to the published version of the manuscript.

Funding: This work was funded by the Shenzhen Science and Technology Program (KCXFZ20201221173608022) and the Laboratory of Lingnan Modern Agriculture Project (NZ2021022).

Institutional Review Board Statement: Not applicable.

Informed Consent Statement: Not applicable.

Data Availability Statement: The data presented in this study are available on request from the corresponding author.

Acknowledgments: The authors wish to thank Yu Peng and Yuhang Zhou, Sendelta International Academy, for their support. Yiyang Wu thanks Yuqi Zhang and Boxiang Wang for their suggestions.

Conflicts of Interest: All authors declare that there are no conflict of interest.

References

1. Baumann, P.; Baumann, L.; Lai, C.; Rouhbakhsh, D.; Moran, N.; Clark, M. Genetics, physiology, and evolutionary relationships of the genus *Buchnera*: Intracellular symbionts of aphids. *Annu. Rev. Microbiol.* **1995**, *49*, 55–94. [CrossRef]
2. Kikuchi, Y. Endosymbiotic bacteria in insects: Their diversity and culturability. *Microbes. Environ.* **2009**, *24*, 195–204. [CrossRef]
3. Schäfer, A.; Konrad, R.; Kuhnigk, T. Hemicellulose-degrading bacteria and yeasts from the termite gut. *J. Appl. Bacteriol.* **1996**, *5*, 471–478. [CrossRef]

4. Bignell, D.E.; Oskarsson, H.; Anderson, J.M. Association of actinomycete-lik bacteria with soil-feeding termites (Termitidae, Termitinae). *Appl. Environ. Microbiol.* **1979**, *2*, 339–342. [CrossRef]
5. Currie, C.R.; Scott, J.A.; Summerbell, R.C.; Malloch, D. Fungus-growing ants use antibiotic-producing bacteria to control garden parasites. *Nature* **2003**, *398*, 701–704. [CrossRef]
6. Guo, Z.K.; Zhang, G.F.; Jiao, R.H.; Shen, Y.; Xu, Q.; Tan, R.X.; Ge, H.M. Actinotetraoses A–H: Tetrasaccharide derivatives from a grasshopper-associated *Amycolatopsis* sp. HCal. *Planta Med.* **2012**, *78*, 988–994. [CrossRef] [PubMed]
7. Lu, Y.H.; Li, S.; Zhou, D.X.; Zhang, Y.L. Isolation and identification of termination antagonistic actinomycetes BYC 01 and its active metabolites. *Acta Microbiol. Sinica.* **2014**, *7*, 754–759. [CrossRef]
8. Mueller, U.G.; Dash, D.; Rabeling, C.; Rodrigues, A. Coevolution between attine ants and actinomycete bacteria: A reevaluation. *Evolution* **2008**, *62*, 2894–2912. [CrossRef]
9. Chevrette, M.G.; Carlson, C.M.; Ortega, H.E.; Thomas, C.; Ananiev, G.E.; Barns, K.J.; Book, A.J.; Cagnazzo, J.; Carlos, C.; Flanigan, W.; et al. The antimicrobial potential of *Streptomyces* from insect microbiomes. *Nat. Commun.* **2019**, *10*, 516. [CrossRef] [PubMed]
10. Blodgett, J.A.V.; Oh, D.C.; Cao, S.G.; Currie, C.R.; Kolter, R.; Clardy, J. Common biosynthetic origins for polycyclic tetramate macrolactams from phylogenetically diverse bacteria. *Proc. Natl. Acad. Sci. USA* **2010**, *107*, 11692–11697. [CrossRef] [PubMed]
11. Scott, J.J.; Oh, D.C.; Yuceer, M.C.; Klepzig, K.D.; Clardy, J.; Currie, C.R. Bacterial protection of beetle-fungus mutualism. *Science* **2008**, *322*, 63. [CrossRef] [PubMed]
12. Poulsen, M.; Oh, D.C.; Clardy, J.; Currie, C.R. Chemical analyses of wasp-associated *Streptomyces* bacteria reveal a prolific potential for natural products discovery. *PLoS ONE* **2011**, *6*, e16763. [CrossRef] [PubMed]
13. Weber, N.A. Fungus-growing ants: A symbiotic relationship exists between an insect and a plant, involving an effective culturing technique. *Science* **1966**, *3736*, 587–604. [CrossRef]
14. Guo, L.F. Diversity of Symbiotic Actinomycetes from Camponotus janponicus Mayr and Their Antibacterial Activity. Master's Thesis, Northeast Agricultural University, Harbin, China, 2016.
15. Zhang, X.Y.; Zhao, X.F.; Xu, Y.J.; Cheng, D.F.; Lu, Y.Y. Identification and culture of *Streptomyces* sp. DF-5 isolated from *Solenopsis invicta* and the antifungal activity of its fermentation broth to plant pathogenic fungi. *Acta J. Environ. Entomol.* **2018**, *4*, 917–924.
16. Oh, D.C.; Poulsen, M.; Currie, C.R.; Clardy, J. Dentigerumycin: A bacterial mediator of an ant-fungus symbiosis. *Nat. Chem. Biol.* **2009**, *5*, 391–393. [CrossRef]
17. Baranova, A.A.; Chistov, A.A.; Tyurin, A.P.; Prokhorenko, I.A.; Korshun, V.A.; Biryukov, M.V.; Alferova, V.A.; Zakalyukina, Y.V. Chemical ecology of *Streptomyces albidoflavus* strain A10 associated with carpenter ant *Camponotus vagus*. *Microorganisms* **2020**, *8*, 1948. [CrossRef]
18. Akalyukina, Y.V.; Birykov, M.V.; Lukianov, D.A.; Shiriaev, D.I.; Komarova, E.S.; Skvortsov, D.A.; Kostyukevich, Y.; Tashlitsky, V.N.; Polshakov, V.I.; Nikolaev, E.; et al. Nybomycin-producing *Streptomyces* isolated from carpenter ant *Camponotus vagus*. *Biochimie* **2019**, *160*, 93–99. [CrossRef]
19. Osterman, I.A.; Wieland, M.; Maviza, T.P.; Lashkevich, K.A.; Lukianov, D.A.; Komarova, E.S.; Zakalyukina, Y.V.; Buschauer, R.; Shiriaev, D.I.; Leyn, S.A.; et al. Tetracenomycin X inhibits translation by binding within the ribosomal exit tunnel. *Nature Chem. Biol.* **2020**, *16*, 1071–1077. [CrossRef]
20. Tarsa, C.; McMillan, A.; Warren, R.J. Plant pathogenic fungi decrease in soil inhabited by seed-dispersing ants. *Insect. Soc.* **2018**, *65*, 315–321. [CrossRef]
21. MacMahon, J.A.; Mull, J.F.; Crist, T.O. Harvester ants (*Pogonomyrmex* spp.): Their community and ecosystem influences. *Annu. Rev. Ecol. Syst.* **2000**, *31*, 265–291. [CrossRef]
22. Kang, Z.S. Current status and development strategy for research on plant fungal diseases in China. *Plant Prot.* **2010**, *3*, 9–12. [CrossRef]
23. Li, B.D.; Shen, C.R. The fungicide resistance of plant pathogens and solutions. *Acta Phytopathol. Sinica* **1994**, *24*, 193–196. [CrossRef]
24. Goodfellow, M.; Williams, S.T. Ecology of Actinomycetes. *Ann. Rev. Microbiol.* **1984**, *1*, 189–216. [CrossRef]
25. Solecka, J.; Zajko, J.; Postek, M.; Rajnisz, A. Biologically active secondary metabolites from *Actinomycetes*. *Cent. Euro. J.Bio.* **2012**, *7*, 373–390. [CrossRef]
26. Zuo, S.M.; Wang, Z.B.; Chen, X.J.; Gu, F.; Zhang, Y.F.; Chen, Z.X.; Pang, X.B. Evaluation of resistance of a novel rice germplasm YSBR1 to sheath blight. *Acta Agron. Sin.* **2009**, *04*, 608–614. [CrossRef]
27. Kong, G.H.; Feng, D.N.; Li, W.; Lian, S.L.; Xi, P.G.; Jiang, Z.D. Research progress in studies on the downy blight disease in litchi. *J. Fruit Sci.* **2021**, *38*, 603–612. [CrossRef]
28. Mannaa, M.; Kim, K.D. Biocontrol activity of volatile-producing *Bacillus megaterium* and *Pseudomonas protegens* against *Aspergillus* and *Penicillium* spp. Predominant in stored rice grains: Study II. *Mycobiology* **2018**, *46*, 52–63. [CrossRef]
29. Savi, D.C.; Shaaban, K.A.; Gos, F.M.W.; Thorson, J.S.; Glienke, C.; Rohr, J. Secondary metabolites produced by *Microbacterium* sp. LGMB471 with antifungal activity against the phytopathogen *Phyllosticta citricarpa*. *Folia Microbiol.* **2019**, *64*, 453–460. [CrossRef] [PubMed]
30. Lopez, S.M.Y.; Pastorino, G.N.; Balatti, P.A. Volatile organic compounds profile synthesized and released by endophytes of tomato (*Solanum lycopersici* L.) and their antagonistic role. *Arch. Microbiol.* **2021**, *203*, 1383–1397. [CrossRef]
31. Lapidus, A.; Pukall, R.; LaButtii, K.; Copeland, A.; Del Rio, T.G.; Nolan, M.; Chen, F.; Lucas, S.; Tice, H.; Cheng, J.F.; et al. Complete genome sequence of Brachybacterium faecium type strain (Schefferle 6–10(T)). *Stand. Genom. Sci.* **2009**, *1*, 3–11. [CrossRef]

32. Ravi, S.; Sevugapperumal, N.; Nallusamy, S.; Shanmugam, H.; Mathiyazhagan, K.; Rangasamy, A.; Subbiah, K.A.; Ganesan, M.V. Differential bacterial endophytome in Foc-resistant banana cultivar displays enhanced antagonistic activity against *Fusarium oxysporum* f.sp. *cubense* (Foc). *Environ. Microbiol.* **2022**, *24*, 2701–2715. [CrossRef]
33. Schultz, T.R. In search of ant ancestors. *Proc. Natl. Acad. Sci. USA* **2000**, *97*, 14028–14029. [CrossRef]
34. Carr, G.; Derbyshire, E.R.; Caldera, E.; Currie, C.R.; Clardy, J. Antibiotic and antimalarial quinones from fungus-growing ant-associated *Pseudonocardia* sp. *J. Nat. Prod.* **2012**, *75*, 1806–1809. [CrossRef]
35. Caldera, E.J.; Poulsen, M.; Suen, G.; Currie, C.R. Insect symbioses: A case study of past, present, and future fungus-growing ant research. *Env. Entomol.* **2009**, *38*, 78–92. [CrossRef]
36. Schoenian, I.; Spiteller, M.; Ghaste, M.; Wirth, R.; Herz, H.; Spiteller, D. Chemical basis of the synergism and antagonism in microbial communities in the nests of leaf-cutting ants. *Proc. Natl. Acad. Sci. USA* **2011**, *108*, 1955–1960. [CrossRef]
37. Haeder, S.; Wirth, R.; Herz, H.; Spiteller, D. Candicidin-producing Streptomyces support leaf-cutting ants to protect their fungus garden against the pathogenic fungus Escovopsis. *Proc. Natl. Acad. Sci. USA* **2009**, *106*, 4742–4746. [CrossRef]
38. Ronque, M.U.V.; Lyra, M.L.; Migliorini, G.H.; Bacci, M., Jr.; Oliveira, P.S. Symbiotic bacterial communities in rainforest fungus-farming ants: Evidence for species and colony specificity. *Sci. Rep.* **2020**, *10*, 10172. [CrossRef]
39. Seipke, R.F.; Barke, J.; Brearley, C.; Hill, L.; Yu, D.W.; Goss, R.J.; Hutchings, M.I. A single Streptomyces symbiont makes multiple antifungals to support the fungus farming ant Acromyrmex octospinosus. *PLoS ONE* **2011**, *6*, e22028. [CrossRef]
40. Zani, R.D.A.; Ferro, M.; Bacci, M. Three phylogenetically distinct and culturable diazotrophs are perennial symbionts of leaf-cutting ants. *Ecol. Evol.* **2021**, *11*, 17686–17699. [CrossRef]

insects

MDPI

Article

A New Albomycin-Producing Strain of *Streptomyces globisporus* subsp. *globisporus* May Provide Protection for Ants *Messor structor*

Yuliya V. Zakalyukina [1,2,*], Nikolay A. Pavlov [3], Dmitrii A. Lukianov [3,4], Valeria I. Marina [3,4], Olga A. Belozerova [5], Vadim N. Tashlitsky [4], Elena B. Guglya [5,6], Ilya A. Osterman [1,3,4] and Mikhail V. Biryukov [1,3,7]

1 Center for Translational Medicine, Sirius University of Science and Technology, Olympic Avenue 1, 354340 Sochi, Russia
2 Department of Soil Science, Lomonosov Moscow State University, Leninskie Gory 1, 119991 Moscow, Russia
3 Center of Life Sciences, Skolkovo Institute of Science and Technology, Bolshoy Boulevard 30, Bld. 1, 121205 Moscow, Russia
4 Department of Chemistry, Lomonosov Moscow State University, Leninskie Gory 1, 119991 Moscow, Russia
5 Shemyakin-Ovchinnikov Institute of Bioorganic Chemistry of the Russian Academy of Sciences, Miklukho-Maklaya st. 16/10, 117997 Moscow, Russia
6 Institute of Translational Medicine, Pirogov Russian National Research Medical University, Ostrovityanova st. 1, 117997 Moscow, Russia
7 Department of Biology, Lomonosov Moscow State University, Leninskie Gory 1, 119991 Moscow, Russia
* Correspondence: juline@soil.msu.ru; Tel.: +7-9175548004

Simple Summary: Insects are the most numerous and diverse animals on our planet. They have mastered different habitats and are able to resist many external threats. Many of them have long concluded mutually beneficial alliances with microorganisms that are capable of producing biologically active substances. We found that *Messor structor* ants have a very common actinobacteria that secretes albomycin, an antibiotic capable of suppressing the growth of enthomopathogenic bacteria in the smallest concentrations. Perhaps this is one of the secrets of ants' resistance to external factors and their successful evolutionary development.

Abstract: There are several well-studied examples of protective symbiosis between insect host and symbiotic actinobacteria, producing antimicrobial metabolites to inhibit host pathogens. These mutualistic relationships are best described for some wasps and leaf-cutting ants, while a huge variety of insect species still remain poorly explored. For the first time, we isolated actinobacteria from the harvester ant *Messor structor* and evaluated the isolates' potential as antimicrobial producers. All isolates could be divided into two morphotypes of single and mycelial cells. We found that the most common mycelial morphotype was observed among soldiers and least common among larvae in the studied laboratory colony. The representative of this morphotype was identified as *Streptomyces globisporus* subsp. *globisporus* 4-3 by a polyphasic approach. It was established using a *E. coli* JW5503 pDualRep2 system that crude broths of mycelial isolates inhibited protein synthesis in reporter strains, but it did not disrupt the in vitro synthesis of proteins in cell-free extracts. An active compound was extracted, purified and identified as albomycin δ2. The pronounced ability of albomycin to inhibit the growth of entomopathogens suggests that *Streptomyces globisporus* subsp. *globisporus* may be involved in defensive symbiosis with the *Messor structor* ant against infections.

Keywords: actinobacteria; *Streptomyces globisporus* subsp. *globisporus*; albomycin; defensive symbiosis; ants; *Messor structor*

Citation: Zakalyukina, Y.V.; Pavlov, N.A.; Lukianov, D.A.; Marina, V.I.; Belozerova, O.A.; Tashlitsky, V.N.; Guglya, E.B.; Osterman, I.A.; Biryukov, M.V. A New Albomycin-Producing Strain of *Streptomyces globisporus* subsp. *globisporus* May Provide Protection for Ants *Messor structor*. *Insects* **2022**, *13*, 1042. https://doi.org/10.3390/insects13111042

Academic Editors: Hongyu Zhang, Xiaoxue Li and Yin Wang

Received: 11 October 2022
Accepted: 7 November 2022
Published: 11 November 2022

Publisher's Note: MDPI stays neutral with regard to jurisdictional claims in published maps and institutional affiliations.

Copyright: © 2022 by the authors. Licensee MDPI, Basel, Switzerland. This article is an open access article distributed under the terms and conditions of the Creative Commons Attribution (CC BY) license (https://creativecommons.org/licenses/by/4.0/).

1. Introduction

The widespread use of antimicrobial compounds in medicine and agriculture has led to the emergence of multidrug-resistant pathogens, recognized now as a significant threat to human health [1,2]. The search for novel compounds possessing antimicrobial properties is still one of the ways we could overcome this global challenge, and microorganisms are the main source in this research [3–5]. Recently great attention was attracted by microbes that form symbioses with higher organisms, in particular, plants and animals [6]. The main reason for that is the mutual evolutionary path, wherein microbes have proved their usefulness to the host [7,8].

Some prokaryotes, principally the phylum Actinomycetota [9], are involved in the formation of so-called "defensive (or protective) symbioses" [10–12] with many eukaryotic organisms. The Insecta class, with the largest number of species, is remarkable among them for these interactions [13,14]. Through the release of various antibiotic compounds, actinobacteria protect insects, their brood and food substrate from potential pathogens and parasites [7,15–18]. Well known in this respect are leaf-cutting ants (Atta, Acromyrmex) of the subfamily Myrmicinae. Their existence, and in particular their feeding and development, depends entirely on the symbiotic actinobacteria of the genus *Pseudonocardia* [19], localized on the insect cuticle.

However, the specificity of "defensive symbioses" in other species of Formicidae remains unclear, including one of the dominant ants of the steppe zone, *Messor structor* (the steppe harvester ant). In this paper, we report on the isolation of actinobacteria from a laboratory colony of *Messor structor*, the identification of the produced antimicrobial compound (albomycin δ2), and its high activity against individual entomopathogens.

2. Materials and Methods

2.1. Ant Colony Rearing and Microbial Isolation

Prior to rearing an ant colony, a special incubator was designed; it consisted of a glass tube with sterile water and poppy seeds as the main nutrient substrate. The mated queen was maintained there during a 3-month period, until the adult ant quantity was sufficient to place the colony in a specific formicarium. The formicarium is made from an acrylic plastic with two main chambers—an arena, through which the seeds are supplemented, and a system of chambers, where ants raise their brood. In the center of the formicarium, a specific watering cell is present, which maintains the humidity and water level inside the nest. Humidity is maintained between 70 and 90% and temperature at 24 °C respectively, without direct sunlight on the formicarium.

The queen was collected during the mating season in July 2017 in the Astrakhan region, Russia (46°51′13.5″ N 47°59′06.2″ E). The isolation of actinobacteria strains was performed only after the stabilization of a population number of at least 50 specimens.

Actinobacterial strains were isolated from the bodies of larvae, pupae and imago worker and soldier castes of *Messor structor*. A total of 14 individuals from each group were examined. Every specimen was washed three times in sterile distilled water and then crushed by a tissue microhomogenizer with sterile saline solution. Aliquots of this mixture and their 10-fold dilutions were spread over mineral agar 1 [20] and organic agar 79 [21] supplemented nystatin and nalidixic acid at final concentrations 250 µg/mL and 10 µg/mL, accordingly, and incubated for 14 days at 28°C [22]. Actinobacteria isolates were purified and maintained on ISP 3 slants [23] and preserved as a suspension of mycelial fragments and spores in 20% glycerol at −20 °C.

2.2. 16S rRNA Phylogeny of Isolated Strains

The extraction of the genomic DNA of isolates and PCR amplification were achieved using procedures described elsewhere [24]. Both the pair of universal primers F27 (5′-AGAGTTTGATCMTGGCTCAG-3′) and R1492 (5′-TACGGYTACCTTGTTACGACTT-3′) and actinobacterial primers 243F (5′-GGATGAGCCCGCGGCCTA-3′) and A3R (5′-CC AGCCCCACCTTCGAC-3′) were used. The amplicons were purified and sequenced using a commercial service (EvroGen). All sequences were identified by searching close relatives with the BLAST service (https://blast.ncbi.nlm.nih.gov/Blast.cgi, accessed on 1 October 2022) and were submitted in GenBank with the assignment of access numbers.

2.3. Genome Features and Phylogenomic Analysis

Genome of strain 4-3 was sequenced de novo by Skoltech Genomics Core Facility, using the Illumina HiSeq 4000 platform (Illumina, San Diego, CA, USA). The quality control and adapter trimming was done by the bbDuk tool from the BBMap suite v38.42 (https://sourceforge.net/projects/bbmap/, accessed on 1 February 2022). Genome assembly was performed by SPAdes v3.13.0 [25]. The genome was annotated using RASTtk pipeline implemented on the PATRIC web service [26]. Assembly is available in the European Nucleotide Archive with project accession PRJEB51905.

Values of average nucleotide identity (ANI), genome completeness and quality were evaluated using a web service MiGA (http://microbial-genomes.org/, accessed on 1 October 2022), and in silico digital DNA:DNA hybridization (DDH) values were calculated by using the GGDC method, with the recommended formula 2, available at the TYGS web service (https://tygs.dsmz.de/, accessed on 1 October 2022) [27].

Phylogenomic analysis was performed using Type (Strain) Genome Server (https://tygs.dsmz.de/, accessed on 1 October 2022). The phylogenomic tree inferred with FastME 2.1.6.1 [28] from GBDP distances calculated from genome sequences. The branch lengths are scaled in terms of GBDP distance formula d5.

The full-length 16S rRNA gene sequences of strain 4-3 was extracted from the whole genome sequence (PRJEB51905) and was compared to sequences of related *Streptomyces globisporus* subsp. *globisporus* and some actinobacteria species, firstly isolated from insects. Evolutionary trees based on 16S rRNA gene sequences were inferred with the neighbor-joining [29], maximum-parsimony [30] and maximum-likelihood [31] treemaking algorithms after CLUSTAL W alignment by using MEGA software version X [32].

2.4. Analysis of Bioactive Compound Biosynthetic Gene Clusters

Secondary metabolite biosynthetic gene clusters in complete genome strain 4-3 and its neighbors were identified with the bacterial version of antiSMASH 6.1.0 (https://antismash.secondarymetabolites.org/, accessed on 1 October 2022). Homologous regions on each genome were identified using NCBI Blastn (https://blast.ncbi.nlm.nih.gov/, accessed on 1 October 2022).

2.5. Phenotypic Characterization

Cultural characteristics of strain 4-3 were observed in ISP 2–ISP 7 media [23] after cultivation for up to 14 days at 28 °C. The RAL Classic Standard was used to determine the designations of colony colors. The shape of spore chains and the spore surface of strain 4-3 on ISP 3 after cultivation at 28 °C for 14 days were studied using light microscopy (Fisherbrand AX-502, Fisher Scientific, Merelbeke, Belgium) and scanning electron microscopy (JSM-6380LA, JEOL, Tokyo, Japan).

Carbon source utilization was assessed on basal medium ISP 9 [23] with the addition of 0.04% solution of bromocresol purple at 28 °C for 14 days. Enzyme activities were estimated using a paper indicator system (NPO Microgen, Nizhny Novgorod, Russia) according to the manufacturer's recommendations at 28 °C for 7 days. The degradation of casein, starch and cellulose was estimated on the clearing of the insoluble compounds around areas of growth [33].

2.6. Biological Activity Testing

2.6.1. Screening of the Antimicrobial Potential

The ability of actinobacteria isolates to inhibit bacterial growth was assessed by the agar diffusion method. The isolates were challenged against different clinically significant microorganisms: *Staphylococcus aureus* ATCC 25923, *Bacillus subtilis* ATCC 6633, *Candida albicans* CBS 8836 and *Aspergillus niger* INA 00760. Anti-entomopathogenic activity was also investigated against: *Bacillus thuringiensis* VKM B-6650, *Paenibacillus alvei* VKM B-502, *Beauveria bassiana* VKM F-1357 and *Entomophthora coronata* VKM F-1359.

Test bacteria, yeast and fungi strains were individually inoculated in Luria–Bertani agar, dextrose–peptone-yeast agar and potato dextrose agar, accordingly. Agar plugs of each actinobacterial isolate (2-week-old cultures in ISP6) were placed on the surface of the inoculated media. The plates were incubated at 37 °C, and after 24–48 h, the inhibition zones were checked.

2.6.2. Reporter Assays on Agar Plates

The two *E. coli* reporter strains: BW25113 wild-type pDualRep2 and JW5503 ΔtolC pDualRep2 [34] were used in this work as previously described [35]. Briefly, 100 μL of cultural broth were placed into wells in agar that had the lawn of a reporter strain. Two control antibiotics, erythromycin (Ery, 2 μg) and levofloxacin (Lev, 0.05 μg), were additionally applied to an agar plate. Plates were incubated at 37 °C overnight and then scanned by ChemiDoc (Bio-Rad) in the modes 'Cy3-blot' for RFP and 'Cy5-blot' for Katushka2S. The expression of the rfp gene occurred in the case of the activation of the SOS-response system of the cell and katushka2S, in the case of a violation of translation, when the ribosome was stalled on the mRNA template. When scanning, the signal from two black and white images was superimposed on each other, with the assignment of green for the signal from the RFP protein and red for Katushka2S.

2.6.3. Determination of Minimal Inhibitory Concentration (MIC)

Overnight cultures of *E. coli* ΔtolC, *Bacillus thuringiensis* and *Paenibacillus alvei* were diluted 1:1000 in LB medium. A sterile 96-well plate was then loaded with 200 mL of the diluted cultural media, with the initial row having 400 mL prior serial dilution. A stock solution of a HPLC-purified 4-3 sample was variously seeded in the initial row, along with erythromycin (Ery), which was used as a control for the experiment. Other wells were left without antibiotic but with diluted LB-culture media, while the rest were left with LB media only as additional controls. A two-fold serial dilution was then carried out, with gentle mixing in each row. The plates were then incubated overnight at 37 °C with shaking at 200 rpm. Cell growth was measured at 590 nm using a microplate reader (VICTOR X5 Light Plate Reader, PerkinElmer, Waltham, MA, USA). UV absorption of 4-3 was measured with a spectrophotometer (NanoPhotometer™ NP80, Implen, München, Germany), and the concentration was calculated using the known coefficients of extinction at 306 and 425 nm.

2.6.4. Cell-free Translation

Cell-free (in vitro) translation reactions were performed in the presence of HPLC 4-3 fraction (1/10 of final volume) in 5 μL using the PURExpress®In Vitro system (NEB, Ipswich, MA, USA) supplemented with 100 ng Fluc mRNA and 0.05 mM D-luciferin. Chemiluminescence was recorded with VICTOR X5 Light Plate Reader. The Fluc mRNA obtained byMEGAscript™ T7 Transcription Kit (ThermoFisher, Carlsbad, CA, USA) from the circular DNA template.

2.7. Purification and Identification of Albomycin

To obtain a sufficient amount of the active compound for detailed bioactivity studies, strain 4-3 was cultured in four 750 mL Erlenmeyer flasks with 250 mL of liquid ISP 6 at 28 °C for 14 days under static conditions. One liter of fermentation broth 4-3 was subjected to gravity-flow reverse-phase chromatography on the sorbent LPS500H (polyvinylbenzene, pore size 50–1000 Å) (LLC "Technosorbent", Moscow, Russia). The fermentation broth was applied to chromatographic media, which then was eluted consistently with 10, 20, 30, 40, 50, 75 and 100% solutions of acetonitrile in water. The pDualrep2 double reporter system was used to analyze the activity of the collected fractions. The most active fractions were eluted by 10 and 20% acetonitrile, they induced the expression of reporter protein Katushka2S. The 10% acetonitrile fraction was further purified by high-performance liquid chromatography (HPLC) (Agilent 1260, isocratic elution 4% of MeCN, 10 mM AcONH4, 1 mL/min, 25 °C) using a Phenomenex HPLC column (Luna 5 μm C18

(2) 100 Å, 4.6 × 250 mm), and the collected fractions were analyzed using the reporter pDualrep2.

Fractions with antibacterial activity corresponding to an individual peak on chromatograms were collected, and the active compound was identified using ultra-high-performance liquid chromatography–electrospray ionization–high-resolution mass spectrometry (UPLC–ESI–HRMS). Analysis was carried out on an Ultimate 3000 RSLCnano HPLC system connected to an Orbitrap Fusion Lumos mass spectrometer (ThermoFisher Scientific). A sample of the active compound was separated on Luna Omega C18 100 × 2.1 mm 1.6 µm columns at a 0.2 mL/min flow rate and at RT. Separation was done by a gradient elution in a two-component mixture from the initial 5% to 20% of component B for 10 min. Component A was 0.1% formic acid plus 10 mM formate ammonium in water, and component B was a mixture of 0.1% formic acid in 100% MeCN and 10% 10 mM formate ammonium in water in ratio 9:1. UV data were registered at 290 nm. MS1 and MS2 spectra were collected in positive ion mode and recorded at 30 K and 15 K resolution, respectively, with HCD fragmentation.

3. Results

3.1. Isolation of Actinobacteria Strains, Associated with Messor Structor Ants

All bacterial strains isolated from *Messor structor* individuals were divided into two morphotypes: one of which formed beige branching mycelium, dark-pigmented and straight spore chains (later labeled as 4-3), and the other, formed rough colorless colonies from Gram-positive cocci (L1). Results demonstrated that strains of 4-3-morphotype had more association with the caste of soldiers with 89% frequency among this group, while workers' caste, pupae and larvae showed less specific results—50%, 21%, and 7% respectively. On the contrary, L1-bacteria were found in the vast majority of individuals from these groups besides soldiers (Table S1).

A comparison with the GenBank database demonstrated that all strains assigned to the 4-3-morphotype, in addition to their phenotypical similarity, had an identical sequence of the 16S rRNA genes, indicating their belonging to the *Streptomyces* genus. The closest strains were *Streptomyces globisporus* subsp. *globisporus* DSM 40136 (formerly a type strain of *Streptomyces albovinaceus*), *Streptomyces globisporus* subsp. *globisporus* DSM 40199^T, *Streptomyces rubiginosohelvolus* DSM 40176^T and *Streptomyces pluricolorescens* DSM 40019^T. The partial 16S rRNA sequences of L1-morphotype strains showed 100% similarity with *Staphylococcus gallinarum* DSM 20610^T. Among the representatives of the mycelial morphotype, strain 4-3 was selected for a more detailed study of the genome features and antagonistic activity.

3.2. Genome Features and Phylogenomic Analysis of Streptomyces sp. Strain 4-3

Phylogenomic analysis based on whole-genome sequences showed that strain 4-3 formed a well-supported monophyletic clade with *S. globisporus* subsp. *globisporus* DSM 40199^T and *S. globisporus* subsp. *globisporus* DSM 40136 with 95% bootstrap value (Figure 1).

The complete genome size of strain 4-3 was 7,941,828 bp with DNA G + C content of 71.6%, which was consistent with the G + C content of the genus *Streptomyces* [36]. The closest neighbors *S. globisporus* subp. *globisporus* DSM 40199^T and *S. globisporus* subp. *globisporus* DSM 40136 are characterized by similar genome size and G + C content (Table S2).

The ANI and in silico dDDH values between strains 4-3 and *S. globisporus* subp. *globisporus* DSM 40136 and *S. globisporus* subp. *globisporus* DSM 40199^T were above the recommended threshold of 96% and 70% (Table 1) needed for species separation [27,37]. Based on this, strain 4-3 most likely belongs to the species *Streptomyces globisporus* subp. *globisporus*.

Neighbor-joining phylogenetic analysis demonstrated that 4-3 was most closely related to type and not type strains of *Streptomyces globisporus* subsp. *globisporus*: DSM 40199^T, DSM 40139, C-1027, TFH56, as well as to *S. rubiginosohelvolus* DSM 40176^T, *S. pluricolorescens* DSM 40019^T, *S. sindenensis* DSM 40255^T, *S. anulatus* DSM 40361^T and *S. griseus* subsp. *griseus* ATCC 13273 and formed a share clade with 100% bootstrap value (Figure S1).

However, type strains of actinobacterial species, first isolated from ants and other insects, did not form well-supported clades with 4-3. This relationship was also supported in the phylogenetic trees generated with maximum-parsimony and maximum-likelihood methods (Figures S2 and S3, available in Supplementary Materials).

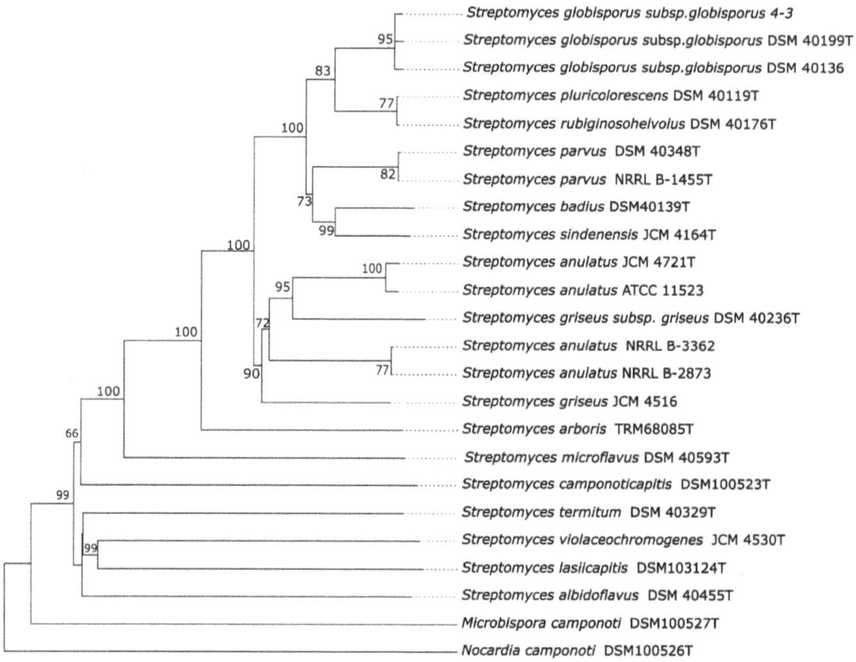

Figure 1. Phylogenetic tree based on whole-genome sequences from 4-3, related type strains and actinobacteria isolated from ants. The branch lengths are scaled in terms of GBDP distance formula d5. Numbers above branches are GBDP pseudo-bootstrap support values > 60% from 100 replications, with an average branch support of 81.4%. *Nocardia camponoti* DSM 100526^{T} as outgroup [27].

Table 1. Genome relatedness of 4-3 and *Streptomyces* type-strains.

Subject Strain	ANI,%	dDDH (in %)	G + C Content Difference (in %)
Streptomyces globisporus subsp. *globisporus* DSM 40136	99.3	96.3	0.14
Streptomyces globisporus subsp. *globisporus* DSM 40199^{T}	99.4	95.9	0.12
Streptomyces rubiginosohelvolus DSM 40176^{T}	96.3	66.2	0.18
Streptomyces pluricolorescens DSM 40019^{T}	96.2	66.1	0.14
Streptomyces parvus NRRL B-1455^{T}	94.8	56.2	0.01
Streptomyces parvus JCM 4069^{T}	94.7	55.7	0.05
Streptomyces sindenensis JCM 4164^{T}	94.2	52.9	0.25
Streptomyces badius JCM 4350^{T}	94.2	52.5	0.11
Streptomyces anulatus JCM 4721^{T}	92.1	41.8	0.15

3.3. Phenotypic Characterization of *Streptomyces sp.* Strain 4-3

To further evaluate the features of the 4-3 strain using a polyphasic taxonomy approach, the cultural, morphological and physiological properties of 4-3 were compared with ones of the type strains of *Streptomyces globisporus* subsp. *globisporus*. Results demonstrated the identity of these organisms in morphology—the shape of sporophores and spore surface (Table S3, Figure 2) and the high similarity of their cultural characteristics on the series of ISP media (Table 2, Figure S4).

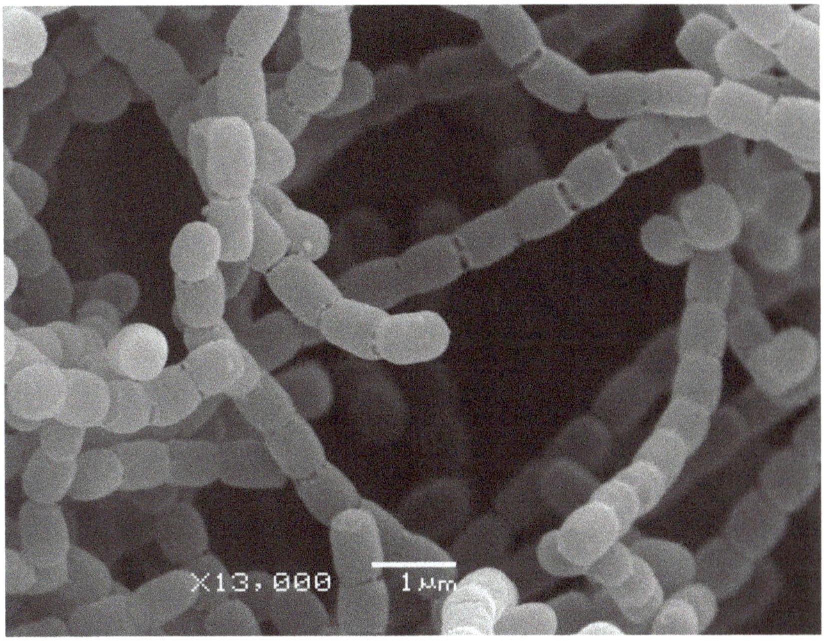

Figure 2. Scanning electron micrograph of the strain *Streptomyces globisporus* subsp. *globisporus* 4-3, showing the spore surface after incubation on ISP 3 medium at 28 °C for 14 days.

However, some differences should be noted in the biochemical and physiological properties of strain 4-3 and the closest type strains of *S. globisporus* subsp. *globisporus* (Table S3). For example, the production acid from glucose and xylose was positive in strain 4-3 (Figure S5), whereas the other type strains showed negative results. In the decomposition of polymers, strain 4-3 was unable to use cellulose as a sole carbon source, whereas the other type strains utilized it. Furthermore, the enzyme assay of 4-3 was negative for L-ornithine decarboxylase and L-arginine decarboxylase; in contrast, DSM 40199^T and DSM 40136 demonstrated positive responses (Table S3). However, most of the biochemical tests showed similar results.

According to the obtained data, we may conclude that strain 4-3 isolated from *Messor structor* ants can be classified as *Streptomyces globisporus* subsp. *globisporus.*

3.4. Analysis of 4-3 Bioactive Compound Biosynthetic Gene Clusters

The bioinformatics analysis of *Streptomyces globisporus* subsp. *globisporus* 4-3 genome revealed a biosynthetic gene cluster of albomycins, consisting of 18 genes from *abm*A to *abm*R, completely identical to that of *Streptomyces* sp. ATCC 700974 (Figure 3), described in detail earlier [38]. The presence of *abm*K, participating directly in the formation of SB-217452 (the active seryl-tRNA synthetase inhibitor component of albomycin) [39] and also providing self-resistance to albomycins [40], indicates the ability of *Streptomyces globisporus* subsp. *globisporus* 4-3 to actively produce albomycin δ2.

Table 2. Cultural characteristics of strain 4-3 (1) and closely related *Streptomyces globisporus* subsp. *globisporus* DSM 40199^T (2) and DSM 40136 (3).

Media	1	2	3
Yeast extract-malt extract (ISP 2)			
Growth	good	good	good
Aerial spore-mass color	oyster white	oster white	cream
Substrate mycelial color	beige	ochre yellow	beige
Soluble pigment	none	none	none
Oatmeal (ISP 3)			
Growth	good	good	good
Aerial spore-mass color	oyster white	oster white	cream
Substrate mycelial color	brown beige	ivory	beige
Soluble pigment	brown beige	none	none
Inorganic salts-starch (ISP 4)			
Growth	good	good	good
Aerial spore-mass color	white	light gray	sparse
Substrate mycelial color	colorless	green brown	beige
Soluble pigment	none	none	none
Glycerol-asparagine (ISP 5)			
Growth	good	good	good
Aerial spore-mass color	light olive	none	none
Substrate mycelial color	sand yellow	ivory	beige
Soluble pigment	sand yellow	none	none
Peptone-yeast extract iron (ISP 6)			
Growth	good	good	good
Aerial spore-mass color	white	none	none
Substrate mycelial color	beige	sand yellow	beige
Soluble pigment	none	none	none
Tyrosine (ISP 7)			
Growth	weak	good	good
Aerial spore-mass color	ivory	none	cream
Substrate mycelial color	yellow–red	beige	beige
Soluble pigment	none	none	none

Data for *Streptomyces globisporus* subsp. *globisporus* DSM 40199^T and DSM 40136 are from DSMZ catalogue (https://www.dsmz.de/collection/catalogue/microorganisms/catalogue, accessed 1 October 2022).

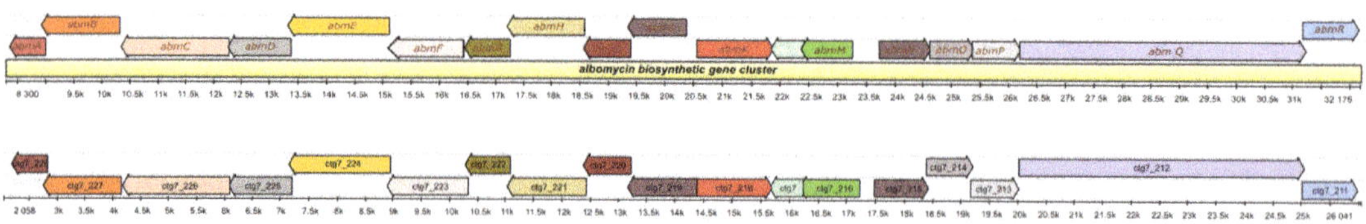

Figure 3. Biosynthetic gene clusters of albomycins: genetic organization of the albomycin (abm) gene cluster in *Streptomyces* sp. ATCC 700974 (**top**) and gene cluster in *Streptomyces globisporus* subsp. *globisporus* 4-3 (**bottom**). The homologous abm and ctg genes are filled with the same colors.

Furthermore, the 4-3 strain genome contains a number of second metabolic biosynthesis gene clusters (SMBGCs), coded production of antimicrobial compounds (streptophenazines B/C/E/H/G, mayamycins), odor substances (geosmin), pigments (melanin, isorenieratene), siderophores (streptobactin, coelichelin), cytoprotectants (ectoine) and others (Table S4).

3.5. Screening Antimicrobial Activity

All isolated strains were initially tested for the ability to secrete antimicrobial substances. It was noticed that all phenotypically similar mycelial strains showed the same activity pattern, so we focused on the study of strain 4-3. Analysis of antimicrobial activity demonstrated that the 4-3 strain noticeably inhibited the growth of various pathogenic microorganisms (Table S5): bacteria (*Bacillus subtilis, Staphylococcus aureus*) and fungi (*Aspergillus niger*), but it is especially active on entomopathogenic microorganisms (*Bacillus thuringiensis, Paenibacillus alvei, Beauveria bassiana, Entomophthora coronata*). To evaluate the MIC of the 4-3 compound, we chose strains that were the most susceptible during the screening procedure. The estimated concentration of the HPLC-purified sample was 0.6 µg/mL. In the prepared series of microdilutions, the test strains demonstrated high sensitivity to the 4-3 compound (Table S5), including entomopathogenic bacteria.

The agar plugs and cultural broth aliquots of 4-3 demonstrated prominent antibiotic activity in tests on the reporter strains (Figure 4) and exhibited strong Katushka2S reporter induction, indicating that the active compound produced by the isolate functions as an inhibitor of protein biosynthesis. To further evaluate the possible mechanism of action, the in vitro translation analysis was performed.

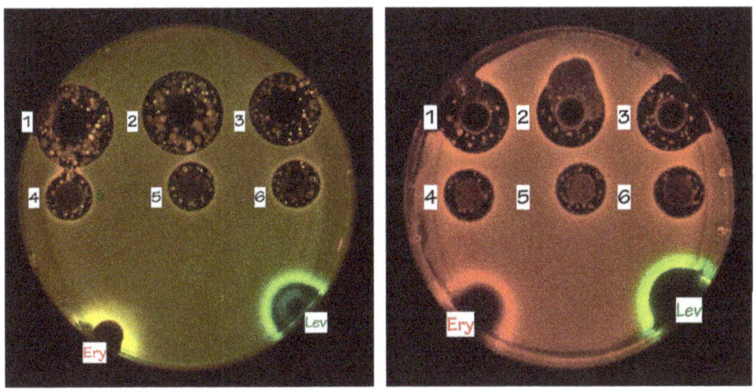

(**a**) (**b**)

Figure 4. In Vitro testing of ant-associated strains activity using: (**a**) *E. coli* BW25113 wild-type pDualRep2 reporter strain; (**b**) *E. coli* JW5503 ΔtolC pDualRep2 reporter strain. 1, 2, 3—crude broth aliquots and 4, 5, 6—agar plugs of X2, X1 and 4-3, accordingly. The agar plates were spotted with erythromycin, 5 µg/mL (Ery) and levofloxacin, 2 µg/mL (Lev).

3.6. Analysis of Bioactive Compounds: Cell-free Translation

It was decided to test the HPLC-purified sample 4-3 in in vitro translation procedure to completely evaluate its translation inhibitory activity. Despite induction of the Katushka2S, indicating the inhibition of protein synthesis in the cells of the reporter strains (Figure 4), the active compound of the strain 4-3 did not suppress translation in the cell-free system (Figure 5).

3.7. Purification and Identification of Albomycin

The liquid culture of *Streptomyces globisporus* subsp. *globisporus* 4-3 was preconcentrated and purified by solid-phase extraction (SPE) on LPS500H resin. It was binded on LPS500H, washed with water and the active compound was eluted with 10% MeCN solution in water. Further purification of active SPE fraction was carried out by RP HPLC on a C18 column in isocratic mode with aqueous solution of ammonium acetate—MeCN as eluent. Thus, a pure compound was isolated with a UV maxima at 283 and 424 nm, and the activity testing confirmed that it corresponded to an active metabolite (Figure S6).

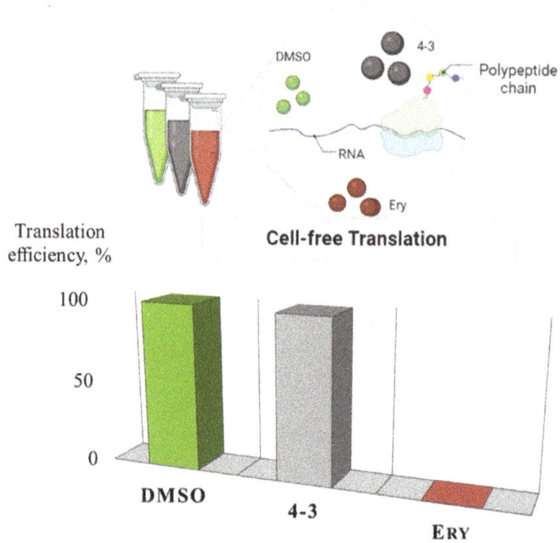

Figure 5. General scheme of cell-free translation assay (**top**) and translation effectiveness (**bottom**) in the presence of sample 4-3, DMSO 1% (negative control) and erythromycin (Ery, 5 μg/mL) as positive control.

The metabolite identification was carried out using LC-HRMS/MS analysis (Figure S7). The observed exact masses 1046.3102 of molecular ion $[M + H]^+$ of the compound and characteristic isotope distribution corresponded to the composition $C_{37}H_{57}FeN_{12}O_{18}S$ (calculated exact mass 1046.3057). The main fragmentary ion in the MS2 spectrum due to the loss of the cytosine part was observed at m/z 878. MS1–MS2 raw data were analyzed in Compound Discoverer 3.2 software (Thermo Fisher Scientific). Peak annotation was performed with ChemSpider, Natural Product Atlas 2020 and COCONUT databases using the mass spectra information with 5 ppm mass accuracy, isotopic distribution $\geq$ 50% and match score $\geq$ 85%. The result of the analysis allowed us to conclude that the active compound is a known inhibitor of bacterial seryl-tRNA synthetase, albomycin δ2 (Figure 6) [41].

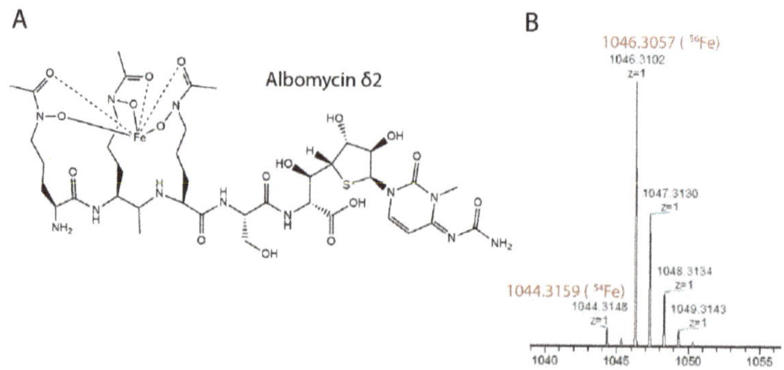

Figure 6. *Cont.*

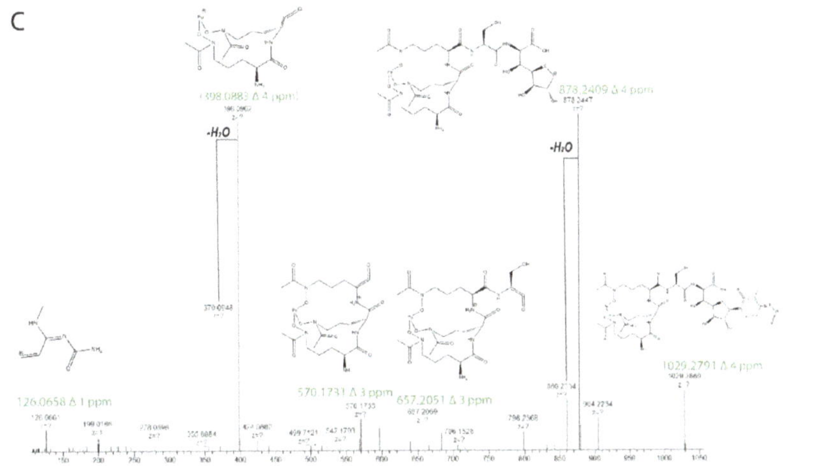

Figure 6. Identification of albomycin δ2. (**A**) Molecular structure of albomycin δ2. (**B**) Albomycin δ2 molecular ion $[M + H]^+$ observed at 1046.3102 and its isotopic distribution (calculated mass are signed as red). (**C**) The positive-mode HCD mass spectra of the parent ion at m/z 1046.3102 with the molecular fragment structures (calculated mass for the fragments are signed as green).

4. Discussion

Albomycin δ2 belongs to the group of sideromycins, antibiotics covalently bound to siderophore fragments and penetrating into the cell through siderophore absorption pathways, implementing the so-called "Trojan horse" strategy [42].

Albomycin, originally reported as grisein, were first isolated from soil from the soil-dwelling *Streptomyces griseus* in 1947 by S. Waksman and colleagues. It was also identified in *Streptomyces subtropicus* (previously known as *Actinomyces subtropicus*) by Gause and Brazhnikova in 1951, and their identity was confirmed later [43]. It is noteworthy that albomycin was also known as alveomycin, antibiotics A 1787, LA 5352 and LA 5937, and Ro 5-2667 in the literature [44].

Albomycins have attracted significant attention due to their potent antibacterial activities against both Gram-negative and Gram-positive bacteria, including multi-drug-resistant strains [45]. Moreover, no toxicity was observed during in vivo studies of albomycins, and it was well tolerated and safe up to a maximum dose evaluated in mice [41].

The structure of albomycin and related compounds (δ1, δ2, and ε) was fully established more than 30 years after the initial discovery [46]. Albomycins have a thioribosyl nucleoside moiety linked to an iron-chelating ferrichrome-type siderophore through a serine residue. The L-serine-thioheptose dipeptide partial structure, known as SB-217452, has been found to be the active seryl-tRNA synthetase inhibitor [39].

The iron-chelator portion serves as a vehicle for the active delivery of the albomycin warhead inside both Gram-positive and Gram-negative bacterial cells through the ferrichrome-specific transporter system. The formation of the active inhibitory form of albomycin, SB-217452, occurs intracellularly under the action of PepN peptidase (*E. coli*), which cleaves off the siderophore part. As a result, the toxic nucleoside part is accumulated in the cytoplasm of *E. coli* in ~500-fold excess over the concentration of the antibiotic in the medium [42]. This explains the fact that albomycin does not act on translation in a cell-free extract—in such a system, there are no enzymes that cleave off the siderophore (Figure 4). Likewise, added directly to bacterial culture, the nucleoside portion of albomycin does not inhibit cell growth [42] since it cannot get inside the cells without the siderophore part.

In the crops of *Messor structor* ants contained in laboratory conditions, we constantly observed mycelial bacteria of a certain phenotype, most closely associated with individuals from the soldier caste. A representative of this phenotype, strain 4-3, was identified using a polyphase approach as *Streptomyces globisporus* subsp. *globisporus*.

The members of *Streptomyces* are widely known for their ability to produce various antibiotic compounds and often in association with insects: ants, wasps, beetles [7]. According to the literature, *Streptomyces globisporus* was most often isolated from soils, plants and so on [47] but rarely from insects. There is an example of the *Streptomyces globisporus* SP6C4, which plays a significant role in the mutualistic relationship between pineapple strawberries (*Fragaria ananassa*) and bees (Apidae), protecting both the plant and the insect from pathogenic microorganisms, including the phytopathogenic fungus *Botrytis cinerea* [48]. In addition, it is reported about *Streptomyces globisporus* WA5-2-37, isolated from the intestinal tract of the American cockroach (*Periplaneta americana*), produced actinomycin X2 and collismycin A, which showed great activity against MRSA ATCC 43300 [49]. This is the first reported naturally occurring strain of *S. globisporus* subsp. *globisporus* isolated from Formicidae.

The genus *Messor* (Forel, 1890) is a moderately large genus, with more than 126 species worldwide recognized; they are mainly distributed in the Palearctic, Afrotropical and Oriental regions. Messor species are granivorous and play an important role in ecosystem maintenance and plant-seed dispersal. The harvesting ant, *Messor structor* (Latreille, 1798), is an ecosystem engineer in many dry biocenoses [50].

The main food resource for harvester ants are grains of cereals and oilseed plants. Such seeds have a solid endosperm and require considerable effort to grind them. Representatives of the soldier caste have a large head—the result of the development of massive occipital muscles responsible for the work of the lower jaw—and powerful mandibles. They initially grind the seeds, and then smaller worker ants process the prepared pieces of seeds, since it requires less effort; turn them into flour; moisten them with saliva and uses them as food for the colony. Their saliva is dominated by amylase enzymes that break down starch [51].

The greatest abundance of actinobacteria, isolated from soldiers and workers, which represent a conveyor for the production of food for the colony, may indicate a possible symbiotic relationship between ants and streptomycetes. Actinobacteria receive food and shelter, and in return produce a substance with a wide spectrum of action that protects the food resource of ants from spoilage. The ability of associated streptomycetes to synthesize albomycin can be extremely useful for hosts—since harvester ants contact with soils and plants and encounter a large number of microorganisms, Gram-positive and Gram-negative, as well as fungi.

As is known, albomycin δ2 is characterized by surprisingly low inhibitory concentrations for many pathogenic microorganisms: minimum inhibitory concentrations (MICs) as low as 5 ng/mL against *Escherichia coli* and 10 ng/mL against *Streptococcus pneumoniae* [52]. In our in vitro experiments, *Streptomyces globisporus* subsp. *globisporus* 4-3 very actively suppressed the growth of various entomopathogens: *Paenibacillus alvei* VKM B-502, *Bacillus thuringiensis* VKM B-6650, *Beauveria bassiana* VKM F-1357 and *Entomophthora coronata* VKM F-1359 (Table S5), which can provide one of the forms of protection of the ant family health.

A culture-dependent study of the actinobiome of a *Messor structor* colony, living under laboratory conditions, revealed the strains dominant in adult individuals from soldier and worker casts. All mycelial isolates demonstrated the same genotypic and phenotypic properties and were identified as *Streptomyces globisporus* subsp. *globisporus*. They produced protein synthesis inhibitor albomycin δ2, which was active against entomopathogens. The distribution of these actinobacteria among individuals from different castes suggests their essential role in maintaining the health of the ant family. The confirmation of this assumption, as well as the hypothesis about the methods of the transmission of strains between individuals, needs more detailed studies on a wider range of colonies of harvester ants living in formicaria.

Supplementary Materials: The following supporting information can be downloaded at: https://www.mdpi.com/article/10.3390/insects13111042/s1: Table S1: The isolation of Gram-positive bacteria strains from *Messor structor* individuals; Table S2: General features of the genome of strain 4-3 and its closely related strains of *Streptomyces globisporus*; Figure S1: Neighbor-joining phylogenetic tree of strain 4-3; Figure S2: Maximum-parsimony phylogenetic tree of strain 4-3; Figure S3: Maximum-likelihood phylogenetic tree of strain 4-3; Table S3: Morphological, physiological and biochemical characteristics of strain 4-3 and closest *Streptomyces globisporus* subsp. *globisporus* strains; Figure S4: Cultural properties of 4-3 on ISP media after 14 days at 28 °C; Figure S5: Utilization of different sugars (1.0%, w/v) as the sole carbon source by strain 4-3; Figure S6. RP HPLC of an active fraction and UV spectrum of an active metabolite peak; Figure S7. LC-MS data for albomycin δ2; Table S4: Some of the secondary metabolite gene clusters in *Streptomyces globisporus* subsp. *globisporus* 4-3; Table S5: Antimicrobial activity of *Streptomyces globisporus* subsp. *globisporus* 4-3, isolated from *Messor structor* ants.

Author Contributions: Conceptualization, Y.V.Z.; methodology, Y.V.Z., M.V.B., D.A.L. and I.A.O.; investigation, N.A.P., Y.V.Z., E.B.G., O.A.B., V.N.T., D.A.L. and V.I.M.; resources, N.A.P. and M.V.B.; writing—original draft preparation, Y.V.Z. and N.A.P.; data curation, Y.V.Z.; writing—review and editing, D.A.L., V.I.M., E.B.G. and V.N.T.; visualization, Y.V.Z., E.B.G., O.A.B. and V.N.T.; supervision, Y.V.Z. and M.V.B.; project administration, M.V.B. and I.A.O. All authors have read and agreed to the published version of the manuscript.

Funding: The research was funded by Russian Science Foundation (project No. 22-24-00278).

Data Availability Statement: Genomic data of *Streptomyces globisporus* subsp. *globisporus* 4-3 can be found at https://www.ebi.ac.uk/ena/browser/home (accessed on 1 November 2022) with project accession PRJEB51905.

Acknowledgments: SEM studies were carried out at the Shared Research Facility "Electron microscopy in life sciences" at Moscow State University (Unique Equipment "Three-dimensional electron microscopy and spectroscopy"). The authors acknowledge partial support from M.V.Lomonosov Moscow State University Program of Development. This work was partly supported by the Ministry of Health of the Russian Federation project "New antibiotics from marine microorganisms of the Russian Far East: studying structures, mechanisms of action and pharmacological prospectives". We acknowledge the Skoltech Genomic Core Facility. The authors would like to thank Andrei Osterman, Semen Leyn and Nick Wong for technical help and the critical appraisal of the manuscript.

Conflicts of Interest: The authors declare no conflict of interest.

References

1. Levy, S.B.; Marshall, B. Antibacterial resistance worldwide: Causes, challenges and responses. *Nat. Med.* **2004**, *10*, S122–S129. [CrossRef] [PubMed]
2. Larsson, D.G.J.; Flach, C.-F. Antibiotic resistance in the environment. *Nat. Rev. Microbiol.* **2022**, *20*, 257–269. [CrossRef] [PubMed]
3. Abdel-Razek, A.S.; El-Naggar, M.E.; Allam, A.; Morsy, O.M.; Othman, S.I. Microbial natural products in drug discovery. *Processes* **2020**, *8*, 470. [CrossRef]
4. Newman, D.J.; Cragg, G.M. Natural products as sources of new drugs over the nearly four decades from 01/1981 to 09/2019. *J. Nat. Prod.* **2020**, *83*, 770–803. [CrossRef]
5. Dai, J.; Han, R.; Xu, Y.; Li, N.; Wang, J.; Dan, W. Recent progress of antibacterial natural products: Future antibiotics candidates. *Bioorg. Chem.* **2020**, *101*, 103922. [CrossRef]
6. Rosenberg, E.; Zilber-Rosenberg, I. Microbes drive evolution of animals and plants: The hologenome concept. *mBio* **2016**, *7*, e01395-15. [CrossRef]
7. Chevrette, M.G.; Carlson, C.M.; Ortega, H.E.; Thomas, C.; Ananiev, G.E.; Barns, K.J.; Book, A.J.; Cagnazzo, J.; Carlos, C.; Flanigan, W.; et al. The antimicrobial potential of streptomyces from insect microbiomes. *Nat. Commun.* **2019**, *10*, 516. [CrossRef]
8. Wein, T.; Picazo, D.R.; Blow, F.; Woehle, C.; Jami, E.; Reusch, T.B.H.; Martin, W.F.; Dagan, T. Currency, exchange, and inheritance in the evolution of symbiosis. *Trends Microbiol.* **2019**, *27*, 836–849. [CrossRef]
9. Oren, A.; Garrity, G.M.Y. 2021 Valid publication of the names of forty-two phyla of prokaryotes. *Int. J. Syst. Evol. Microbiol.* **2021**, *71*, 005056. [CrossRef]
10. Clay, K. Defensive symbiosis: A microbial perspective. *Funct. Ecol.* **2014**, *28*, 293–298. [CrossRef]
11. Flórez, L.V.; Biedermann, P.H.W.; Engl, T.; Kaltenpoth, M. Defensive symbioses of animals with prokaryotic and eukaryotic microorganisms. *Nat. Prod. Rep.* **2015**, *32*, 904–936. [CrossRef] [PubMed]

12. Hopkins, S.R.; Wojdak, J.M.; Belden, L.K. Defensive symbionts mediate host–parasite interactions at multiple scales. *Trends Parasitol.* **2017**, *33*, 53–64. [CrossRef] [PubMed]
13. Arnam, E.B.V.; Currie, C.R.; Clardy, J. Defense contracts: Molecular protection in insect-microbe symbioses. *Chem. Soc. Rev.* **2018**, *47*, 1638–1651. [CrossRef] [PubMed]
14. Gil, R.; Latorre, A. Unity makes strength: A review on mutualistic symbiosis in representative insect clades. *Life* **2019**, *9*, 21. [CrossRef] [PubMed]
15. Kaltenpoth, M. Actinobacteria as mutualists: General healthcare for insects? *Trends Microbiol.* **2009**, *17*, 529–535. [CrossRef] [PubMed]
16. Seipke, R.F.; Kaltenpoth, M.; Hutchings, M.I. Streptomyces as symbionts: An emerging and widespread theme? *FEMS Microbiol. Rev.* **2012**, *36*, 862–876. [CrossRef]
17. Kaltenpoth, M.; Engl, T. Defensive microbial symbionts in Hymenoptera. *Funct. Ecol.* **2014**, *28*, 315–327. [CrossRef]
18. van der Meij, A.; Worsley, S.F.; Hutchings, M.I.; van Wezel, G.P. Chemical ecology of antibiotic production by Actinomycetes. *FEMS Microbiol. Rev.* **2017**, *41*, 392–416. [CrossRef]
19. Currie, C.R.; Wong, B.; Stuart, A.E.; Schultz, T.R.; Rehner, S.A.; Mueller, U.G.; Sung, G.-H.; Spatafora, J.W.; Straus, N.A. Ancient tripartite coevolution in the Attine ant-microbe symbiosis. *Science* **2003**, *299*, 386–388. [CrossRef]
20. Gause, G.F.; Preobrazhenskaya, T.P.; Sveshnikova, M.A.; Terekhova, L.P.; Maksimova, T.S. *Guide for Determination of Actinomycetes: Genera Streptomyces, Streptoverticillium, and Chainia*; Nauka: Moscow, Russia, 1983.
21. Prauser, H.; Falta, R. Phagensensibilität, zellwand-zusammensetzung und taxonomie von actinomyceten. *J. Basic Microbiol.* **1968**, *8*, 39–46. [CrossRef]
22. Zakalyukina, Y.V.; Osterman, I.A.; Wolf, J.; Neumann-Schaal, M.; Nouioui, I.; Biryukov, M.V. *Amycolatopsis camponoti* sp. nov., new tetracenomycin-producing actinomycete isolated from carpenter ant *Camponotus vagus*. *Antonie Van Leeuwenhoek* **2022**, *115*, 533–544. [CrossRef] [PubMed]
23. Shirling, E.B.; Gottlieb, D. Methods for characterization of *Streptomyces* species. *Int. J. Syst. Bacteriol.* **1966**, *16*, 313–340. [CrossRef]
24. Zakalyukina, Y.V.; Birykov, M.V.; Lukianov, D.A.; Shiriaev, D.I.; Komarova, E.S.; Skvortsov, D.A.; Kostyukevich, Y.; Tashlitsky, V.N.; Polshakov, V.I.; Nikolaev, E.; et al. Nybomycin-producing streptomyces isolated from carpenter ant *Camponotus vagus*. *Biochimie* **2019**, *160*, 93–99. [CrossRef]
25. Prjibelski, A.; Antipov, D.; Meleshko, D.; Lapidus, A.; Korobeynikov, A. Using SPAdes de novo assembler. *Curr. Protoc. Bioinform.* **2020**, *70*, e102. [CrossRef] [PubMed]
26. Brettin, T.; Davis, J.J.; Disz, T.; Edwards, R.A.; Gerdes, S.; Olsen, G.J.; Olson, R.; Overbeek, R.; Parrello, B.; Pusch, G.D.; et al. RASTtk: A modular and extensible implementation of the RAST algorithm for building custom annotation pipelines and annotating batches of genomes. *Sci. Rep.* **2015**, *5*, 8365. [CrossRef] [PubMed]
27. Meier-Kolthoff, J.P.; Göker, M. TYGS is an automated high-throughput platform for state-of-the-art genome-based taxonomy. *Nat. Commun.* **2019**, *10*, 2182. [CrossRef]
28. Lefort, V.; Desper, R.; Gascuel, O. FastME 2.0: A comprehensive, accurate, and fast distance-based phylogeny inference program. *Mol. Biol. Evol.* **2015**, *32*, 2798–2800. [CrossRef]
29. Saitou, N.; Nei, M. The Neighbor-Joining method: A new method for reconstructing phylogenetic trees. *Mol. Biol. Evol.* **1987**, *4*, 406–425. [CrossRef]
30. Nei, M.; Kumar, S. *Molecular Evolution and Phylogenetics*; Oxford University Press: Oxford, UK, 2000; ISBN 0-19-513584-9.
31. Tamura, K.; Nei, M. Estimation of the number of nucleotide substitutions in the control region of mitochondrial DNA in humans and chimpanzees. *Mol. Biol. Evol.* **1993**, *10*, 512–526. [CrossRef]
32. Kumar, S.; Stecher, G.; Li, M.; Knyaz, C.; Tamura, K. MEGA X: Molecular evolutionary genetics analysis across computing platforms. *Mol. Biol. Evol.* **2018**, *35*, 1547–1549. [CrossRef]
33. Volynkina, I.A.; Zakalyukina, Y.V.; Alferova, V.A.; Belik, A.R.; Yagoda, D.K.; Nikandrova, A.A.; Buyuklyan, Y.A.; Udalov, A.V.; Golovin, E.V.; Kryakvin, M.A.; et al. Mechanism-based approach to new antibiotic producers screening among actinomycetes in the course of civil science. *Antibiotics* **2022**, *11*, 1198. [CrossRef] [PubMed]
34. Osterman, I.A.; Komarova, E.S.; Shiryaev, D.I.; Korniltsev, I.A.; Khven, I.M.; Lukyanov, D.A.; Tashlitsky, V.N.; Serebryakova, M.V.; Efremenkova, O.V.; Ivanenkov, Y.A.; et al. Sorting out antibiotics' mechanisms of action: A double fluorescent protein reporter for high-throughput screening of ribosome and DNA biosynthesis inhibitors. *Antimicrob. Agents Chemother.* **2016**, *60*, 7481–7489. [CrossRef] [PubMed]
35. Osterman, I.A.; Wieland, M.; Maviza, T.P.; Lashkevich, K.A.; Lukianov, D.A.; Komarova, E.S.; Zakalyukina, Y.V.; Buschauer, R.; Shiriaev, D.I.; Leyn, S.A.; et al. Tetracenomycin X inhibits translation by binding within the ribosomal exit tunnel. *Nat. Chem. Biol.* **2020**, *16*, 1071–1077. [CrossRef] [PubMed]
36. Nouioui, I.; Carro, L.; García-López, M.; Meier-Kolthoff, J.P.; Woyke, T.; Kyrpides, N.C.; Pukall, R.; Klenk, H.-P.; Goodfellow, M.; Göker, M. Genome-based taxonomic classification of the phylum Actinobacteria. *Front. Microbiol.* **2018**, *9*, 2007. [CrossRef] [PubMed]
37. Ciufo, S.; Kannan, S.; Sharma, S.; Badretdin, A.; Clark, K.; Turner, S.; Brover, S.; Schoch, C.L.; Kimchi, A.; DiCuccio, M. Using average nucleotide identity to improve taxonomic assignments in prokaryotic genomes at the NCBI. *Int. J. Syst. Evol. Microbiol.* **2018**, *68*, 2386–2392. [CrossRef]

38. Zeng, Y.; Kulkarni, A.; Yang, Z.; Patil, P.B.; Zhou, W.; Chi, X.; Van Lanen, S.; Chen, S. Biosynthesis of albomycin δ2 provides a template for assembling siderophore and aminoacyl-tRNA synthetase inhibitor conjugates. *ACS Chem. Biol.* **2012**, *7*, 1565–1575. [CrossRef]
39. Ushimaru, R.; Chen, Z.; Zhao, H.; Fan, P.; Liu, H. Identification of the enzymes mediating the maturation of the seryl-tRNA synthetase inhibitor SB-217452 during the biosynthesis of albomycins. *Angew. Chem. Int. Ed.* **2020**, *59*, 3558–3562. [CrossRef]
40. Zeng, Y.; Roy, H.; Patil, P.B.; Ibba, M.; Chen, S. Characterization of two seryl-tRNA synthetases in albomycin-producing *Streptomyces* sp. strain ATCC 700974. *Antimicrob. Agents Chemother.* **2009**, *53*, 4619–4627. [CrossRef]
41. Gause, G.F. Recent studies on albomycin, a new antibiotic. *Br. Med. J.* **1955**, *2*, 1177–1179. [CrossRef]
42. Travin, D.Y.; Severinov, K.; Dubiley, S. Natural Trojan horse inhibitors of aminoacyl-tRNA synthetases. *RSC Chem. Biol.* **2021**, *2*, 468–485. [CrossRef]
43. Stapley, E.O.; Ormond, R.E. Similarity of albomycin and grisein. *Science* **1957**, *125*, 587–589. [CrossRef] [PubMed]
44. Maehr, H. antibiotics and other naturally occurring hydroxamic acids and hydroxamates. *Pure Appl. Chem.* **1971**, *28*, 603–636. [CrossRef]
45. Lin, Z.; Xu, X.; Zhao, S.; Yang, X.; Guo, J.; Zhang, Q.; Jing, C.; Chen, S.; He, Y. Total synthesis and antimicrobial evaluation of natural albomycins against clinical pathogens. *Nat. Commun.* **2018**, *9*, 3445. [CrossRef]
46. Benz, G.; Schröder, T.D.; Kurz, J.; Wünsche, C.; Karl, W.; Steffens, G.; Pfitzner, J.D.; Schmidt, D. Constitution of the deferriform of the albomycins Δ1, Δ2 and ϵ. *Angew. Chem.* **1982**, *21*, 527–528. [CrossRef]
47. Li, X.; Lei, X.; Zhang, C.; Jiang, Z.; Shi, Y.; Wang, S.; Wang, L.; Hong, B. Complete genome sequence of *Streptomyces globisporus* C-1027, the producer of an enediyne antibiotic lidamycin. *J. Biotechnol.* **2016**, *222*, 9–10. [CrossRef] [PubMed]
48. Kim, D.-R.; Cho, G.; Jeon, C.-W.; Weller, D.M.; Thomashow, L.S.; Paulitz, T.C.; Kwak, Y.-S. A Mutualistic interaction between streptomyces bacteria, strawberry plants and pollinating bees. *Nat. Commun.* **2019**, *10*, 4802. [CrossRef] [PubMed]
49. Chen, Z.; Ou, P.; Liu, L.; Jin, X. Anti-MRSA Activity of Actinomycin X2 and Collismycin a produced by *Streptomyces globisporus* WA5-2-37 from the intestinal tract of american cockroach (*Periplaneta americana*). *Front. Microbiol.* **2020**, *11*, 555. [CrossRef] [PubMed]
50. Arnoldi, K.V. survey of harvester ants of the genus *Messor* (Hymenoptera, Formicidae) of the fauna of the USSR. *Zool. Zh.* **1977**, *56*, 1637–1648. [CrossRef]
51. Kipyatkov, V.E. *The World of Social Insects*, 2nd ed.; Leningrad University Press: Leningrad, Russia, 1991; ISBN 5-288-00376-9.
52. Pramanik, A.; Stroeher, U.H.; Krejci, J.; Standish, A.J.; Bohn, E.; Paton, J.C.; Autenrieth, I.B.; Braun, V. Albomycin is an effective antibiotic, as exemplified with *Yersinia enterocolitica* and *Streptococcus pneumoniae*. *Int. J. Med. Microbiol.* **2007**, *297*, 459–469. [CrossRef]

insects

MDPI

Article

Molecular and Functional Characterization of Peptidoglycan Recognition Proteins OfPGRP-A and OfPGRP-B in *Ostrinia furnacalis* (Lepidoptera: Crambidae)

Zengxia Wang [1,2,*], Wan Zhou [3], Baohong Huang [1], Mengyuan Gao [1], Qianqian Li [1], Yidong Tao [1] and Zhenying Wang [2,*]

1 College of Agriculture, Anhui Science and Technology University, Fengyang 233100, China; bhh826@163.com (B.H.); gaomytryhard@163.com (M.G.); qian_11030112@163.com (Q.L.); yidongtao2022@163.com (Y.T.)
2 State Key Laboratory for Biology of Plant Diseases and Insect Pests, MOA—CABI Joint Laboratory for Bio-Safety, Institute of Plant Protection, Chinese Academy of Agricultural Science, Beijing 100193, China
3 College of Resource and Environment, Anhui Science and Technology University, Fengyang 233100, China; eliauk1221@163.com
* Correspondence: wangzengxia100@163.com (Z.W.); wangzy61@163.com (Z.W.)

Simple Summary: The Asian corn borer, *Ostrinia furnacalis* (Guenée), is the most destructive lepidopteran insect pest of corn (*Zea mays* L.) in China. Pathogenic microorganisms play an important role in the population control of the Asian corn borer. Although microorganisms can cause the death of *O. furnacalis*, an immune response also occurs as an attempt to fight off and eliminate invading pathogens. If the molecular mechanism of interaction between *O. furnacalis* and pathogenic bacteria is clarified, the lethal effect of pathogenic microorganisms can be better exerted by inhibiting the natural immune response of *O. furnacalis*. As an important member of the pattern-recognition receptor family, peptidoglycan recognition protein (PGRP) plays a key role in the insect innate immune response. In this study, we cloned two PGRP genes from *O. furnacalis* and analyzed their spatiotemporal expression. In combination with bacterial induction experiments, we revealed the immune signal recognition pathway involved in the two proteins. The results of this study deepen the understanding of the natural immune response of *O. furnacalis* and provide new ideas for better utilization of pathogenic microorganisms in biological control of the Asian corn borer.

Abstract: Peptidoglycan recognition proteins (PGRPs) are important components of insect immune systems, in which they play key roles. We cloned and sequenced two full-length PGRP, named *OfPGRP-A* and *OfPGRP-B*, from the Asian corn borer, *Ostrinia furnacalis*. These two genes comprise open reading frames of 658 and 759 bp, encoding proteins of 192 and 218 amino acids, respectively. qPCR showed that *OfPGRP-A* and *OfPGRP-B* are prominently expressed in the midgut of *O. furnacalis* fourth instar larvae. After inoculation with *Staphylococcus aureus* and *Bacillus thuringiensis*, the expression of *OfPGRP-A* was significantly upregulated, whereas the expression of *OfPGRP-B* was enhanced after inoculation with Escherichia coli. This suggests that *OfPGRP-A* mainly recognizes Gram-positive bacteria and may participate in the Toll signaling pathways, while *OfPGRP-B* identifies Gram-negative bacteria and may participate in Imd signaling pathways. Our results provide insights into the roles of PGRPs in *O. furnacalis* immune function and a foundation for using pathogens for the biological control of *O. furnacalis*.

Keywords: innate immunity; bacterial infection; *Ostrinia furnacalis*; PGRPs

Citation: Wang, Z.; Zhou, W.; Huang, B.; Gao, M.; Li, Q.; Tao, Y.; Wang, Z. Molecular and Functional Characterization of Peptidoglycan Recognition Proteins OfPGRP-A and OfPGRP-B in *Ostrinia furnacalis* (Lepidoptera: Crambidae). *Insects* **2022**, *13*, 417. https://doi.org/10.3390/insects13050417

Academic Editors: Hongyu Zhang, Yin Wang and Xiaoxue Li

Received: 11 April 2022
Accepted: 26 April 2022
Published: 28 April 2022

Publisher's Note: MDPI stays neutral with regard to jurisdictional claims in published maps and institutional affiliations.

Copyright: © 2022 by the authors. Licensee MDPI, Basel, Switzerland. This article is an open access article distributed under the terms and conditions of the Creative Commons Attribution (CC BY) license (https://creativecommons.org/licenses/by/4.0/).

1. Introduction

Innate immunity is the first defense in insects and mammals against bacterial, fungal, viral, and parasitic infections. Identifying invading pathogens is key to triggering the host immune responses. Peptidoglycan recognition protein (PGRP), a member of the pattern-recognition receptor family, can recognize pathogen-associated molecular patterns present on the pathogen surface. After recognition, the Toll signaling pathway, Imd pathway, JAK–STAT pathway, reactive oxygen metabolism, and a melanization reaction are selectively activated to eliminate pathogens [1–3].

PGRP was first found in the blood of *Bombyx mori*, and then in mammals (humans and mice) [4]. More than 100 PGRP family members have been identified [5]. These proteins are highly conserved, containing at least one conserved PGRP domain, which is similar to *N*-acetylmuramic acid-L-alanine amidase [6]. PGRPs have been detected in some insects, mollusks, and vertebrates but not in plants, nematodes, and aphids [7]. PGRPs can be divided into two categories according to their molecular size: short PGRPs contain a signal peptide but no transmembrane domain, whereas long PGRPs have a signal peptide and predicted transmembrane domain [8,9]. Insect PGRPs are generally expressed in immune organs. Some of these PGRPs are constitutively expressed, whereas other PGRPs are inducible. For example, in *Drosophila melanogaster*, although PGRP-S is secreted in different parts of the body, it is mostly expressed in immune organs, particularly after microbial infection. *PGRP-SB1, SC2,* and *SD* are highly expressed in the fat body, *PGRP-SA* is expressed in the epithelial tissue, and *PGRP-SC* is specifically expressed in viscera [10].

PGRPs play important roles in identifying invading pathogens and regulating the innate immune responses, and they can directly kill invading pathogens. PGRPs can recognize pathogenic microorganisms and activate the Toll signaling pathway and Imd pathway, which induces the synthesis of antimicrobial peptides. These two pathways are activated by Lys-type and DAP-type peptidoglycans (PGNs), respectively [11]. Dm PGRP-SA and Dm PGRP-SD in *D. melanogaster* bind with Gram-negative bacteria binding protein 1 to activate the Toll signaling pathway [12]. Dm PGRP-LC and Dm PGRP-LE of *D. melanogaster* can recognize DAP-PGN on the cell wall of Gram-negative bacteria, and the PGRP functional area outside the cell membrane binds to PGN to activate the Imd signaling pathway [13]. Different PGRPs share partly conserved structures but exhibit varied functions and expression patterns. Numerous studies have demonstrated how PGRPs in the model insects *D. melanogaster* and *B. mori* recognize bacteria, initiate immune responses, and participate in signaling pathways, whereas relatively few studies have focused on other lepidopterans.

The Asian corn borer, *Ostrinia furnacalis* (Guenée), is the most destructive lepidopteran insect pest of corn (*Zea mays* L.) in China [14]. It is widely distributed in the Asia Pacific region, where it has been reported to cause 20% to 80% yield losses in the Philippines, and 10% to 30% yield losses in China [15]. The larvae of *O. furnacalis* can bore into stalks and ear shanks, feeding on whorl leaves and young kernels, thus resulting in yield losses. The use of entomopathogens for controlling *O. furnacalis* has recently received increasing attention [16,17]. However, research on entomopathogens, including their recognition by the host immune system and role in triggering the innate immune response, remains limited.

Initially, two PGRPs were identified by analyzing full-length transcriptome sequences of pathogen-induced *O. furnacalis*. After sequence alignment, these genes were named *OfPGRP-A* (Gene ID:ON152884) and *OfPGRP-B* (Gene ID: ON152885). In this study, we conducted cloning and bioinformatics analysis of these two genes. Based on the tissue expression patterns of *OfPGRP-A* and *OfPGRP-B* after bacterial inoculation, we explored the role of the two genes in the innate immune response. We predicted the functions of *OfPGRP-A* and *OfPGRP-B* proteins and their involvement in immune signal pathways. This study provides a theoretical basis for using microbial pesticides to control Asian corn borer and for developing more efficient and environmentally friendly pesticides.

2. Materials and Methods

2.1. Insect Rearing and Bacteria Culture

O. furnacalis larvae were reared at 26 ± 1°C, RH = 80% ± 5%, and a photoperiod of 16 L: 8 D. The larvae were fed on a semi-artificial diet as described by Zhou [18]. The larval instar was identified according to head width.

The bacteria used in this experiment were the Gram-positive bacteria *Staphylococcusaureus* and *Bacillus thuringiensis* and the Gram-negative bacterium *Escherichia coli*. These three bacterial strains were purchased from the Institute of Microbiology, Chinese Academy of Sciences (Beijing, China). The bacteria were activated and then cultured at 37 °C and 200 rpm until the OD_{600} reached around 1. After centrifugation at 5000× *g* for 5 min at 4 °C, the supernatant was removed. The bacteria were collected, washed three times with sterile phosphate-buffered saline (pH 6.4), and resuspended to the required bacterial concentration. After autoclaving at 121 °C for 20 min, they were kept at −20 °C until use.

2.2. Total RNA Extraction and cDNA Synthesis

The TRIzol method was used to extract total RNA from *O. furnacalis*. First, 1% agarose gel electrophoresis was used to detect the quality of the total RNA, then a Nanodrop 2000 (Thermo Scientific, Waltham, MA, USA) was used to detect the concentration and purity of the total RNA, and an Agilent 2100 bioanalyzer (Agilent Technologies, Santa Clara, CA, USA) was used to determine the RNA integrity to ensure the quality of the RNA samples. Finally, high-quality RNA samples were reverse transcribed to synthesize cDNA using a TransScript One-step gDNA Removal and cDNA Synthesis SuperMix kit (TransGen Biotech, Beijing, China).

2.3. cDNA Cloning of OfPGRP-A and OfPGRP-B

According to the transcriptome sequencing results of *O. furnacalis*, two PGRP gene sequences were obtained. Based on sequence alignment and phylogenetic tree analysis, these genes were named *OfPGRP-A* and *OfPGRP-B*. Based on the cDNA sequences of *OfPGRP-A* and *OfPGRP-B*, primers were designed using Primer Premier 5.0 software (Premier Biosoft, Rockville, MD, USA) and DNAMAN5.0 (LynnonBioSoft, San Ramon, CA, USA), respectively (Table 1). The cDNA that was obtained from the fourth instar larvae at 24 h after injecting bacteria (a mixture of *S. aureus*, *B. thuringiensis*, and *E. coli*.) into *O. furnacalis* was used as a template to amplify the full-length sequence of two PGRPs. The 50-μL reaction system contained 1 μL cDNA template, 1 μL each of upstream and downstream primers (10 mmol/L), 25 μL Taq PCR Master Mixblue dye (2x), and 22 μL ddH_2O. The PCR conditions were as follows: 94 °C for 3 min, 94 °C for 30 s, 60 °C for 30 s, and 72 °C for 1 min for 30 cycles, followed by 72 °C for 10 min. The PCR product was detected by 1% agarose gel electrophoresis. The target band was recovered using an AxyGen DNA gel recovery kit (Union City, CA, USA) and ligated into the pGEM-T Easy vector (Promega, Beijing, China). The ligation product was transformed into E. coli DH5α competent cells, and positive clones were screened using PCR with M13 universal primers. Finally, the samples were sent for sequencing to Sangon Biotech Co., Ltd. (Nanjing, China).

Table 1. Primers used in the experiment.

Gene	Forward Primer	Reverse Primer	Purpose
OfPGRP-A1	TCAGTACCTGCCGAGGCCAGTC	GAAGGAAGAACCAATGTCCCACCAA	Cloning
OfPGRP-B1	TTCATTTCAACAGCGTCAGCCTCG	CGGGTGCGGTGAGTAGTGTTTCC	
OfPGRP-A2	ATGTTCCGAAAGTTGAATATTT	GATCGAGCTGACGTCGTCCATC	Full-length clone
OfPGRP-B2	ATGCCGGGTCCGCTGCCAGTA	TCATCATAAGTTGCATTCCCCCT	
OfPGRP-A3	TGCTGGCCAAAGTCTAGACA	AGTAAGGAACATCGCCCCAA	Real-time PCR
OfPGRP-B3	TGGCCGATGAGAGTGTAGTC	GGATACAGTTTTGCGGTGGG	
RPL18	ACGGAGGTGGTAACCATCAACA	ACGCCTCCTTCTTGGTGTCG	

2.4. Sequence Analysis of OfPGRP-A and OfPGRP-B

We spliced and evaluated the sequenced *O. furnacalis* PGRP genes to obtain complete cDNA sequences. ExPASy (https://web.expasy.org/compute_pi/, accessed on 10 April 2022) was used to predict the molecular weight and isoelectric point of the proteins. SMART (http://smart.embl-heidelberg.de/, accessed on 10 April 2022) was used to analyze functional domains in the protein sequences. The online software SignalP 5.0 (https://services.healthtech.dtu.dk/service.php?SignalP-5.0, accessed on 10 April 2022) was used to analyze signal peptides. TMHMM Server 2.0 (https://services.healthtech.dtu.dk/service.php?TMHMM-2.0, accessed on 10 April 2022) was used to analyze the transmembrane domain. We downloaded the PGRP sequences of other insects from NCBI, used Clustal X2.0 for homology alignment analysis, and used Jalview software 2.10 for multiple sequence alignment editing. We also used MEGA 7.0 software for neighbor-joining analysis to construct a phylogenetic tree, with bootstrapping 1000 times for testing.

2.5. Spatiotemporal Expression Profiles of OfPGRP-A and OfPGRP-B

We collected *O. furnacalis* larvae from the 1st–5th instars and dissected 4th instars to collect samples from different tissues (hemolymph, fat body, midgut, and epidermis). Sixty 4th instar *O. furnacalis* larvae were randomly selected, their stomachs and feet were disinfected with 70% alcohol, and they were placed on ice for 5 min for freezing and anesthesia. Next, we sterilized the larvae's gastropod using a sterilized insect needle, gently squeezed the gastropod to drop the hemolymph on parafilm, and this sample was placed in a 1.5 mL centrifuge tube. The epidermis was dissected, and the midgut, fat body, and body wall were separately collected. The hemolymph was centrifuged at 4 °C and 12,000× *g* for 30 min, and the supernatant was collected. All tissue samples were stored at −80 °C until use.

Primer Premier 5.0 software was used to design quantitative primers (Table 1). A Takara TB Green Premix Ex Taq kit (Shiga, Japan) and ABI ViiA7 real-time fluorescent quantitative PCR machine (Applied Biosystems, Foster City, CA, USA) were used to detect the expression levels of *OfPGRP-A* and *OfPGRP-B* in different stages and different tissues of the larvae. qPCR reaction consists of a system (20 μL): 1 μL cDNA template, 1 μL each upstream and downstream primers (10 mmol/L), 1 μL ROX Reference Dye II, 10 μL TB Green Premix Ex Taq, and 6 μL ddH_2O. The program was set according to the RR420A manual as follows: 95 °C for 30 s, 95 °C for 5 s, and 60 °C for 34 s for 40 cycles, followed by 95 °C for 15 s, 60 °C for 1 min, and 95 °C for 15 s. *Ostrinia furnacalis RPL-18* was used as an internal reference gene [19].

2.6. OfPGRP-A and OfPGRP-B Expression after Bacterial Induction

Fourth instar larvae of *O. furnacalis* of the same size were selected and divided into a control group, phosphate-buffered saline injection group, *S. aureus* injection group, *E. coli* injection group, and *B. thuringiensis* injection group. Thirty larvae were injected for each treatment as described by Sun and Bai [20]. A microinjector was used to draw 5 μL of inactivated bacterial solution (around 3.0×10^6 cells/mL); this solution was injected into the larval gastropod. After injection, the larvae surfaces were disinfected with 70% alcohol. The larvae were then reared at 26 ± 1 °C, RH = 80 ± 5%, and a photoperiod of 16 L: 8 D.

Five larvae were collected for dissection at 2, 4, 8, 12, 24, and 48 h after injecting the bacteria, and different tissue samples were collected for quantitative analysis. Three replicates were used for each treatment. The collection method and qPCR program were the same as those described in Section 2.5, and the sample RNA extraction and cDNA synthesis steps were performed as described in Section 2.2.

2.7. Data Analysis

The relative expression levels of *OfPGRP-A* and *OfPGRP-B* in *O. furnacalis* at different instars, in different tissues, and after induction with different bacteria were calculated using the $2^{-\Delta\Delta Ct}$ method [21]. Differences in expression between samples or treatments were analyzed using analysis of variance (ANOVA). SPSS16.0 software (SPSS, Inc., Chicago,

IL, USA) was used for statistical analysis. Illustrator software Origin 8.0 (OriginLab, Northampton, MA, USA) was used to prepare illustrations.

3. Results

3.1. Molecular Characteristics and Phylogenetic Analysis of OfPGRP-A and OfPGRP-B cDNA Sequence

The full-length sequences of *O. furnacalis OfPGRP-A* and *OfPGRP-B* were obtained by PCR amplification. The complete cDNA of *OfPGRP-A* consisted of 658 bp and encoded a protein of 192 amino acids (Figure 1A). The predicted molecular mass of the *OfPGRP-A* protein was 21.79 kD, with an estimated pI of 8.20, and lacked a transmembrane domain but had a PGRP domain (amino acids 22–164), Ami2 domain (amino acids 33–170), and a *N*-terminal signal peptide comprising 20 amino acids. The open reading frame of *OfPGRP-B* was 759 bp and encoded a protein of 218 amino acids (Figure 1B). The predicted molecular weight of the encoded protein was 24.44kD with a pI of 5.68, and it had no transmembrane domain nor signal peptide but had a PGRP domain (amino acids 17-160) and an Ami2 domain (amino acids 29-166). These structural characteristics indicated that *OfPGRP-A* and *OfPGRP-B* are short-form PGRPs.

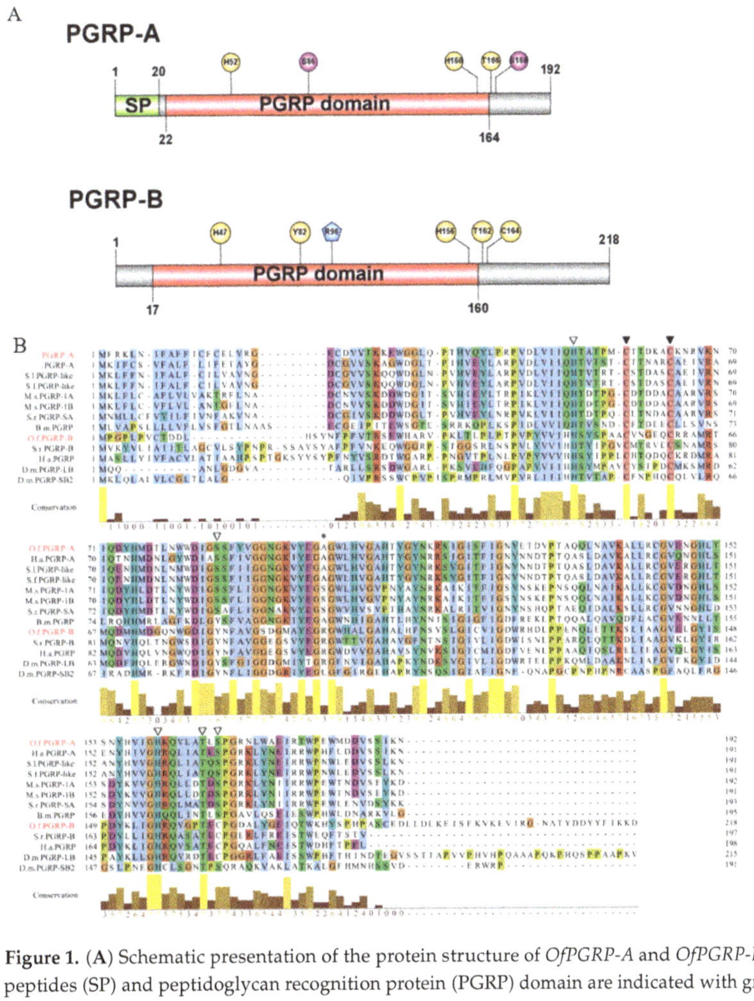

Figure 1. (**A**) Schematic presentation of the protein structure of *OfPGRP-A* and *OfPGRP-B*. Signal peptides (SP) and peptidoglycan recognition protein (PGRP) domain are indicated with green and

red, respectively. Yellow circles and purple circles represent key amino acid sites that determine zinc ion/amidase activity and mutation sites, respectively. Blue pentagon represents key DAP-type amino acid sites. (**B**) Multiple-sequence alignment of *OfPGRP-A* and *OfPGRP-B* with other insect PGRPs. Ha: *Helicoverpaarmigera*; Ms: *Manduca sexta*; Sl: *Spodopteralitura*; Sf: *Spodopterafrugiperda*; Sr: *Samiaricini*; Bm: *Bombyx mori*; Dm: *Drosophila melanogaster*. ▽ Key amino acid sites that determine zinc ion/amidase activity, ▼ conserved cysteine sites, and * key DAP-type amino acid sites.

The PGRPs of *Helicoverpaarmigera, Manduca sexta, Spodopteralitura, Spodopterafrugiperda,* and *Samiaricini* were selected for multiple-sequence alignment with OfPGRP-A and OfPGRP-B amino acid sequences (Figure 1A,B). The conserved regions of insect PGRPs are very consistent, and the domains are highly conserved. OfPGRP-A and OfPGRP-B contain two and three conserved cysteine residues, respectively. Some insects contain the H–Y–H–T–C structure, which is necessary for forming the amidase active sites with amidase activity. The sequence alignment results showed that both *O. furnacalis* OfPGRP-A and OfPGRP-B contained the five key amino acid residues required for PGRP/amidase activity. However, in the sequence of OfPGRP-SA, tyrosine is mutated to serine and cysteine is mutated to serine, suggesting that the protein encoded by OfPGRP-A lacks amidase activity. In contrast, OfPGRP-B retains the five key amidase active sites H–Y–H–T–C, indicating that the encoded protein has amidase activity. Additionally, OfPGRP-B contains arginine residues and a DAP-type PGN recognition site, indicating that OfPGRP-B can recognize DAP-type PGN.

The phylogenetic tree of *O. furnacalis* PGRP and that of other species constructed using MEGA 7.0 software and online website modification is shown in Figure 2. Cluster analysis showed that OfPGRP-A and the *P. xylostella* PGRP are closely related. OfPGRP-B clustered with PGRP-Bs from other insects and is closely related to that in *M. sexta*, with an amino acid sequence conservation of 96%.

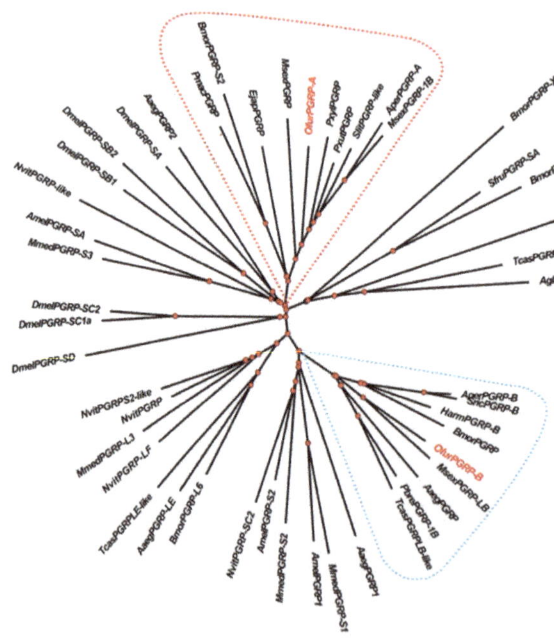

Figure 2. Phylogenetic tree of OfPGRP and PGRPs from other species. Bmor: *Bombyx mori*; Sfru: *Spodopterafrugiperda*; Aaeg: *Aedes aegypti*; Tcas: *Triboliumcastaneum*; Agla: *Anoplophoraglabripennis*;

Aper: *Antheraeapernyi;* Sric: *Samiaricini;* Harm: *Helicoverpaarmigera;* Msex: *Manduca sexta;* Pbre: *Protaetiabrevitarsis;* Mmed: *Microplitis mediator;* Amel: *Apis mellifera;* Nvit: *Nasoniavitripennis;* Dmel: *Drosophila melanogaster;* Pmac: *Papiliomachaon;* Ejap: *Eumeta japonica;* Pxyl: *Plutellaxylostella;* Pxut: *Papilio Xuthus;* Slit: *Spodopteralitura.*

3.2. Spatiotemporal Expression of OfPGRP-A and OfPGRP-B

We used *PRL-18* as an internal reference gene to analyze the expression patterns of PGRP in different stages and different tissues of *O. furnacalis* larvae. The expression levels of *OfPGRP-A* and *OfPGRP-B* were highest in 4th instar larvae, showing significantly higher levels compared with those in other instars ($p < 0.05$). The expression level of *OfPGRP-A* in 4th instar larvae was 55-fold higher than that in 1st instar larvae, and *OfPGRP-B* showed the lowest expression level in 1st instar larvae, with the expression being almost undetectable (Figures 3A and 4A). *OfPGRP-A* was expressed in all larval tissues. Expression was highest in the midgut, which showed significantly higher levels compared with all other tissues ($p < 0.05$), followed by expression in the fat body. Expression was lowest in the hemolymph; expression in the midgut was nearly 140-fold higher than that in the hemolymph. *OfPGRP-B* was also highly expressed in the midgut at significantly higher levels than in the epidermis and hemolymph, indicating that the expression of *OfPGRP* is tissue-specific (Figures 3B and 4B).

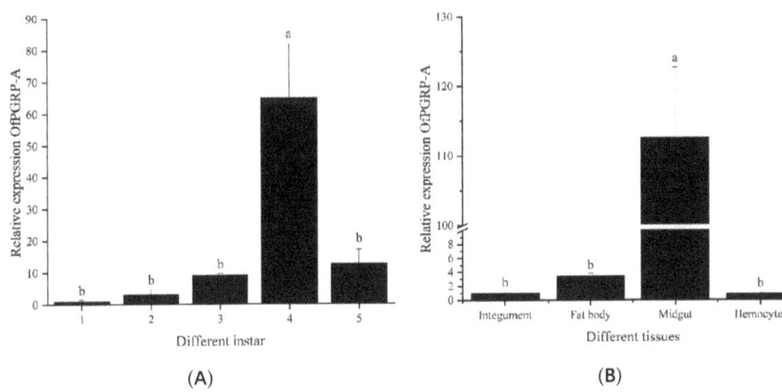

Figure 3. Relative expression levels of *OfPGRP-A* in different developmental stages (**A**) and adult tissues (**B**) of *Ostrinia furnacalis*. The small letters represent a signifiicant difference (one-way ANOVA, followed by Tukey's test as post hoc, $p < 0.05$).

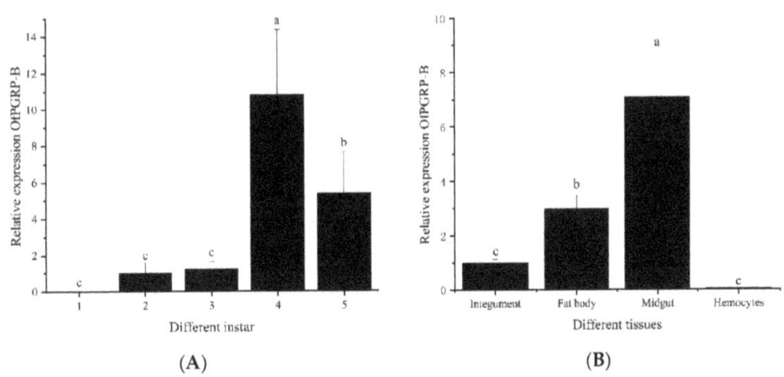

Figure 4. Relative expression levels of *OfPGRP-B* in different developmental stages (**A**) and adult tissues (**B**) of *Ostrinia furnacalis*. The small letters represent a signifiicant difference (one-way ANOVA, followed by Tukey's test as post hoc, $p < 0.05$).

3.3. Expression of OfPGRP-A and OfPGRP-B after Bacterial Inoculation

Inoculation of the three bacteria caused changes in the expression level of *OfPGRP-A* in *O. furnacalis* larvae (Figure 5). In the epidermis, the expression of *OfPGRP-A* was highest at 4 h after *B. thuringiensis* injection (35-fold higher than that in the control ($p < 0.05$)), after which the expression gradually decreased. *OfPGRP-A* was also significantly upregulated at multiple time points after *S. aureus* induction. In the hemolymph, *OfPGRP-A* was significantly activated by *S. aureus*. The expression levels at 2, 4, 8, and 12 h after induction were significantly higher than those in the control and after other bacterial treatments. At 8 h after treatment, the expression of *OfPGRP-A* reached the highest level ($p < 0.05$; 205-fold higher than that in the control). In the fat body, *OfPGRP-A* expression was upregulated at all time points. At 2 and 4 h after *S. aureus* injection, the expression level was significantly higher than in the other groups ($p < 0.05$), after which the expression level decreased. The expression level of *OfPGRP-A* following *B. thuringiensis* induction reached the highest level after 12 h. After induction by *E. coli*, the expression level of *OfPGRP-A* was upregulated, although the increase was not as significant as that observed with the other two bacteria. In the midgut, the expression of *OfPGRP-A* reached the highest level at 12 h after each of the three bacterium injections, with the expression being significantly higher than that in the control ($p < 0.05$). The most significant increase in *OfPGRP-A* expression was induced by *S. aureus*. Some PGRPs act as pattern recognition receptors that recognize and bind to foreign pathogens and then activate the Toll and Imd pathways. Lysine-type PGN mainly activates the Toll pathway, while the DAP-type PGN activates the Imd pathway [22,23]. According to our results, *OfPGRP-A* expression was significantly increased after induction by the Gram-positive bacteria *S. aureus* and *B. thuringiensis*. This gene may be involved in activating the Toll pathway.

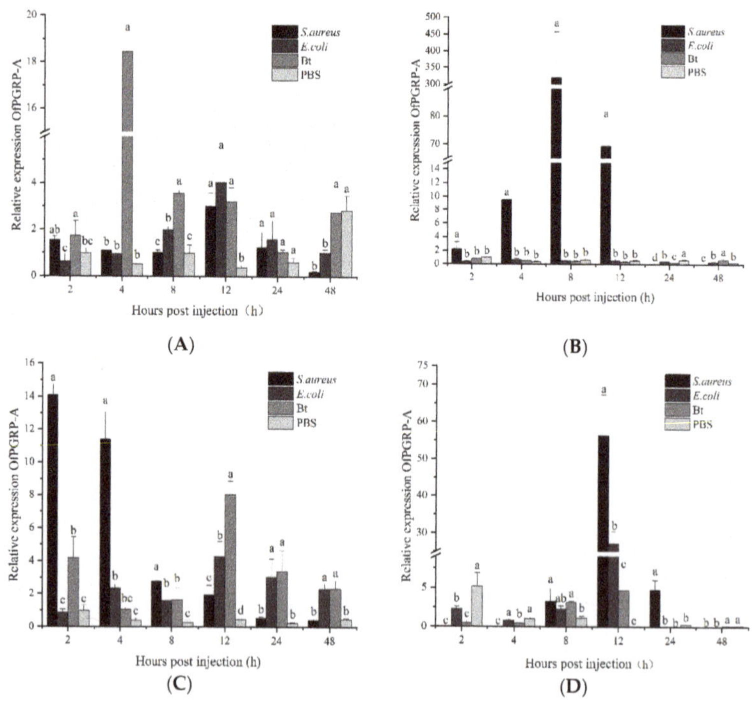

Figure 5. Expression levels of *OfPGRP-A* in different tissues of 4th instar larvae of *Ostrinia furnacalis* after injection with different pathogens: (**A**) epidermis; (**B**) hemolymph; (**C**) fat body; (**D**) midgut.

PBS: PBS buffer; Bt: *Bacillus thuringiensis*; Staph: *Staphylococcus aureus*; RIL: *Escherichia coli*. The small letters represent a signifiicant difference (one-way ANOVA, followed by Tukey's test as post hoc, $p < 0.05$).

The expression level of the *OfPGRP-B* gene in *O. furnacalis* larvae was higher than that in the control group and other bacterial treatment groups at various time points and in different tissues after *E. coli* injection (Figure 6). In the epidermis, the expression of *OfPGRP-B* reached the highest level at 2 h after *E. coli* induction, showing significantly higher expression than that in the control group ($p < 0.05$), after which it decreased. In the hemolymph, after *E. coli* induction, the expression of *OfPGRP-B* increased, reaching the highest expression level at 8 h, and then decreased. Compared with *S. aureus* and *B. thuringiensis*, there is a significant difference in expression ($p < 0.05$). *OfPGRP-B* expression was significantly upregulated in the fat body after *E. coli* induction compared with that in the control group at all time points ($p < 0.05$). The fold-increase was particularly significant in the first eight hours. *S. aureus* and *B. thuringiensis* could also increase *OfPGRP-B* expression, but to lower levels than that induced by *E. coli* injection. In the midgut, after injection of the three bacteria, the induction effect of *E. coli* was the most significant, followed by *S. aureus*. Compared with the control, the expression of *OfPGRP-B* significantly differed at 4, 8, and 12 h after induction, reaching the highest level at 8 h. According to these results, the expression of *OfPGRP-B* increased significantly after induction by the Gram-negative bacteria *E. coli*. This gene may activate the Imd pathway.

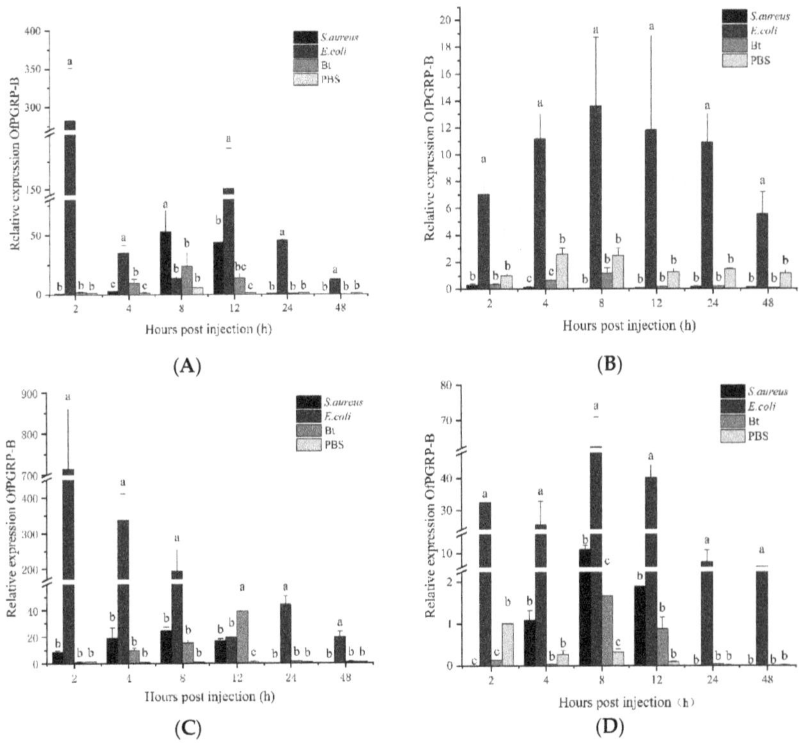

Figure 6. Expression levels of *OfPGRP-B* in different tissues of 4th instar larvae of *Ostrinia furnacalis* after injection with different pathogens: (**A**) epidermis; (**B**) hemolymph; (**C**) fat body; (**D**) midgut. PBS: PBS buffer; Bt: *Bacillus thuringiensis*; Staph: *Staphylococcus aureus*; RIL: *Escherichia coli*. The small letters represent a signifiicant difference (one-way ANOVA, followed by Tukey's test as post hoc, $p < 0.05$).

4. Discussion

Insects possess innate immunity against invading microbial pathogens. PGRPs in insects play important roles in recognizing bacterial infections. These proteins can specifically recognize PGNs on bacterial surfaces, which activate the Toll and Imd pathways as well as downstream immune responses [7,12]. In recent years, the functions of various members of the PGRP family have been determined in insects; however, these studies mainly focused on model insects, such as *D. melanogaster* and *B. mori*. Two PGRP gene sequences (*OfPGRP-A* and *OfPGRP-B*) were screened from the sequence data of the transcriptome of *O. furnacalis* larvae by sequence comparison and evolutionary analysis. In this study, the full-length cDNA sequences of *OfPGRP-A* and *OfPGRP-B* were successfully cloned, the sequence characteristics of the genes were analyzed using bioinformatics software, and their spatiotemporal and pathogen-induced expression patterns were detected using qPCR.

Insect PGRP can be divided into short PGRP-S and long PGRP-L according to the molecular size and characteristic structure. For example, 13 PGRPs were found in *Drosophila melanogaster*, of which seven (SA, SB1, SB2, SC1a, SC1b, SC2, and SD) were type PGRP-S and contained signal peptides but no transmembrane domains. The other six were PGRP-L, which are divided into two categories: those in one category contain signal peptides and transmembrane domains, whereas members of the other contain no signal peptides nor transmembrane domains [24]. Sequence analysis revealed that *OfPGRP-A* and *OfPGRP-B* have conserved PGRP domains. *OfPGRP-A* has a signal peptide but no transmembrane domain, whereas *OfPGRP-B* has neither a transmembrane domain nor a signal peptide. We predicted that both genes belong to the short form of PGRP, which is a secreted extracellular protein. For most PGRPs that have amidase activity, this activity is contained in the H–Y–H–T–C structure [25]. Sequence comparison showed that in the *OfPGRP-A* sequence, tyrosine is mutated to serine and cysteine is mutated to serine, suggesting that *OfPGRP-A* lacks amidase activity. *OfPGRP-B* retains five key amidase active sites of H–Y–H–T–C. It also has arginine residues in the DAP-type PGN recognition site, indicating that this gene has amidase activity. *OfPGRP-B* may be involved in recognizing Gram-negative bacteria, leading to activation of the Imd signaling pathway. *OfPGRP-A* and *OfPGRP-B* clustered with the PGRP genes of *P. xylostella* and *M. sexta*, respectively, indicating that PGRP has relatively conserved evolutionary characteristics in lepidopterans.

PGRPs show significantly different expression levels at various developmental stages as well as tissue-specific expression. The expression levels of *BdPGRP-LCa* and *BdPGRP-LCb* of *Bactrocera dorsalis* increase from the mature larva to early pupal stage and from the late pupal to early adult stage. Both genes are highly expressed in the larval midgut and fat bodies but lower in the gonads. *BdPGRP-SA* exhibits the highest expression in adults, lowest expression in the egg and early larval stages, and high expression in the fat body and midgut [26,27]. *PxPGRP-S1* shows the highest expression in the 4th instar larval stage and pupal stage of *Plutellaxylostella*; lowest expression in the egg, 2nd instar, and adult stages; and highest expression in the fat body among tissues [28]. In this study, *OfPGRP-A* and *OfPGRP-B* showed the highest expression in 4th instar larvae. Before the 4th instar, the expression level gradually increased with age. As the larvae grew, immune function gradually enhanced, and the ability to resist pathogens was strongest in the 4th instar larvae. The midgut is an essential immune response center in the intestine, whereas the fat body synthesizes, stores, detoxifies, and metabolizes toxic substances. Both the midgut and fat body are important immune organs that defend against pathogens [29]. In this study, the expression levels of *OfPGRP-A* and *OfPGRP-B* were significantly higher in the midgut than in other tissues. After inoculation by different bacteria, *OfPGRP-A* and *OfPGRP-B* expression changed significantly in each tissue to different extents. Thus, *OfPGRP* expression may differ between tissues.

Insect PGRPs can recognize PGN in the bacterial cell wall and activate the Toll and Imd pathways [30]. PGRP activates the Toll signaling pathway upon recognition of PGN from Gram-positive bacteria but activates the Imd signaling pathway upon recognition of PGN from Gram-negative bacteria. In *Drosophila*, PGRP-SA can bind to lysine-type PGN and activate the Toll pathway [31], whereas PGRP-LE and PGRP-LC cooperate to activate the Imd pathway [32]. In this study, *OfPGRP-A* expression increased significantly after injection of *S. aureus* and *B. thuringiensis*, and *OfPGRP-B* expression increased significantly after injection of *E. coli*. *S. aureus* and *B. thuringiensis* are Gram-positive bacteria with lysine-type PGN, whereas *E. coli* are Gram-negative bacteria with DAP-type PGN. The change in PGRP gene expression after injection with different bacteria indicates that PGRP regulates different signaling pathways. Combined with previous analysis of the gene characteristics, *OfPGRP-A* may be involved in activation of the Toll pathway, and *OfPGRP-B* may be involved in activation of the Imd pathway. In further studies, we will obtain the recombinant proteins of OfPGRP-A and OfPGRP-B using a prokaryotic expression system and perform in vitro experiments to explore the functions of OfPGRP-A and OfPGRP-B proteins, thus determining the possible immune signal pathways in which they participate. This study lays a foundation for further research on the function of PGRP in *O. furnacalis* and will help facilitate the use of entomopathogens for control of this insect pest.

Author Contributions: Conceptualization: Z.W. (Zengxia Wang) and B.H.; methodology: W.Z., M.G., Q.L. and Y.T.; data curation: Z.W. (Zengxia Wang); formal analysis: Q.L. and Y.T.; investigation: M.G. and W.Z.; project administration: Z.W. (Zengxia Wang); supervision: B.H. and Z.W. (Zhenying Wang); validation: Z.W. (Zengxia Wang); visualization: B.H.; writing—original draft: Z.W. (Zengxia Wang) and W.Z.; funding acquisition: Z.W. (Zengxia Wang); writing—review and editing: Z.W. (Zhenying Wang) and B.H. All authors have read and agreed to the published version of the manuscript.

Funding: This research was partially supported by the following grants: Natural Science Foundation of Anhui Province (1908085QC136); Major science and technology project of Anhui Province (201903a06020027) and the Research Start-up Fund for the New Doctoral Staff by Anhui Science and Technology University (NXYJ201701).

Institutional Review Board Statement: Not applicable.

Informed Consent Statement: Not applicable.

Data Availability Statement: The data presented in this study are available on reasonable request from the corresponding author.

Acknowledgments: We would like to thank Xing Ge for English language editing and useful comments on the manuscript.

Conflicts of Interest: The authors declare no conflict of interest.

References

1. Huang, W.; Xu, X.; Freed, S.; Zheng, Z.; Wang, S.; Ren, S.; Jin, F. Molecular cloning and charac-terization of a β-1,3-glucan recognition protein from *Plutella xylostella* (L.). *New Biotechnol.* **2015**, *32*, 290–299. [CrossRef]
2. Jiang, H.; Vilcinskas, A.; Kanost, M.R. Immunity in lepidopteran insects. *Adv. Exp. Med. Biol.* **2010**, *708*, 181–204. [PubMed]
3. Lu, Y.; Su, F.; Li, Q.; Zhang, J.; Yu, X.Q. Pattern recognition receptors in *Drosophila* immune responses. *Dev. Comp. Immunol.* **2019**, *102*, 103468. [CrossRef] [PubMed]
4. Lee, W.J.; Lee, J.D.; Kravchenko, V.V.; Ulevitch, R.J.; Brey, P.T. Purification and molecular cloning of an inducible gram-negative bacteria-binding protein from the silkworm, *Bombyx mori*. *Proc. Natl. Acad. Sci. USA* **1996**, *93*, 7888–7893. [CrossRef] [PubMed]
5. Gerardo, N.M.; Altincicek, B.; Anselme, C.; Atamian, H.; Barribeau, S.M.; Vos, M.D.; Duncan, E.J.; Evans, J.D.; Gabaldón, T.; Ghanim, M.; et al. Immunity and other defenses in pea aphids, *Acyrthosiphon pisum*. *Genome Biol.* **2010**, *11*, R21. [CrossRef]
6. Xiong, G.H.; Xing, L.S.; Lin, Z.; Saha, T.T.; Wang, C.; Jiang, H.; Zou, Z. High throughput profiling of the cotton bollworm *Helicoverpa armigera* immunotranscriptome during the fungal and bacterial infections. *BMC Genom.* **2015**, *16*, 321. [CrossRef] [PubMed]
7. Royet, J.; Gupta, D.; Dziarski, R. Peptidoglycan recognition proteins: Modulators of the microbiome and inflammation. *Nat. Rev. Immunol.* **2011**, *11*, 837–851. [CrossRef]
8. Lemaitre, B.; Hoffmann, J. The host defense of *Drosophila melanogaster*. *Ann. Rev. Immunol.* **2007**, *25*, 697–743. [CrossRef]
9. Tanaka, H.; Ishibashi, J.; Fujita, K.; Nakajima, Y.; Sagisaka, A.; Tomimoto, K.; Suzuki, N.; Yoshiyama, M.; Kaneko, Y.; Iwasaki, T.; et al. A genome-wide analysis of genes and gene families involved in innate immunity of *Bombyx mori*. *Insect Biochem. Mol. Biol.* **2008**, *38*, 1087–1110. [CrossRef]

10. Gottar, M.; Gobert, V.; Michel, T.; Belvin, M.; Duyk, G.; Hoffmann, J.A.; Ferrandon, D.; Royet, J. The *Drosophila* immune response against gram-negative bacteria is mediated by a peptidoglycan recognition protein. *Nature* **2002**, *416*, 640–644. [CrossRef] [PubMed]
11. Steiner, H.; Hultmark, D.; Engstrom, A.; Bennich, H.; Boman, H.G. Sequence and specificity of two antibacterial proteins involved in insect immunity. *Nature* **1981**, *292*, 246–248. [CrossRef] [PubMed]
12. Wang, Q.; Ren, M.J.; Liu, X.Y.; Xia, H.C.; Chen, K.P. Peptidoglycan recognition proteins in insect immunity. *Mol. Immunol.* **2019**, *106*, 69–76. [CrossRef] [PubMed]
13. Kurata, S. Peptidoglycan recognition proteins in *Drosophila* immunity. *Dev. Comp. Immunol.* **2014**, *42*, 36–41. [CrossRef] [PubMed]
14. Wang, Z.Y.; Lu, X.; He, K.L.; Zhou, D.R. Review of history, present situation and prospect of the Asian maize borer research in China. *J. Shenyang Agric. Univ.* **2000**, *31*, 402–412.
15. Schreiner, I.H.; Nafus, D.M. Detasselling and insecticides for control of *Ostrinia furnacalis* (Lepidoptera: Pyralidae) on sweet corn. *J. Econ. Entomol.* **1987**, *80*, 263–267. [CrossRef]
16. Duan, X.L.; He, K.L.; Wang, Z.Y.; Wang, X.Y.; Li, Q. Investigation on the main parasitoids and case fatality rate of the overwintering larvae of the Asian corn borer, *Ostrinia furnacalis*. *Chin. J. Biol. Control* **2014**, *6*, 823–827.
17. Feng, S.D.; Li, X.H.; Wang, Y.Z.; Zhang, J.; Xu, W.J.; Zhang, Z.K.; Wang, D.; Li, Q. Ecological control effects of two mating types of *Beauveria bassiana* on *Ostrinia furnacalis*. *Acta Ecol. Sin.* **2017**, *37*, 650–658.
18. Zhou, D.R.; Wang, Y.Y.; Liu, B.L.; Ju, Z.L. Studies on the artificial mass propagation of the corn borer, *Ostrinia furnacalis*: I. A semi-artificial feed and its improvement. *J. Plant Prot.* **1980**, *7*, 113–122.
19. Wu, T.Y.; Zhao, Y.; Wang, Z.Y.; Song, Q.S.; Wang, Z.X.; Xu, Q.W.; Wang, Y.; Wang, L.; Zhang, Y.; Feng, C. β-1,3-glucan recognition protein 3 activates the prophenoloxidase system in response to bacterial infection in *Ostrinia furnacalis* guenée. *Dev. Comp. Immunol.* **2018**, *79*, 31–43. [CrossRef]
20. Sun, J.; Bai, Y. Predator-induced stress influences fall armyworm immune response to infecting bacteria. *J. Invertebr. Pathol.* **2020**, *172*, 107352. [CrossRef]
21. Livak, K.J.; Schmittgen, T.D. Analysis of relative gene expression data using real-time quantitative PCR. *Methods* **2002**, *25*, 402–408. [CrossRef] [PubMed]
22. Kaneko, T.; Goldman, W.E.; Mellroth, P.; Steiner, H.; Fukase, K.; Kusumoto, S.; Harley, W.; Fox, A.; Golenbock, D.; Silverman, N. Monomeric and polymeric gram-negative peptidoglycan but not purified LPS stimulate the *Drosophila* IMD pathway. *Immunity* **2004**, *20*, 637–649. [CrossRef]
23. Leulier, F.; Parquet, C.; Pili-Floury, S.; Ryu, J.H.; Lemaitre, B. The *Drosophila* immune system detects bacteria through specific peptidoglycan recognition. *Nat. Immunol.* **2003**, *4*, 478–484. [CrossRef] [PubMed]
24. Thomas, W.; Liu, G. A family of peptidoglycan recognition proteins in the fruit fly *Drosophila melanogaster*. *Proc. Natl. Acad. Sci. USA* **2001**, *97*, 13772–13777.
25. Liu, F.; Li, H.; Yang, P.; Rao, X. Structure-function analysis of PGRP-S1 from the oriental armyworm, *Mythimna separata*. *Arch. Insect Biochem. Physiol.* **2021**, *106*, e21763. [CrossRef]
26. Wei, D.; Liu, Y.W.; Zhang, Y.X.; Wang, J.J. Characterization and function of two short peptidoglycan recognition proteins involved in the immunity of *Bactrocera dorsalis* (Hendel). *Insects* **2019**, *10*, 79. [CrossRef]
27. Wei, D.; Wang, Z.; Xu, H.Q.; Niu, J.Z.; Wang, J.J. Cloning and functional characterization of two peptidoglycan recognition protein isoforms (PGRP-LC) in *Bactrocera dorsalis* (Diptera: Tephritidae). *J. Integr. Agric.* **2020**, *12*, 3025–3034. [CrossRef]
28. Zhang, Z.; Kong, J.; Mandal, S.D.; Li, S.; Zheng, Z.; Jin, F.; Xu, X. An immune-responsive PGRP-S1 regulates the expression of antibacterial peptide genes in diamondback moth, *Plutella xylostella* (L.). *Int. J. Biol. Macromol.* **2020**, *142*, 114–124. [CrossRef]
29. Zhao, G.; Guo, H.; Zhang, H.; Zhang, X.; Xu, A. Effects of pyriproxyfen exposure on immune signaling pathway and transcription of detoxification enzyme genes in fat body of silkworm, *Bombyx mori*. *Pestic. Biochem. Physiol.* **2020**, *168*, 104621. [CrossRef] [PubMed]
30. Hetru, C.; Hoffmann, J.A. NF-kB in the immune response of *Drosophila*. *Cold Spring Harb. Perspect. Biol.* **2009**, *6*, a000232.
31. Werner, T.; Borge-Renberg, K.; Mellroth, P.; Steiner, H.; Dan, H. Functional diversity of the *Drosophila* PGRP-LC gene cluster in the response to lipopolysaccharide and peptidoglycan. *J. Biol. Chem.* **2003**, *278*, 26319–26322. [CrossRef] [PubMed]
32. Kaneko, T.; Yano, T.; Aggarwal, K.; Lim, J.H.; Ueda, K.; Oshima, Y.; Peach, C.; Erturk-Hasdemir, D.; Goldman, W.E.; Oh, B.H.; et al. PGRP-LC and PGRP-LE have essential yet distinct functions in the *Drosophila* immune response to monomeric DAP-type peptidoglycan. *Nat. Immunol.* **2006**, *7*, 715–723. [CrossRef] [PubMed]

insects

MDPI

Article

BdNub Is Essential for Maintaining gut Immunity and Microbiome Homeostasis in *Bactrocera dorsalis*

Jian Gu †, Ping Zhang †, Zhichao Yao, Xiaoxue Li * and Hongyu Zhang *

National Key Laboratory for Germplasm Innovation and Utilization for Fruit and Vegetable Horticultural Crops, Hubei Hongshan Laboratory, Institute of Urban and Horticultural Entomology, College of Plant Science and Technology, Huazhong Agricultural University, Wuhan 430070, China

* Correspondence: xiaoxueli@mail.hzau.edu.cn (X.L.); hongyu.zhang@mail.hzau.edu.cn (H.Z.)

† These authors contributed equally to this work.

Simple Summary: The innate immune system of insects can recognize various pathogens that invade insects and make rapid immune responses. However, excessive immune activation is detrimental to the survival of insects. *Nub* gene of the OCT/POU family plays an important role in regulating the intestinal IMD pathway. In this study, an important horticultural pest, *Bactrocera dorsalis,* was adopted to study its high adaptability in complex habitats. Through NCBI database analysis, we found that the *BdNub* gene of *B. dorsalis* produced two transcription isoforms, *BdNubX1* and *BdNubX2*. After Gram-negative bacterium *Escherichia coli* with system infection, the immunoeffector genes of Imd signaling pathway, antimicrobial peptides Diptcin (Dpt), Cecropin (Cec), AttcinA (AttA), AttcinB (AttB) and AttcinC (AttC) were significantly up-regulated. The expression levels of antimicrobial peptide genes *Dpt, Cec, AttA, AttB,* and *AttC* were significantly up-regulated at 6 h and 9 h after intestinal infection with the Gram-negative bacterium *Providencia rettgeri*. RNAi showed that the silencing of the *BdNubX1* and *BdNubX2* genes could make the gut more sensitive to *Providencia rettgeri* infection, reduce the survival rate significantly, and cause changes in the gut microbiota's structure. These results suggest that the maintenance of immune balance plays an important role in *B. dorsalis* high invasiveness.

Abstract: Insects face immune challenges posed by invading and indigenous bacteria. They rely on the immune system to clear these microorganisms. However, the immune response can be harmful to the host. Therefore, fine-tuning the immune response to maintain tissue homeostasis is of great importance to the survival of insects. The *Nub* gene of the OCT/POU family regulates the intestinal IMD pathway. However, the role of the *Nub* gene in regulating host microbiota remains unstudied. Here, a combination of bioinformatic tools, RNA interference, and qPCR methods were adopted to study *BdNub* gene function in *Bactrocera dorsalis* gut immune system. It's found that *BdNubX1*, *BdNubX2,* and antimicrobial peptides (AMPs), including *Diptcin* (*Dpt*), *Cecropin* (*Cec*), *AttcinA* (*Att A*), *AttcinB* (*Att B*) and *AttcinC* (*Att C*) are significantly up-regulated in Tephritidae fruit fly *Bactrocera dorsalis* after gut infection. Silencing *BdNubX1* leads to down-regulated AMPs expression, while *BdNubX2* RNAi leads to increased expression of AMPs. These results indicate that *BdNubX1* is a positive regulatory gene of the IMD pathway, while *BdNubX2* negatively regulates IMD pathway activity. Further studies also revealed that *BdNubX1* and *BdNubX2* are associated with gut microbiota composition, possibly through regulation of IMD pathway activity. Our results prove that the *Nub* gene is evolutionarily conserved and participates in maintaining gut microbiota homeostasis.

Keywords: *Nub*; *Bactrocera dorsalis*; the antibacterial peptide; gut immunity; gut microbes; IMD pathway

Citation: Gu, J.; Zhang, P.; Yao, Z.; Li, X.; Zhang, H. *BdNub* Is Essential for Maintaining gut Immunity and Microbiome Homeostasis in *Bactrocera dorsalis*. *Insects* **2023**, *14*, 178. https://doi.org/10.3390/insects14020178

Academic Editor: Bessem Chouaia

Received: 13 December 2022
Revised: 7 February 2023
Accepted: 8 February 2023
Published: 10 February 2023

Copyright: © 2023 by the authors. Licensee MDPI, Basel, Switzerland. This article is an open access article distributed under the terms and conditions of the Creative Commons Attribution (CC BY) license (https://creativecommons.org/licenses/by/4.0/).

1. Introduction

Insects, composed of over 5 million different species, are the most abundant species on earth and can survive in all kinds of complex environments [1]. These highly diverse habitats also exert tremendous survival pressure on insects. Infection by environmental microorganisms and colonization by indigenous bacteria can be detrimental to insects [2]. During their long-term evolution, insects, like all other animals, developed efficient immune defense systems. Although lacking an adaptive immune response, insects can resist bacteria, fungi, viruses, and nematodes using their innate immunity. The IMD pathway plays a vital role in defense against Gram-negative bacteria. This process is marked by antimicrobial peptides (AMPs) production and phagocytosis of bacteria accomplished by blood cells [3–5]. Insects also synthesize and accumulate many immune effectors, which are released into the hemolymph and play a role in the immune response, which is called the humoral response [6].

In insects, humoral immune effectors are mainly AMPs. About 20 inducible antimicrobial peptides have been identified in *Drosophila melanogaster*, showing a broad spectrum of antimicrobial effects [7]. AMPs can act either in a specific or synergistic way [8]. In *Drosophila* and honeybee, *Diptericin*, *Drosocin*, and *Attacin* play essential roles in the defense against Gram-negative bacteria [7,9]. *Defensin* mainly resists Gram-positive bacteria, while *Drosomycin* and *Metchnikowin* are the main active antifungal substances. Cecropin plays an essential role in both anti-bacterial and anti-fungal processes [10], and the constitutively expressed antimicrobial peptide *Andropin* is continuously expressed in male reproductive organs for defense [11].

The *Nub* gene of the POU/OCT family was early discovered in *D. melanogaster* [12,13]. The POU/OCT family regulates key regulators of metabolism, immunity, and cancer [14]. The *Nub* gene encodes two distinct proteins with independent functions, *Nub-PB* and *Nub-PD*. These two proteins have similar expression patterns but perform different functions. The difference between *Nub-PB* and *Nub-PD* proteins' N-terminal sequence leads to a different regulatory mechanism. *Nub-PD* is a transcriptional repressor of the antimicrobial peptide gene, while *Nub-PB* is a transcriptional activator of the antimicrobial peptide gene in *Drosophila* [14,15]. In *Drosophila*, overexpressing *Nub-PB* results in increased AMPs abundance [15]. On the other side, overexpressing *Nub-PD* results in reduced AMP abundance. Furthermore, co-overexpressing *Nub-PB* and *Nub-PD* does not induce changes in AMPs gene expression [14,15].

Microorganisms are found in many plants, animals, and other organisms [16]. Insects host probably the largest group of commensal bacteria [2]. These bacteria are abundant in the gut, body cavity, and specific cells of insects [2]. They have a wide range of functions, including contributions to host growth and development, nutrient acquisition, and resistance to pathogens [17–20]. On the other hand, dysregulation of gut microbiota can also be harmful to the host [21]. Therefore, the gut microbiota needs to be tightly controlled. In *Drosophila*, proper gut microbiota composition, density, and localization were altered in IMD-deficient flies, suggesting IMD's prominent role in gut microbiota control [22]. Maintaining gut microbiota homeostasis and eliminating pathogenic bacteria are essential to the host's health. In a recent study, compartmentalized IMD pathway receptors *PGRP-LC* and *PGRP-SB*, *PGRP-LB* expression act to eliminate pathogenic bacteria and protect symbiotic bacteria in *Bactrocera dorsalis* [23].

B. dorsalis is a major horticultural and agricultural pest. It damages more than 250 kinds of fruits and vegetables, causing substantial economic loss. Its larva lives inside rotten fruits and faces threats posed by bacteria. Therefore, its immune system, especially the gut immune system, must be precisely regulated to ensure its survival [21,24,25]. Since Nub gene is essential for AMPs gene expression regulation, we expect it also plays an indispensable role in *B. doraslis* gut immunity. In this paper, we also aim to decipher *BdNub* function in *B. dorsalis* microbiota homeostasis. Gut microbiota has been proven to be necessary for *B. dorsalis* overall fitness [20,24]. Our findings suggest that *BdNub* could be a novel target for developing pest control strategies targeting gut homeostasis.

2. Materials and Methods

2.1. Insects Rearing

The experimental insects were collected from Guangdong Province, China using protein bait and maintained in the Institute of Urban and Horticultural Insects, Huazhong Agricultural University, Wuhan, Hubei, China. The photoperiod of the insect-rearing room was 12 h:12 h. The room's relative humidity was 70–80%, and the temperature was maintained at 28 ± 1 °C. Larvae were raised on larval food (wheat bran 80 g, corn flour 40 g, sucrose 40 g, yeast powder 15 g, water 200 mL). After eclosion, adult flies were moved to 30 cm × 30 cm × 30 cm cages. Adult flies were raised on a sucrose and yeast mix at a ratio of 3:1.

2.2. *BdNub* Identification

We blasted the *Drosophila Nub* protein (NCBI REFSEQ: accession NM_001103683.2) sequence against NCBI. *BdNub* BLAST results showed that *BdNub* has two different transcripts. The online analysis tool SPLIGN was used to identify *BdNub* gene introns and exons. The Neighbor-Joining phylogenetic tree was built using MEGA7. DNAMAN was used to perform amino acid homology analysis. We used SnapGene to analyze nucleotide sequence homology. The online tool SMART was used to predict and analyze the conserved protein domains. The protein secondary structure was analyzed using the online tool SOPMA to submit amino acid sequences to the SOPMA working page. The results show an alpha helix (blue), a beta turn (green), a random coil (yellow), and an extended strand (red). SWISS-MODEL was deployed to predict protein structures, input the target amino acid sequence, and build a model. A simple way to evaluate the quality of a model is to look at the GMQE value (Global Model Quality Estimate), which is between 0 and 1. The closer to 1, the better the quality of the model.

2.3. *BdNub* Spatial and Temporal Expression Profiles

For spatial analysis, adult flies were dissected. Tissue samples were collected from the head, gut, Malpighian tubules, ovary, testis, and fat body. The samples were stored at −80 °C for further use. For temporal analysis, whole eggs or insects were collected from different developmental stages, including eggs, first instar larvae, second instar larvae, third instar larvae, one-day-old pupa, nine-day-old pupa, one-day-old adult flies (male and female apart), and 14-day-old mature adult flies. Samples from different developmental stages were rinsed once in 75% alcohol, followed by two rinses in PBS.

2.4. RNA Extraction and cDNA Synthesis

For spatial expression profiles, total RNA was isolated from 30 different tissue samples per biological replicate; three biological replicates were conducted. For temporal expression profiles, total RNA was isolated from 20 different stage of development samples per biological replicate; three biological replicates were conducted. Samples were placed into a 1.5 mL RNase-free centrifuge tube containing ground beads and homogenized (Jinxin, Shanghai, China). Total RNA was extracted using Trizol (TaKaRa, Otsu, Shiga, Japan) following the manufacturer's instructions. RNA quality and concentration were determined by electrophoresis (Liuyi Biotechnology, Beijing, China) and a Nanodrop spectrophotometer (Thermo Fisher Scientific Inc., Waltham, MA, USA). RNA was stored at −80 °C for later use. The first-strand complementary DNA (cDNA) of each pool was synthesized from 1 μg of total RNA using the PrimeScriptTM RT reagent kit (Takara, Otsu, Shiga, Japan) with a gDNA eraser to remove residual DNA contamination.

2.5. Real-Time PCR

Real-time PCR was performed using the Hieff UNICON® qPCR SYBR Green Master Mix No Rox kit (Bio-Rad, Hercules, CA, USA)on a real-time Bio-Rad CFX96 (Bio-Rad, Hercules, CA, USA) PCR instrument. Real-time PCR was performed using a Bio-Rad CFX Connect system with the following protocol: initial denaturation at 95 °C for 30 s, followed by 45 cycles of 95 °C for 15 s and 60 °C for 30 s. Melting curve analysis was performed at the end of each amplification run to confirm the presence of a single peak with the following protocol: 55 °C for 60 s, followed by 81 cycles starting at 55 °C for 10 s with a 0.5 °C increase each cycle. Real-time PCR results were relatively quantified using the $2^{-\Delta\Delta}$CT method [26]. Relative mRNA abundance was normalized using the *RPL32* set as a reference gene. Real-time PCR primers> are listed in Table 1. Each sample was set in a triplicate, and the corresponding blank control was set as required. The total PCR reaction volume was 20 μL, including 10 μL SYBR Green mix, primers 1.6 μL, RNA-free water 6.4 μL, and cDNA 2 μL. The data was analyzed and exported using Graphpad 7.0.

Table 1. The primers used for Quantitative Real-time PCR.

Primer Name	Sequence	Target Gene	Amplicon Size (bp)
QRpl32 F	5′-CCCGTCATATGCTGCCAACT-3′	*Rpl32*	148 bp
QRpl32 R	5′-GCGCGCTCAACAATTTCCTT-3′		
QBdNubX1 F	5′-GCAGTAATGTGCCCCAGAAG-3′	*NubX1*	115 bp
QBdNubX1 R	5′-AACGCAGACGTAGCGGTAAC-3′		
QBdNubX2 F	5′-GTCGAGCATCGAGGTGTTTT-3′	*NubX2*	105 bp
QBdNubX2 R	5′-AGTGTCTGAGCGCTTGTGTG-3′		
QBdDpt F	5′-GCATAGATTTGAGCCTTGACACAC-3′	*Diptcin*	110 bp
QBdDpt R	5′-GCCATATCGTCCGCCCAAAT-3′		
QBdCec F	5′-GGCAAGAAAATTGAGCGGGT-3′	*Cecropin*	100 bp
QBdCec R	5′-CCTTCAATGTTGCTGCCACA-3′		
QBdAttA F	5′-GTGGCAACCTTAATTGGGCG-3′	*Attcin A*	106 bp
QBdAttA R	5′-AGATTGGAACTTGCGCCGTA-3′		
QBdAttB F	5′-ACACGCTTGGACTTGACAGG-3′	*Attcin B*	93 bp
QBdAttB R	5′-ATGAGTCAATCCCAAGCCGG-3′		
QBdAttC F	5′-GAGTTGGCCGGTAGAGCAAA-3′	*Attcin C*	104 bp
QBdAttC R	5′-GTAGTCGCGTTGTCCACTCA-3′		
QBdDef F	5′-CTGGAAAAGTCAATGGGCCG-3′	*Defensin*	105 bp
QBdDef R	5′-AAGCGATACAATGGACAGCG-3′		
QBdactinF	5′-TCGATCATGAAGTGCGATGT-3	*β-actin*	101 bp
QBdactinR	5′-ATCAGCAATACCGGGGTACA-3		

2.6. DsRNA Synthesis and RNA Interference

BdNubX1 and *BdNubX2* specific primers were designed using Prime5 software (Table 2). For *BdNubX1* dsRNA synthesis, the T7 polymerase recognition sequence (GGATCCTAATACGACTCACTATAGGN) was added to the 5′ of the primers. *BdNubX1* dsRNA was synthesized using the T7 Ribomax express RNAi system (Promega, Madison, WI, USA). *Egfp* dsRNA was synthesized as a control. Our initial assessment showed that *BdNubX2* RNAi could not be achieved by dsRNA injection. Therefore, we choose siRNA for *BdNubX2* RNAi. *BdNubX2* siRNA was synthesized using specific primers of *BdNubX2* (RiboBio, Guangzhou, China).

DsRNA integrity and concentration were monitored by 1% agarose gel electrophoresis and a NanoDrop 2000 spectrophotometer (Thermo Fisher Scientific Inc., Waltham, MA, USA). Microinjection was performed using the FemtoJet 5247 micromanipulation system (Microinjector for cell biology, FemtoJet 5247, Hamburg, Germany) with a Pi of 300 hpa and a Ti of 0.3 s. One microliter of 2 μg/μL dsRNA was injected into the adult flies′

abdomen [23]. For ds-*egfp*, ds-*BdNubX1*, si-*egfp*, and si-*BdNubX2* treatments, we used 100 flies (age: 2 days after emergence) for injection.

Table 2. The primers used for synthesis of dsRNA.

*BdNub*X1 T7F	5′-GGATCCTAATACGACTCACTATAGGACCAGGCATTTTGAACCCA-3′
*BdNub*X1 T7R	5′-GGATCCTAATACGACTCACTATAGGTGATCCGCTGACTCCGTCT-3′
Egfp T7F	5′-GGATCCTAATACGACTCACTATAGGACGTAAACGGCCACAAGTTC-3
Egfp T7R	5′-GGATCCTAATACGACTCACTATAGGAAGTCGTGCTGCTTCATGTG-3′
Si-BdNubX2	5′-GTGGGCACATAATGCAGAA-3′

2.7. Bacterial Infection and Survival Assay

Escherichia coli was used for systemic infection. *E. coli* culture was left to grow in 200 mL LB broth for 14 h at 220 rpm at 37 °C. *E. coli* was harvested by centrifuging at 3600 g for 5 min. The final concentration of the *E. coli* solution was adjusted to OD_{600} = 400(~10^{11} cfu/mL). For pricking, we used flies 5 days after RNAi treatment. Briefly, a clean insect needle was first surface sterilized by ethanol and then dipped into the bacterial pellet. Flies were then pricked in the abdomen. For oral infections, we used the gram-negative bacteria *P. rettgeri*. The *P. rettgeri* strain used in this study was isolated from the *B. dorsalis* gut. It could induce a strong immune response through oral infection in adult *B. dorsalis*. The culture method was the same as previously described. The final bacteria concentration for oral infection was OD_{600} = 50(~10^{10} cfu/mL). Adult flies were starved and dehydrated for 24 h without food or water supplies. For infection, flies were then fed an artificial diet with 5% sucrose containing the concentrated microbe solution [23]. The control group was fed only 5% sucrose.

For the survival assay, *BdNubX1* and *BdNubX2* were silenced by injecting corresponding dsRNA and siRNA, and we used 30 flies (age: 2 days after emergence) for injection per biological replicate, and three biological replicates were conducted, respectively. The ds*Egfp*-injected flies were used as the control group. We fed the flies with *P. rettgeri* for 24 h. Next, the infected flies were switched to the normal adult diet. Survival was recorded every day.

2.8. Gut Bacterial DNA Extraction

BdNubX1 RNAi, *BdNubX2* RNAi, and the control flies (age: 2 days after emergence) were surface sterilized using 75% ethanol and then transferred to PBS (pH = 7.2). DNA was extracted from 30 gut regions of flies (age: *BdNubX1* RNAi; *BdNubX2* RNAi after 3 d, 5 d, and 7 d) per biological replicate; three biological replicates were conducted. According to the manufacturer's instructions, total DNA was extracted using an E.Z.N.A. Soil DNA kit (Omega, Norcross, GA, USA).

2.9. Quantification of Bacterial Species or Group by qPCR

For bacteria quantification, qPCR was carried out in a 20 mL reaction volume that included 10 mL of SYBR Green Mix (Bio-Rad, Hercules, CA, USA), 200 nM of each primer, and 5 ng of DNA. Real-time PCR was performed using a Bio-Rad CFX Connect system with the following protocol: (1) preincubation at 50 °C for 2 min and 95 °C for 10 min (2) 45 cycles of denaturation at 95 °C for 15 s and annealing at 60 °C for 1 min; and (3) one cycle at 95 °C for 15 s, 53 °C for 15 s, and 95 °C for 15 s. Real-time quantitative PCR (qPCR) was performed and normalized to the host *β-actin* gene (Table 2). In this study, bacterial primers are listed in (Table 3).

2.10. Statistics and Analysis

Student's *t*-test was used for two independent samples to compare the mean values. For multiple comparisons among samples, the Tukey's test in one-way ANOVA was used, and the significance level was set at $p < 0.05$. An analysis of variance was completed with SPSS 18 software. GraphPad Prism 7.0 and Excel software were used to plot and analyze the experimental data.

Table 3. The primers used for different bacteria genera.

Target Bacteria	Primer Name	Sequence	Resources	Target Gene	Amplicon Size (bp)
Total bacteria	*Tol F*	5′-TCCTACGGGAGGCAGCAGT-3′	(Guo et al., 2008) [23,27]	16S RNA	466 bp
	Tol R	5′-GGACTACCAGGGTATCTATCCTGTT-3′			
Enterobacteriaceae	*Ent F*	5′-CATTGACGTTACCCGCAGAAGAAGC-3′	(Bartosch et al., 2004) [23,28]	16S RNA	195 bp
	Ent R	5′-CTCTACGAGACTCAAGCTTGC-3′			
Enterococcus	*Eco F*	5′-CCCTTATTGTTAGTTGCCATCATT-3	(Rinttilä et al., 2004) [29]	16S RNA	144 bp
	Eco R	5′-ACTCGTTGTACTTCCCATTGT-3′			
Lactobacillus	*Lac F*	5′-AGCAGTAGGGAATCTTCCA-3′	(Rinttilä et al., 2004) [29]	16S RNA	341 bp
	Lac R	5′-CACCGCTACACATGGAG-3′			
Flavobacterium	*Flavo F*	5′-ATTGGGTTTAAAGGGTCC-3′	(Abell et al., 2005) [30]	16SRNA	349 bp
	Flavo R	5′-CCGTCAATTCCTTTGAGTTT-3′			
Salmonella	*Salmon F*	5′-ACAGTGCTCGTTTACGACCTGAAT-3′	(Ahmed et al., 2008) [31]	*invA*	244 bp
	Salmon R	5′-AGACGACTGGTACTGATCGATAAT-3′			
Pseudomonas	*Pseudo F*	5′-CAAAACTACTGAGCTAGAGTACG-3′	(Matsuda et al., 2009) [32]	16S RNA	215 bp
	Pseudo R	5′-TAAGATCTCAAGGATCCCAACGGCT-3′			
Clostridium	*Cclos F*	5′-AATCTTGATTGACTGAGTGGCGGAC-3′	(Bartosch et al., 2004) [28]	16S RNA	148 bp
	Cclos R	5′-CCATCTCACACTACCGGAGTTTTTC-3′			

3. Results

3.1. Identification of Nub Gene in B. dorsalis

We found only one *Nub* gene (accession number: NW_011876379) in *B. dorsalis* by searching the NCBI database based on protein sequence homology using BLASTp. *BdNub* has two different transcripts, which we named *BdNubX1* and *BdNubX2*, respectively. *BdNub* consists of 6 exons. *BdNubX1* contains five exons, including transcript-specific exons 1 and 2, which are missing in *BdNubX2*, while *BdNubX2* has four exons, including transcript-specific exon 3 (Figure 1A). *BdNubX1* encodes 833 amino acids, and *BdNubX2* encodes 567 amino acids. The two transcripts share the same 529 amino acids. Sequence alignment showed that *BdNubX1* has 304 transcript-specific amino acids, while *BdNubX2* has 38 transcript-specific amino acids. The *Nub* belongs to the OCT/POU family genes, evolutionarily conserved from arthropods to mammals. We performed the phylogenetic analysis on insects, including *B. dorsalis*, *D. melanogaster*, *Musca domestica*, *Ceratitis capitata*, *Bombyx mori*, and *Aedes aegypti*. The results confirmed that the *Nub* gene is conserved in insects. The results also revealed that despite sequence differences, two *Nub* isoforms clustered in the same branch. *BdNub* was closely related to *Bactrocera capsicum* (Figure 1B).

The *Nub* gene has two POU/OCT family conserved domains: the POU domain and the HOX domain. Although *BdNubX1* and *BdNubX2* have different amino acid sequences, they both have two intact conserved domains. We compared the amino acid sequences of the POU and HOX domains of *BdNubX1* and *BdNubX2* (Figure 1C). The amino acid sequences of *BdNub* POU and HOX are highly similar to the functional domains of *B. capsicum* and *D. melanogaster*, suggesting their functional similarity in vivo. We further use three other algorithms, including SMART, SOPMA, and SWISS, to predict the conserved domain and protein structure of *BdNub*. We also confirmed that *BdNub* is a classical POU/OCT family member (Figure S1).

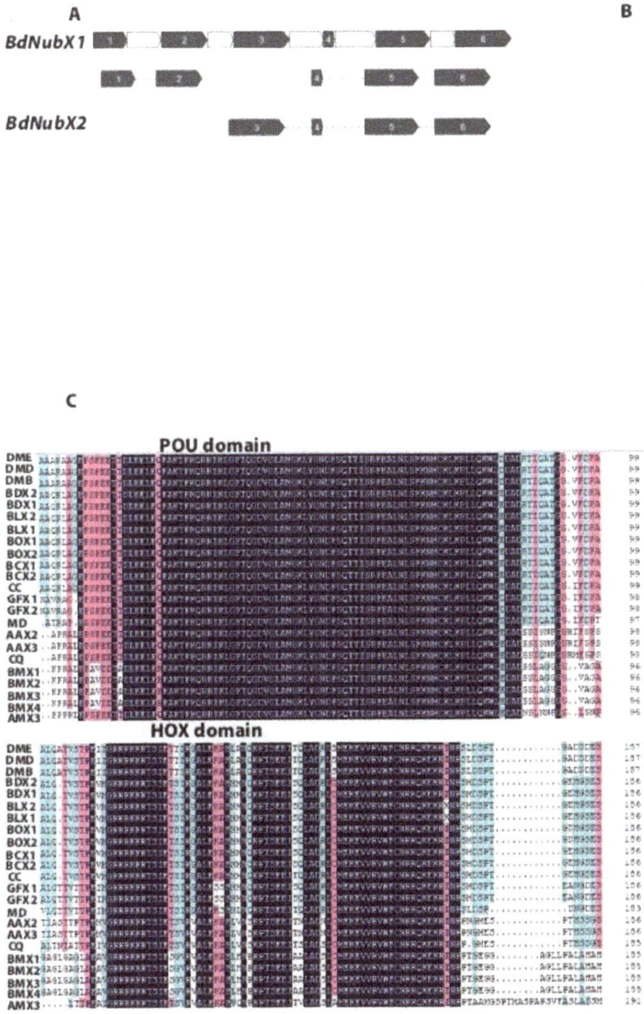

Figure 1. Identification of *BdNub* gene. (**A**): The alternative splicing of *BdNub*. Exons are indicated by grey color, Introns are indicated by white color. (**B**): Phylogenetic analysis of the *Nub* gene, *BdNub* were aligned with *Nub* genes from 11 other insect species. Gene accession numbers were given in parentheses. Black triangle(the location of the target gene). (**C**): Alignment of POU and HOX domain amino acid sequence. Identical sequences were shown in black. 75% conserved amino acids were shown on pink background, and 50% conserved amino acids were shown on blue background. (**B**,**C**). BDX1, *Bactrocera dorsalis nubbin X1* (XP 011207745.1); BDX2, *Bactrocera dorsalis nubbin X2* (XP 011207746.1); BLX1, *Bactrocera latifrons nubbin X1* (XP 018783925.1); BLX2, *Bactrocera latifrons nubbin X* 2(XP 018783926.1); BCX2, *Bactrocera cucurbitae nubbin X2*(XP 011178505.1) ;BCX1, *Bactrocera cucurbitae nubbin X1* (XP 011178504.1); BOX2, *Bactrocera oleae nubbin X2* (XP 014086366.1); BOX1, *Bactrocera oleae nubbin X2;* CC, *Ceratitis capitata nubbin* (XP 004530324.1); MD, *Musca domestica nubbin* (XP 019892278.1); GFX2, *Glossina fuscipes nubbin* X2 (XP 037882400.1); GFX1, *Glossina fuscipes nubbin X1* (XP 037882399.1); DMB, *Drosophila melanogaster nubbin B* (NP 001097153.1); DMD, *Drosophila melanogaster nubbin D* (NP 476659.1) ; DME, *Drosophila melanogaster nubbin E*(NP 001285876.1) ; CQ, *Culex quinquefasciatus nubbin* (XP 001844054.1); AAX3, *Aedes aegypti nubbin X3*(XP 021704008.1); AAX2, *Aedes aegypti nubbin X2* (XP

021704007.1); AMX3, *Apis mellifera nubbin X3* (XP 006558737.1); BMX2, *Bombyx mori nubbin X2* (XP 037870381.1) ; BMX4, *Bombyx mori nubbin X4*(XP 037870383.1); BMX3, *Bombyx mori nubbin X3* (XP 037870382.1); BMX1, *Bombyx mori nubbin X1*(XP 037870380.1).

3.2. Temporal and Spatial Expression Patterns of the BdNub Gene

We analyzed the spatial and temporal expression patterns of *BdNub* gene transcripts at different periods and in different tissues. The results showed that *BdNubX1* was highly expressed in 9-day-old pupa (the late-stage pupa) and the freshly emerged adults, and the expression was low in the egg, larva, and sexually mature adult stages (Figure 2A). Similarly, *BdNubX2* was also highly expressed in the late pupal stage and the newly emerged adult flies (Figure 2B). Spatial analysis revealed that *BdNubX1* expression was highest in the midgut, followed by the foregut and testis (Figure 2C). *BdNubX2* showed a slightly different expression pattern. Although it was highly expressed in the foregut and midgut, it was much lower in the testis compared with *BdNubX1* (Figure 2D). These results suggested that two *BdNub* transcript isoforms have similar expression patterns with high abundance in the late pupal stage, newly emerged adult flies, and in the adult gut.

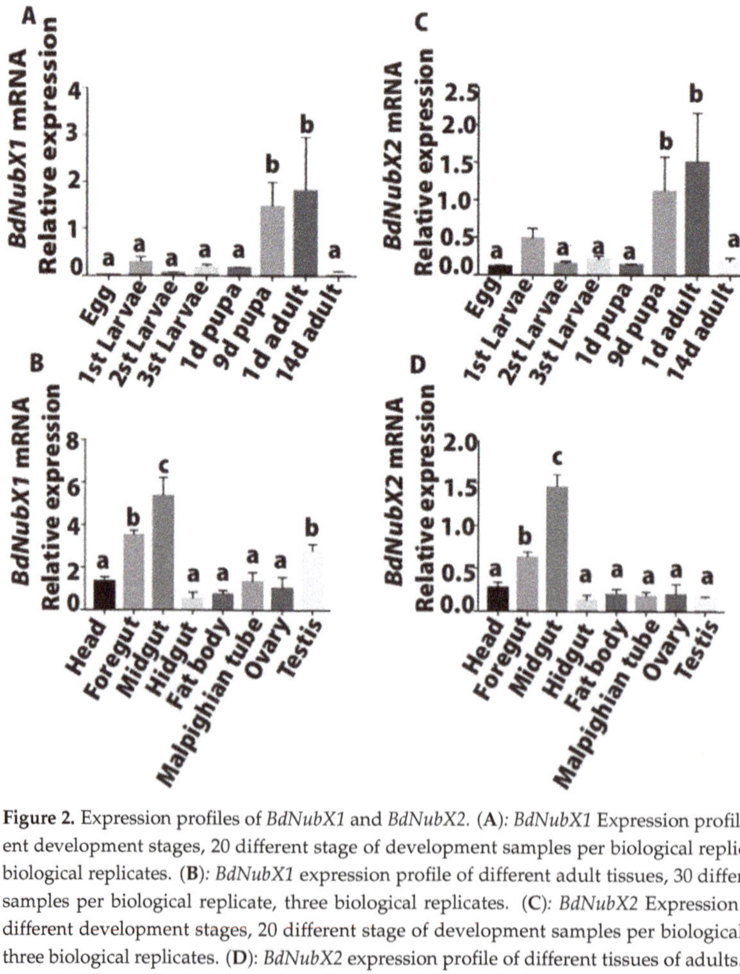

Figure 2. Expression profiles of *BdNubX1* and *BdNubX2*. (**A**): *BdNubX1* Expression profile of different development stages, 20 different stage of development samples per biological replicate, three biological replicates. (**B**): *BdNubX1* expression profile of different adult tissues, 30 different tissue samples per biological replicate, three biological replicates. (**C**): *BdNubX2* Expression profile of different development stages, 20 different stage of development samples per biological replicate, three biological replicates. (**D**): *BdNubX2* expression profile of different tissues of adults. Different letters indicate statistically significant differences in *BdNub* isoforms expression, 30 different tissue samples per biological replicate, three biological replicates. $p < 0.05$, Tukey's test, One way ANOVA.

3.3. BdNub Does Not Participate in Systemic Infection of the IMD Pathway

Next, we examined *BdNubX1* and *BdNubX2* expression at different time points after *E. coli* systemic infection. *E. coli* is a gram-negative pathogen that can elicit a robust immune response in *B.dorsalis*. The results showed that the *BdNubX1* transcript had no significant increase after systemic infection with *E. coli* (Figure 3A). Similarly, the *BdNubX2* transcript showed no significant change after *E. coli* infection (Figure 3B). These results indicated that *BdNub* did not play a role in the systemic infection of Gram-negative *E. coli*. To strengthen our conclusion, we further detected the main immune effectors of the IMD pathway, including AMPs *Dpt*, *Cec*, *AttA*, *AttB*, and *AttC*. The results showed that all five AMPs were significantly up-regulated at 3 h, 6 h, and 12 h after *E. coli* infection, proving that systemic infection works well in *B. dorsalis* (Figure 3C–E).

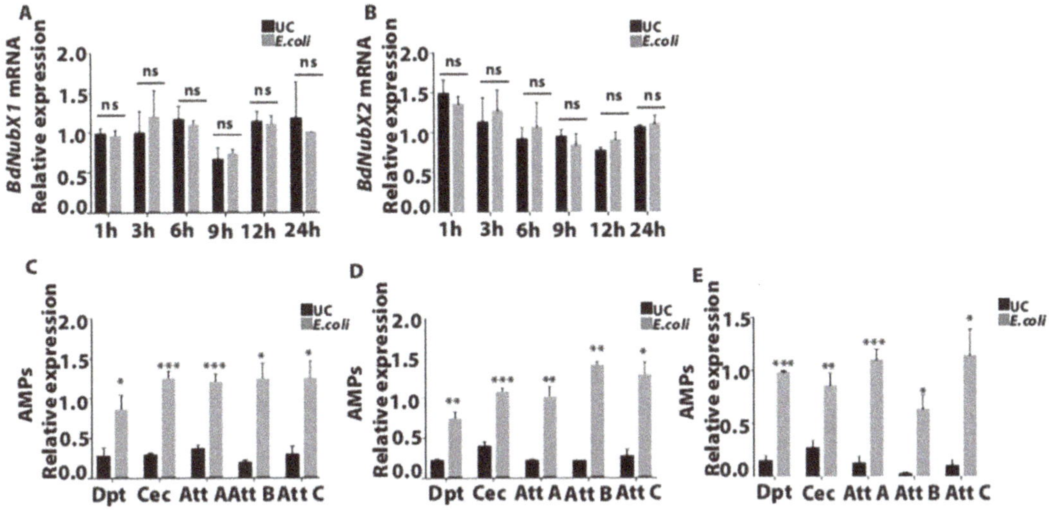

Figure 3. The immune response of *BdNub* after *E. coli* systemic infection. (**A**) The relative expression level of *BdNubX1* after *E. coli* systemic infection. (**B**) The relative expression level of *BdNubX2* after *E. coli* systemic infection. (**C–E**) The relative expression of AMPs genes at 3 h, 6 h, 12 h after *E. coli* systemic infection, respectively, (**A–E**) 30 flies samples per biological replicate, three biological replicates. Gene expressions were normalized to the reference gene RP49. UC: unchallenged flies. * $p < 0.05$, ** $p < 0.01$, *** $p < 0.001$, Student's *t*-test.

3.4. BdNub Regulates the Expression of Gut AMPs after Oral Infection

Previous data in our lab suggests that *E. coli* could not induce a strong and consistent gut immune response in *B. dorsalis*. Our screening found that *P. rettgeri* is a potent inducer of the gut immune response. Our results showed that *BdNubX1* was significantly up-regulated at 6 h post oral infection of *P. rettgeri*, while there was no significant up-regulation at other time points (Figure 4A). This suggested that *BdNubX1* is only transiently activated during the gut immune response. *BdNubX2* showed a somewhat different expression pattern. It was significantly up-regulated at 6 h and 9 h after *P. rettgeri* oral infection, suggesting it played a prolonged role in gut immunity (Figure 4B). Next, to reaffirm that *P. rettgeri* could induce a gut immune response, we further examined AMP expression at different time points after infection. The results showed the immune effector AMPs' expression, including *Dpt*, *Cec*, *AttA*, *AttB*, and *AttC*, increased at 6 h and 9 h after *P. rettgeri* oral infection. The results are as follows: at 6 h after oral infection, all five AMPs genes were significantly up-regulated (Figure 4C), while at 9 h after oral infection, only *Dpt* and *Cec* were significantly up-regulated (Figure 4D).

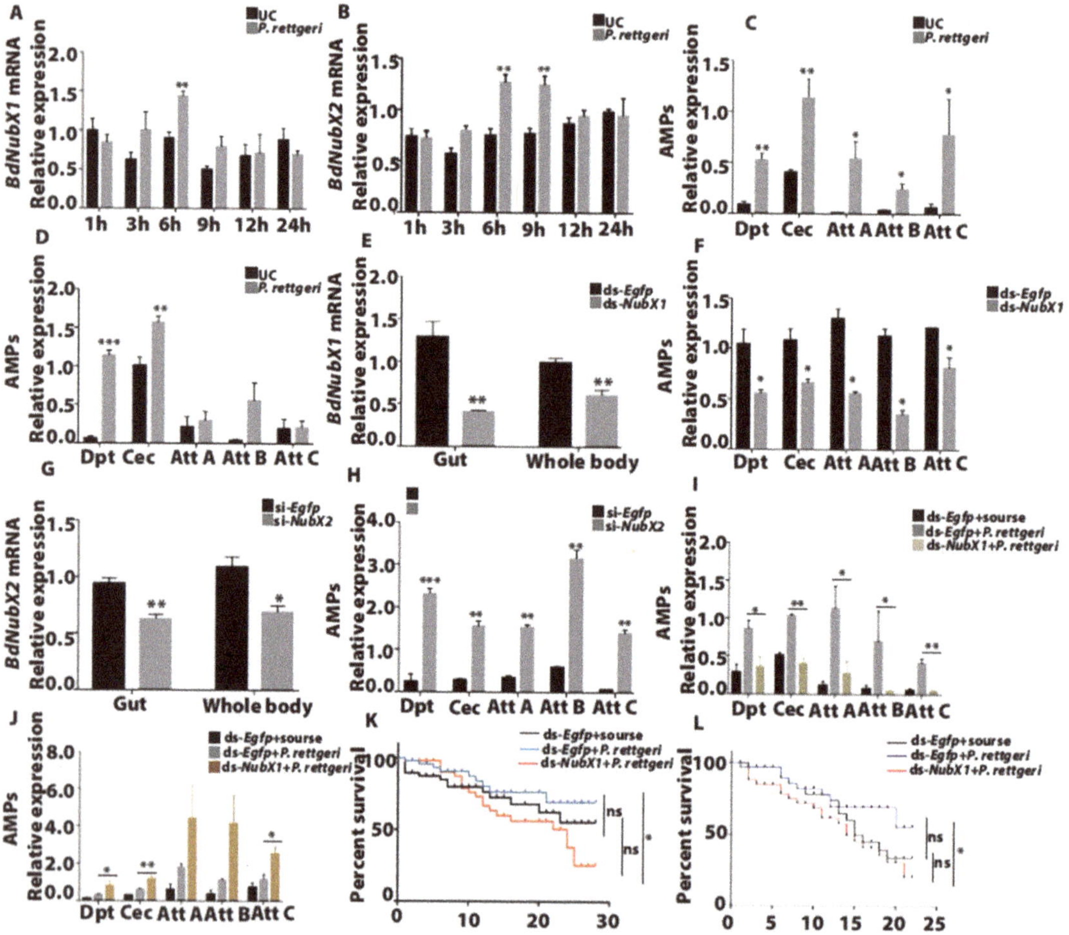

Figure 4. The immune response of *BdNub* after *P. rettgeri* oral infection. (**A**,**B**) The relative expression level of *BdNubX1* and *BdNubX2* after *P. rettgeri* oral infection. (**C**,**D**) The relative expression of the AMPs genes at 6h and 9h after oral infection, respectively. (**E**) The RNAi effect of *BdNubX1* dsRNA injection. (**F**) The expression levels of AMPs genes in *BdNubX1* RNAi flies. (**G**) The RNAi effect of *BdNubX1* siRNA injection. (**H**) The expression levels of AMPs genes in *BdNubX2* RNAi flies. (**I**,**J**) The AMPs genes expressions at 6 h after *P. rettgeri* oral infection in *BdNubX1* RNAi flies (I) and *BdNubX2* RNAi flies (J). ds*Egfp* treated group was used as the control for RNAi. (**K**,**L**) The survival of *BdNubX1* RNAi and *BdNubX2* RNAi flies after rate after *P. rettgeri* oral infection. (**A**–**L**) 30 flies samples per biological replicate, three biological replicates. * $p < 0.05$, ** $p < 0.01$, *** $p < 0.001$, Student's t-test. For survival assay, ns, non-significance, Log-rank (Mantel-Cox) test.

Next, we performed RNAi experiments to elucidate the function of the *BdNub* genes. The results showed that *BdNubX1* RNAi down-regulated IMD target AMP genes, including *Dpt*, *Cec*, *AttA*, *AttB*, and *AttC* (Figure 4E,F). It indicated that *BdNubX1* has a positive regulatory function on the IMD pathway. On the contrary, *BdNubX2* RNAi leads to a significant up-regulation of the IMD-regulated AMPs expression (Figure 4G,H), suggesting that *BdNubX2* has a negative regulatory function on the IMD pathway (Figure 4G,H). In order to further explore the function of the *BdNub* gene in gut immunity, we performed gut infection after RNAi of *BdNubX1* and *BdNubX2*, respectively. The results showed that the expression levels of AMPs in *BdNubX1*-silenced flies after gut infection were significantly down-regulated compared with those in control flies after gut infection, and the expression levels of AMPs were similar to those of non-gut infection (Figure 4I). On the contrary, the expression levels of AMPs were significantly up-regulated in the *BdNubX2*-silenced flies after gut infection compared with the control flies after gut infection (Figure 4J).

Furthermore, to determine whether *BdNubX1* and *BdNubX2* RNAi affect *B. dorsalis* survival, we examined the survival rate of adult flies after *P. rettgeri* oral infection. The results showed that *BdNubX1* RNAi flies fed on *P. rettgeri* died faster than the control group, which was provided with only sucrose. This demonstrated that *P. rettgeri* is indeed a pathogenic bacterium in *B. dorsalis*. On day 20, *BdNubX1* RNAi flies showed a much lower survival rate than the *Egfp* RNAi control group (Figure 4K). Similarly, the results showed that *BdNubX2* RNAi flies fed on *P. rettgeri* died faster than the control group feeding only sucrose. On day 19, *BdNubX2* RNAi flies showed a lower survival rate than the *Egfp* RNAi control group (Figure 4L).

3.5. BdNub Is Necessary to Maintain Gut Microbiota Composition and Structure

The IMD pathway is involved in insect gut microbiota regulation [2,22]. To determine the function of *BdNubX1* and *BdNubX2* isoforms in microbiota regulation, we examined microbiota composition using qRT-PCR in *BdNubX1* and *BdNubX2* RNAi flies. Our results showed that *BdNubX1* RNAi disturbed gut microbiota composition and decreased microbiota abundance (Figure 5A,C,E,G). The bacterial loads of different genera have changed significantly, with strains like *Lactobacillus* and *Enterococcus* significantly down-regulated, and *Pseudomonas* and *Salmonella* significantly increased (Figure 5D,F,H). This result indicated that *BdNubX1* RNAi changed gut microbiota composition and quantity. Similarly, *BdNubX2* RNAi also caused significant changes in gut microbiota. The total intestinal bacteria were down-regulated on day 5 after the dsRNA injection (Figure 5B,I,K,M). Among them, the *Enterobacteriaceae* were significantly down-regulated on days 3 and 5. The abundance of *Pseudomonas, Flavobacterium,* and *Salmonella* also changed at different time points after RNAi (Figure 5J,L,N).

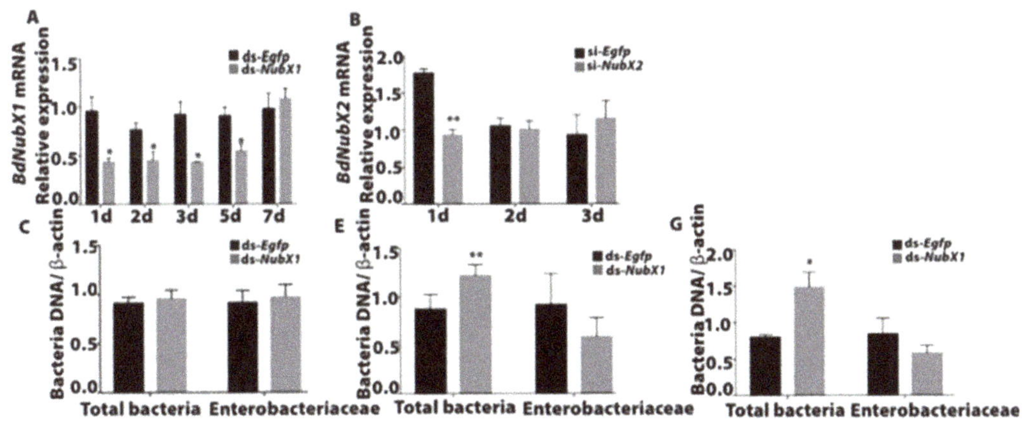

Figure 5. *Cont.*

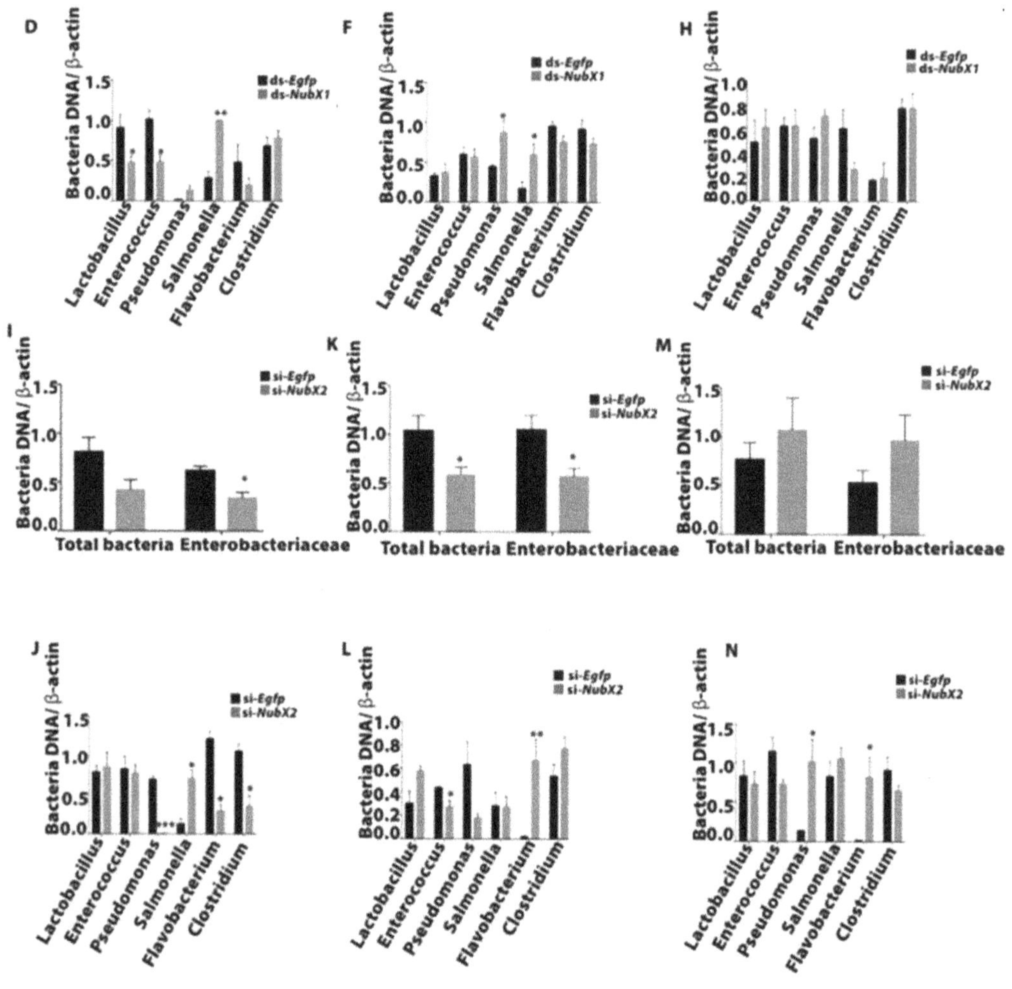

Figure 5. Effect of *BdNubX1* and *BdNubX2* RNAi on gut microbiota. (**A**,**B**) *BdNubX1* and *BdNubX2* expression levels after dsRNA and siRNA injection, respectively. (**C**) The gut total bacteria and *Enterobacteriaceae* abundance at 3d in *BdNubX1* RNAi flies. (**D**) The different genus bacteria abundance at 3 d in *BdNubX1* RNAi flies. (**E**) The gut total bacteria and *Enterobacteriaceae* abundance at 5d in *BdNubX1* RNAi flies. (**F**) The different genus bacteria abundance at 5 d in *BdNubX1* RNAi flies. (**G**) The gut total bacteria and *Enterobacteriaceae* abundance at 7 d in *BdNubX1* RNAi flies. (**H**) The different genus bacteria abundance at 7d in *BdNubX1* RNAi flies. (**I**) The gut total bacteria and *Enterobacteriaceae* abundance at 3 d in *BdNubX2* RNAi flies. (**J**) The different genus bacteria abundance at 3 d in *BdNubX2* RNAi flies. (**K**) The gut total bacteria and *Enterobacteriaceae* abundance at 5 d in *BdNubX2* RNAi flies. (**L**) The different genus bacteria abundance at 5 d in *BdNubX2* RNAi flies. (**M**) The gut total bacteria and *Enterobacteriaceae* abundance at 7 d in *BdNubX2* RNAi flies. (**N**) The different genus bacteria abundance at 7 d in *BdNubX2* RNAi flies, (**A–N**) 30 flies guts samples per biological replicate, three biological replicates. * $p < 0.05$, ** $p < 0.01$, *** $p < 0.001$, Student's *t*-test.

4. Discussion

In this study, we report that *BdNub* encodes two isoforms, *BdNubX1* and *BdNubX2*. They are involved in the gut immune response, not systemic immunity, possibly by regulating the gut IMD pathway. Our results also show that *BdNub* is essential for maintaining gut microbiota.

We identified a putative *B. dorsalis Nub* gene, *BdNub*. Similar to *Drosophila*, it also produces two distinct *BdNub* isoforms [15]. Phylogenetic tree analysis shows that the *BdNub* gene is closely related to *B. capsicum*, and the *Nub* gene is highly conserved in insects. This is also confirmed with the protein alignment of conserved domains, POU and HOX [33]. There are five different POU proteins in the *Drosophila* genome [34], regulating embryonic development and differentiation [12], immune function, and tissue homeostasis [14,15]. In addition, pdm1, a POU family gene, acts as proximal-distal growth of the wing, which has a similar function to *Nub* [35]. Moreover, the HOX domain regulates muscle and wing development [36,37], specifying the anterior posterior axis in all bilaterians [38]. So, this indicates the high expression of *NUB* genes in the old pupa stage and the day 1 adult flies.

Spatial and temporal expression pattern analysis showed that *BdNubX1* and *BdNubX2* were mainly expressed in the late-stage pupa, which may be related to Nub gene function in wing development [39,40]. In *Drosophila*, *Nub* mutant flies showed a severe wing size reduction [35,40–42], indicating that *BdNub* may also have an indispensable role in *B. dorsalis* wing formation. This might relate to the HOX domain of *BdNub*. Furthermore, we also observed high *BdNub* expression in newly emerged adults and guts. *B. dorsalis* must crawl out of the soil to accomplish eclosion. Thus, newly merged adult flies are exposed to various microorganisms from the soil and environment. High *BdNub* expression at this stage suggested its role in regulating gut immune balance in this process [43].

Our results showed that *E. coli* systemic infection induced the AMP genes' expression but not *BdNubX1* and *BdNubX2*. It suggests that the *BdNub* gene is not involved in *E. coli*-induced systemic immunity. This result is consistent with their high expression in the gut and low expression in the fat body, the major systemic immune organ. However, immunostaining reveals that *Nub* protein is present in the fat body of *Drosophila* [14]. This suggests *Nub* could be functional in regulating *Drosophila's* systemic immune response.

In contrast, Gram-negative bacteria *P. rettgeri* oral infection induced a strong immune response and a strong *BdNubX1* and *BdNubX2* up-regulation, suggesting that *BdNub* was involved in the gut immune response. Furthermore, we showed that *BdNubX1* positively regulated gut AMP expression, while *BdNubX2* inhibited AMP expression. The fact that *BdNubX2* is an immunosuppressor of IMD pathway activity is also in line with the *Nub* gene function in the *Drosophila* gut [7,43]. In *Drosophila*, *Nub-PD* RNAi increased AMP gene expression. Similarly, *BdNubX2* RNAi also increased AMP gene expression. Apart from *Nub*, the IMD signaling pathway is regulated by many other factors. For example, the *Pirk* gene encodes the protein binding *PGRP-LC*. It is regulated by the IMD pathway itself. Nevertheless, it establishes a negative feedback loop adjusting IMD pathway activity [44]. *PGRP-SB* and *PGRP-LB* are secreted proteins with an amidase activity that scavenges DAP-type peptidoglycan [6]. They negatively regulate the IMD pathway. A recent study also shows that *PGRP-SB* and *PGRP-LB* are negative regulators of the gut IMD pathway in *B. dorsalis* [23,45,46]. There may also be other unknown factors that regulate the gut IMD pathway activity that have yet to be identified.

In addition, both *BdNubX1* and *BdNubX2* RNAi could make the flies more sensitive to *P. rettgeri* infection. Nevertheless, there was no significant difference in the mortality rate, which was consistent with the results of *Drosophila* [15]. This may be due to the fact that RNAi could not achieve a stable and long-lasting silence effect in *B. dorsalis*. Furthermore, the efficiency of RNAi varies considerably among insects, although RNAi can reach 90% efficacy in *Coleopterans* [47]. *Dipteran* species are not very sensitive to RNAi [48]. Studies using null mutants should be carried out to further elucidate the *Nub* function in gut immunity in the future. Moreover, unlike *Drosophila*, our screening did not find any strong lethal pathogenic bacteria for the gut infection. Although *P. rettgeri* could induce a strong gut immune response, it kills the wild-type flies very slowly, which might be an immune tolerance phenotype rather than an immune resistance phenotype. It indicates that the high adaptability of *B. dorsalis* may be related to its strong immune system.

Since *BdNubX1* and *BdNubX2* transcript isomers play an important role in intestinal immunity, it is plausible that they also participate in controlling the microbiota. Many reports show that changes in immune-related genes will lead to changes in intestinal microbial community structure [21,23]. *BdNubX1* knockdown leads to significantly up-regulated total bacterial abundance, with increased *Pseudomonas* and *Salmonella*. This indicates *BdNubX1* partially regulates gut microbiota composition and abundance through the IMD pathway. The IMD pathway is an important part of the insect gut microbiota regulation mechanism [49]. For example, *PGRP-LB* and *PGRP-SB* have high expression levels in the anterior and middle midgut, which is associated with gut commensal bacteria distribution in *B. dorsalis* [23]. In *Drosophila*, IMD-deficient flies showed a dysregulated gut microbiota and disturbed gut homeostasis [22]. Moreover, we cannot exclude the possibility that *BdNubX1* also interacts with other gut immune mechanisms. For example, *BdDuox* regulates gut microbiota through the production of ROS [21]. Therefore, it is possible that *BdNubX1* might regulate microbiota by affecting ROS production and scavenging.

On the other side, *BdNubX2* knockdown leads to decreased gut microbiota and *Enterobacteriaceae*. It also causes changes in *Pseudomonas*, *Flavobacterium*, and *Salmonella* abundance. In *B. dorsalis*, *Enterobacteriaceae* bacteria account for a large proportion in the gut and are the dominant bacterial group in gut microbiota [50]. In fact, the key ROS production enzyme, *Bdduox* RNAi, induces a similar changes in microbiota composition, with decreased *Enterobacteriaceae* and a rise in secondary microbiota abundance [21]. Several studies have shown that *Enterobacteriaceae* bacteria are beneficial to the host [20,24]. Therefore, decreased *Enterobacteriaceae* in *B. dorsalis* is possibly detrimental to the host. Our results suggest that *BdNubX2* could also be essential to host development and homeostasis through regulating *Enterobacteriaceae* bacteria, which could contribute to the early death of *BdNubX2* RNAi flies. Actually, *Drosophila Caudal* mutants show constitutively activated AMP genes, leading to an increase in *Gluconobacter sp.*, causing an early death of the hosts [49]. Altogether, we proved that the *BdNub* gene regulates the IMD pathway to maintain intestinal microbial homeostasis. In conclusion, *BdNub* plays an important role in the regulation of intestinal immunity, decreasing the host's sensitivity to intestinal opportunistic pathogens, and regulating gut microbiota.

Supplementary Materials: The following supporting information can be downloaded at: https://www.mdpi.com/article/10.3390/insects14020178/s1, Figure S1: Protein functional domain analysis of BdNubX1 and BdNubX2.

Author Contributions: Conceptualization, H.Z., X.L. and Z.Y.; Methodology, J.G. and P.Z.; Formal analysis, J.G. and P.Z.; Investigation, J.G. and P.Z.; Data Curation, J.G. and P.Z.; Visualization, J.G. and P.Z.; Writing—original draft, J.G. and X.L.; Supervision, X.L., Z.Y. and H.Z.; Funding acquisition, H.Z.; Project administration, H.Z. All authors have read and agreed to the published version of the manuscript.

Funding: This study was supported by National Key R&D Program of China (No. 2021YFC2600400), The Major Special Science and Technology Project of Yunnan Province (NO. 202102AE090054),The Earmarked Fund for CARS (CARS-26) and Hubei Hongshan Laboratory.

Data Availability Statement: The data presented in this study are available on request from the corresponding author.

Acknowledgments: Special thanks to Qiongke Ma for insects rearing.

Conflicts of Interest: The authors declare no conflict of interest.

References

1. Basset, Y.; Cizek, L.; Cuénoud, P.; Didham, R.K.; Guilhaumon, F.; Missa, O.; Novotny, V.; Ødegaard, F.; Roslin, T.; Schmidl, J.; et al. Arthropod Diversity in a Tropical Forest. *Science* **2012**, *338*, 1481–1484. [CrossRef] [PubMed]
2. Engel, P.; Moran, N.A. The Gut Microbiota of Insects—Diversity in Structure and Function. *FEMS Microbiol. Rev.* **2013**, *37*, 699–735. [CrossRef] [PubMed]

3. Gottar, M.; Gobert, V.; Michel, T.; Belvin, M.; Duyk, G.; Hoffmann, J.A.; Ferrandon, D.; Royet, J. The Drosophila Immune Response against Gram-Negative Bacteria Is Mediated by a Peptidoglycan Recognition Protein. *Nature* **2002**, *416*, 640–644. [CrossRef] [PubMed]
4. Brennan, C.A.; Anderson, K.V. Drosophila: The Genetics of Innate Immune Recognition and Response. *Annu. Rev. Immunol.* **2004**, *22*, 457–483. [CrossRef]
5. Royet, J.; Dziarski, R. Peptidoglycan Recognition Proteins: Pleiotropic Sensors and Effectors of Antimicrobial Defences. *Nat. Rev. Microbiol.* **2007**, *5*, 264–277. [CrossRef]
6. Lemaitre, B.; Hoffmann, J. The Host Defense of Drosophila Melanogaster. *Annu. Rev. Immunol.* **2007**, *25*, 697–743. [CrossRef]
7. Imler, J.-L.; Bulet, P. Antimicrobial Peptides in Drosophila: Structures, Activities and Gene Regulation. *Mech. Epithel. Def.* **2005**, *86*, 1–21. [CrossRef]
8. Hanson, M.A.; Dostálová, A.; Ceroni, C.; Poidevin, M.; Kondo, S.; Lemaitre, B. Synergy and Remarkable Specificity of Antimicrobial Peptides in Vivo Using a Systematic Knockout Approach. *eLife* **2019**, *8*, e44341. [CrossRef]
9. Horak, R.D.; Leonard, S.P.; Moran, N.A. Symbionts Shape Host Innate Immunity in Honeybees. *Proc. R. Soc. B* **2020**, *287*, 20201184. [CrossRef]
10. Zhang, L.; Gallo, R.L. Antimicrobial Peptides. *Curr. Biol.* **2016**, *26*, R14–R19. [CrossRef]
11. Davis, M.M.; Engström, Y. Immune Response in the Barrier Epithelia: Lessons from the Fruit Fly *Drosophila Melanogaster*. *J. Innate Immun.* **2012**, *4*, 273–283. [CrossRef]
12. Tantin, D. Oct Transcription Factors in Development and Stem Cells: Insights and Mechanisms. *Development* **2013**, *140*, 2857–2866. [CrossRef]
13. Dantoft, W.; Lundin, D.; Esfahani, S.S.; Engström, Y. The POU/Oct Transcription Factor Pdm1/Nub Is Necessary for a Beneficial Gut Microbiota and Normal Lifespan of *Drosophila*. *J. Innate Immun.* **2016**, *8*, 412–426. [CrossRef]
14. Dantoft, W.; Davis, M.M.; Lindvall, J.M.; Tang, X.; Uvell, H.; Junell, A.; Beskow, A.; Engström, Y. The Oct1 Homolog Nubbin Is a Repressor of NF-KB-Dependent Immune Gene Expression That Increases the Tolerance to Gut Microbiota. *BMC Biol.* **2013**, *11*, 99. [CrossRef]
15. Lindberg, B.G.; Tang, X.; Dantoft, W.; Gohel, P.; Seyedoleslami Esfahani, S.; Lindvall, J.M.; Engström, Y. Nubbin Isoform Antagonism Governs Drosophila Intestinal Immune Homeostasis. *PLoS Pathog.* **2018**, *14*, e1006936. [CrossRef]
16. Dillon, R.J.; Dillon, V.M. THE GUT B ACTERIA OF INSECTS: Nonpathogenic Interactions. *Annu. Rev. Entomol.* **2004**, *49*, 71–92. [CrossRef]
17. Consuegra, J.; Grenier, T.; Baa-Puyoulet, P.; Rahioui, I.; Akherraz, H.; Gervais, H.; Parisot, N.; da Silva, P.; Charles, H.; Calevro, F.; et al. Drosophila-Associated Bacteria Differentially Shape the Nutritional Requirements of Their Host during Juvenile Growth. *PLoS Biol.* **2020**, *18*, e3000681. [CrossRef]
18. Storelli, G.; Strigini, M.; Grenier, T.; Bozonnet, L.; Schwarzer, M.; Daniel, C.; Matos, R.; Leulier, F. Drosophila Perpetuates Nutritional Mutualism by Promoting the Fitness of Its Intestinal Symbiont Lactobacillus Plantarum. *Cell Metab.* **2018**, *27*, 362–377.e8. [CrossRef]
19. Flórez, L.V.; Scherlach, K.; Gaube, P.; Ross, C.; Sitte, E.; Hermes, C.; Rodrigues, A.; Hertweck, C.; Kaltenpoth, M. Antibiotic-Producing Symbionts Dynamically Transition between Plant Pathogenicity and Insect-Defensive Mutualism. *Nat. Commun.* **2017**, *8*, 15172. [CrossRef]
20. Cai, Z.; Yao, Z.; Li, Y.; Xi, Z.; Bourtzis, K.; Zhao, Z.; Bai, S.; Zhang, H. Intestinal Probiotics Restore the Ecological Fitness Decline of *Bactrocera Dorsalis* by Irradiation. *Evol. Appl.* **2018**, *11*, 1946–1963. [CrossRef]
21. Yao, Z.; Wang, A.; Li, Y.; Cai, Z.; Lemaitre, B.; Zhang, H. The Dual Oxidase Gene BdDuox Regulates the Intestinal Bacterial Community Homeostasis of Bactrocera Dorsalis. *ISME J.* **2016**, *10*, 1037–1050. [CrossRef] [PubMed]
22. Broderick, N.A.; Buchon, N.; Lemaitre, B. Microbiota-Induced Changes in Drosophila Melanogaster Host Gene Expression and Gut Morphology. *mBio* **2014**, *5*, e01117-14. [CrossRef] [PubMed]
23. Yao, Z.; Cai, Z.; Ma, Q.; Bai, S.; Wang, Y.; Zhang, P.; Guo, Q.; Gu, J.; Lemaitre, B.; Zhang, H. Compartmentalized PGRP Expression along the Dipteran Bactrocera Dorsalis Gut Forms a Zone of Protection for Symbiotic Bacteria. *Cell Rep.* **2022**, *41*, 111523. [CrossRef] [PubMed]
24. Raza, M.F.; Wang, Y.; Cai, Z.; Bai, S.; Yao, Z.; Awan, U.A.; Zhang, Z.; Zheng, W.; Zhang, H. Gut Microbiota Promotes Host Resistance to Low-Temperature Stress by Stimulating Its Arginine and Proline Metabolism Pathway in Adult Bactrocera Dorsalis. *PLoS Pathog.* **2020**, *16*, e1008441. [CrossRef] [PubMed]
25. Guo, Q.; Yao, Z.; Cai, Z.; Bai, S.; Zhang, H. Gut Fungal Community and Its Probiotic Effect on *Bactrocera Dorsalis*. *Insect Sci.* **2022**, *29*, 1145–1158. [CrossRef]
26. Livak, K.J.; Schmittgen, T.D. Analysis of Relative Gene Expression Data Using Real-Time Quantitative PCR and the 2\textbackslashtextminusΔΔCT Method. *Methods* **2001**, *25*, 402–408. [CrossRef]
27. Guo, X.; Xia, X.; Tang, R.; Wang, K. Real-Time PCR Quantification of the Predominant Bacterial Divisions in the Distal Gut of Meishan and Landrace Pigs. *Anaerobe* **2008**, *14*, 224–228. [CrossRef]
28. Bartosch, S.; Fite, A.; Macfarlane, G.T.; McMurdo, M.E.T. Characterization of Bacterial Communities in Feces from Healthy Elderly Volunteers and Hospitalized Elderly Patients by Using Real-Time PCR and Effects of Antibiotic Treatment on the Fecal Microbiota. *Appl. Environ. Microbiol.* **2004**, *70*, 3575–3581. [CrossRef]

29. Rinttila, T.; Kassinen, A.; Malinen, E.; Krogius, L.; Palva, A. Development of an Extensive Set of 16S RDNA-Targeted Primers for Quantification of Pathogenic and Indigenous Bacteria in Faecal Samples by Real-Time PCR. *J. Appl. Microbiol.* **2004**, *97*, 1166–1177. [CrossRef]
30. Abell, G.C.J.; Bowman, J.P. Ecological and Biogeographic Relationships of Class Flavobacteria in the Southern Ocean. *Fems Microbiol. Ecol.* **2005**, *51*, 265–277. [CrossRef]
31. Ahmed, W.; Huygens, F.; Goonetilleke, A.; Gardner, T. Real-Time PCR Detection of Pathogenic Microorganisms in Roof-Harvested Rainwater in Southeast Queensland, Australia. *Appl. Environ. Microbiol.* **2008**, *74*, 5490–5496. [CrossRef]
32. Matsuda, K.; Tsuji, H.; Asahara, T.; Matsumoto, K.; Takada, T.; Nomoto, K. Establishment of an Analytical System for the Human Fecal Microbiota, Based on Reverse Transcription-Quantitative PCR Targeting of Multicopy RRNA Molecules. *Appl. Environ. Microbiol.* **2009**, *75*, 1961–1969. [CrossRef]
33. Bürglin, T.R.; Affolter, M. Homeodomain Proteins: An Update. *Chromosoma* **2016**, *125*, 497–521. [CrossRef]
34. Tang, X.; Engström, Y. Regulation of Immune and Tissue Homeostasis by Drosophila POU Factors. *Insect Biochem. Mol. Biol.* **2019**, *109*, 24–30. [CrossRef]
35. Cifuentes, F.J.; García-Bellido, A. Proximo–Distal Specification in the Wing Disc of *Drosophila* by the Nubbin Gene. *Proc. Natl. Acad. Sci. USA* **1997**, *94*, 11405–11410. [CrossRef]
36. Poliacikova, G.; Maurel-Zaffran, C.; Graba, Y.; Saurin, A.J. Hox Proteins in the Regulation of Muscle Development. *Front. Cell Dev. Biol.* **2021**, *9*, 731996. [CrossRef]
37. Paul, R.; Giraud, G.; Domsch, K.; Duffraisse, M.; Marmigère, F.; Khan, S.; Vanderperre, S.; Lohmann, I.; Stoks, R.; Shashidhara, L.S.; et al. Hox Dosage Contributes to Flight Appendage Morphology in Drosophila. *Nat. Commun.* **2021**, *12*, 2892. [CrossRef]
38. Mallo, M. Reassessing the Role of Hox Genes during Vertebrate Development and Evolution. *Trends Genet.* **2018**, *34*, 209–217. [CrossRef]
39. Fernandez-Nicolas, A.; Ventos-Alfonso, A.; Kamsoi, O.; Clark-Hachtel, C.; Tomoyasu, Y.; Belles, X. Broad Complex and Wing Development in Cockroaches. *Insect Biochem. Mol. Biol.* **2022**, *147*, 103798. [CrossRef]
40. Ng, M.; Diaz-Benjumea, F.J.; Cohen, S.M. Nubbin Encodes a POU-Domain Protein Required for Proximal-Distal Patterning in the Drosophila Wing. *Development* **1995**, *121*, 589–599. [CrossRef]
41. Zirin, J.D.; Mann, R.S. Nubbin and Teashirt Mark Barriers to Clonal Growth along the Proximal–Distal Axis of the Drosophila Wing. *Dev. Biol.* **2007**, *304*, 745–758. [CrossRef] [PubMed]
42. Neumann, C.J.; Cohen, S.M. Boundary Formation in *Drosophila* Wing: Notch Activity Attenuated by the POU Protein Nubbin. *Science* **1998**, *281*, 409–413. [CrossRef] [PubMed]
43. Buddika, K.; Huang, Y.-T.; Ariyapala, I.S.; Butrum-Griffith, A.; Norrell, S.A.; O'Connor, A.M.; Patel, V.K.; Rector, S.A.; Slovan, M.; Sokolowski, M.; et al. Coordinated Repression of Pro-Differentiation Genes via P-Bodies and Transcription Maintains Drosophila Intestinal Stem Cell Identity. *Curr. Biol.* **2022**, *32*, 386–397.e6. [CrossRef] [PubMed]
44. Kleino, A.; Myllymäki, H.; Kallio, J.; Vanha-aho, L.-M.; Oksanen, K.; Ulvila, J.; Hultmark, D.; Valanne, S.; Rämet, M. Pirk Is a Negative Regulator of the *Drosophila* Imd Pathway. *J. Immunol.* **2008**, *180*, 5413–5422. [CrossRef] [PubMed]
45. Mellroth, P.; Steiner, H. akan PGRP-SB1: An N-Acetylmuramoyl l-Alanine Amidase with Antibacterial Activity. *Biochem. Biophys. Res. Commun.* **2006**, *350*, 994–999. [CrossRef] [PubMed]
46. Zaidman-Rémy, A.; Hervé, M.; Poidevin, M.; Pili-Floury, S.; Kim, M.-S.; Blanot, D.; Oh, B.-H.; Ueda, R.; Mengin-Lecreulx, D.; Lemaitre, B. The Drosophila Amidase PGRP-LB Modulates the Immune Response to Bacterial Infection. *Immunity* **2006**, *24*, 463–473. [CrossRef]
47. Joga, M.R.; Zotti, M.J.; Smagghe, G.; Christiaens, O. RNAi Efficiency, Systemic Properties, and Novel Delivery Methods for Pest Insect Control: What We Know So Far. *Front. Physiol.* **2016**, *7*, 553. [CrossRef]
48. Araujo, R.N.; Santos, A.; Pinto, F.S.; Gontijo, N.F.; Lehane, M.J.; Pereira, M.H. RNA Interference of the Salivary Gland Nitrophorin 2 in the Triatomine Bug Rhodnius Prolixus (Hemiptera: Reduviidae) by DsRNA Ingestion or Injection. *Insect Biochem. Mol. Biol.* **2006**, *36*, 683–693. [CrossRef]
49. Ryu, J.-H.; Kim, S.-H.; Lee, H.-Y.; Bai, J.Y.; Nam, Y.-D.; Bae, J.-W.; Lee, D.G.; Shin, S.C.; Ha, E.-M.; Lee, W.-J. Innate Immune Homeostasis by the Homeobox Gene Caudal and Commensal-Gut Mutualism in Drosophila. *Science* **2008**, *319*, 777–782. [CrossRef]
50. Wang, H.; Jin, L.; Zhang, H. Comparison of the Diversity of the Bacterial Communities in the Intestinal Tract of Adult Bactrocera Dorsalis from Three Different Populations. *J. Appl. Microbiol.* **2011**, *110*, 1390–1401. [CrossRef]

Disclaimer/Publisher's Note: The statements, opinions and data contained in all publications are solely those of the individual author(s) and contributor(s) and not of MDPI and/or the editor(s). MDPI and/or the editor(s) disclaim responsibility for any injury to people or property resulting from any ideas, methods, instructions or products referred to in the content.

MDPI
St. Alban-Anlage 66
4052 Basel
Switzerland
Tel. +41 61 683 77 34
Fax +41 61 302 89 18
www.mdpi.com

Insects Editorial Office
E-mail: insects@mdpi.com
www.mdpi.com/journal/insects

www.ingramcontent.com/pod-product-compliance
Lightning Source LLC
Chambersburg PA
CBHW041947260926
51850CB00018B/1924
* 9 7 8 3 0 3 6 5 6 8 6 8 3 *